Recent Titles in This Series

(Continued in the back of this publication)

D0905220

CONTEMPORARY MATHEMATICS

164

Geometric Topology

Joint U.S.–Israel Workshop on Geometric Topology
June 10–16, 1992
Technion, Haifa, Israel

Cameron Gordon
Yoav Moriah
Bronislaw Wajnryb
Editors

American Mathematical Society
Providence, Rhode Island

Geometric Topology

The Joint U.S.-Israel Workshop on Geometric Topology was held June 10–16, 1992, at the Amado Mathematics Building, Technion, Haifa, Israel, with support provided by the Institute of Advanced Studies in Mathematics of the Technion and the U.S.-Israel Binational Science Foundation.

1991 *Mathematics Subject Classification.* Primary 57M05, 57M07, 57M25, 57M50, 57Q45, 57M60.

Library of Congress Cataloging-in-Publication Data

Joint U.S.-Israel Workshop on Geometric Topology (1992: Haifa, Israel)
    Geometric topology: Workshop on Geometric Topology, June 10–16, 1992, Technion, Haifa, Israel / [edited by] Cameron Gordon, Yoav Moriah, Bronislaw Wajnryb.
        p.   cm. — (Contemporary mathematics; v. 164)
    ISBN 0-8218-5182-9
    1. Low-dimensional topology–Congresses.   I. Gordon, Cameron, 1945–.   II. Moriah, Yoav, 1952–.   III. Wajnryb, Bronislaw, 1946–.   IV. Title.   V. Series: Contemporary mathematics (American Mathematical Society); v. 164.
QA612.14.W67   1992   94-11696
514′.2–dc20   CIP

Contents

Introduction

Geometric topology is an area of topology that has gone through tremendous changes in the last decade or so. In fact a lot of the big questions and problems that were facing mathematicians in the area have been answered, and new directions and problems have arisen. One of the characteristics of geometric topology is that the techniques that are used in the area are very diverse. The purpose of the conference was to bring people working in the different subareas to "touch base" and to learn what other people were doing.

These proceedings contain refereed reports from eighteen lectures given at the conference. Altogether sixty-two people registered for the conference. Our guests included mathematicians from all over the world: 25 from the U.S.A., 17 from Israel, 7 from France, 5 from Japan, 3 from Germany, 2 from the U.K., and 1 each from Canada, China, and Mexico.

The organizers wish to thank the sponsors of the conference, which were the Institute of Advanced Studies in Mathematics of the Technion and the U.S.-Israel Binational Science Foundation. We would also like to thank Sylvia Shur and Hava Harel of the Faculty of Mathematics of the Technion for the outstanding job they did before and during the conference.

Finally, we remember our colleague and outstanding mathematician Professor Boris Moishezon, who passed away unexpectedly in August 1993.

Haifa, October 1993

Cameron Gordon
Yoav Moriah
Bronek Wajnryb

List of Participants

Dr. E. Appelbaum, *Department of Mathematics, Technion-Israel Institute of Technology, 32000 Haifa, Israel.*

Professor L. Birbrair, *Kiryat Arba 7/9, 90100 Jerusalem, Israel.*

Professor J. Birman, *Department of Mathematics, Columbia University, New York, NY 10027, U.S.A.*

Dr. D. Blanc, *Department of Mathematics, Hebrew University of Jerusalem, 91904 Jerusalem, Israel.*

Professor M. Boileau, *Department of Mathematics, University Paul Sabatier, UER MIG, 118 Rte de Narbonne, 31062 Toulouse cedex, France.*

Professor F. Bonahon, *Department of Mathematics, University of Southern California, Los Angeles, CA 90089-1113, U.S.A.*

Professor P. Callahan, *Department of Mathematics, University of Illinois, Urbana, IL 61801, U.S.A.*

Professor A. Casson, *Department of Mathematics, University of California, Berkeley, CA 94720, U.S.A.*

Professor M. Domergue, *Department of Mathematics, University of Provence, Case H, F13331 Marseille 3, France.*

Professor Mario Eudave-Munoz, *Instituto de Matematicas, UNAM, Circuito Exterior, Ciudad Universitaria, Mexico D.F. 04510, Mexico.*

Professor M. Farber, *Department of Mathematics, Tel Aviv University, Ramat Aviv, 69978 Tel Aviv, Israel.*

Professor E. Flapan, *Department of Mathematics, Pomona College, Claremont, CA 91711, U.S.A.*

Professor D. Gabai, *Department of Mathematics, Caltech, Pasedena, CA 91125, U.S.A.*

Professor Rita Gitick, *Department of Mathematics, University of Michigan, Ann Arbor, MI 48109, U.S.A.*

Professor V. Goldstein, *Department of Mathematics, Ben Gurion University of the Negev, P.O. Box 653, 84105 Beersheva, Israel.*

Professor C. Gordon, *Department of Mathematics, University of Texas at Austin, Austin, TX 78712, U.S.A.*

Dr. M. Grossberg, *Department of Mathematics, Hebrew University of Jerusalem, 91904 Jerusalem, Israel.*

Professor J. Hass, *Department of Mathematics, University of California, Davis, CA 95616, U.S.A.*

Professor Andrew Duncan, *Department of Mathematics, Heriot-Watt University, Riccarton, Edinburgh EH14 4AS, Scotland, U.K.*

Professor A. Juhasz, *Department of Mathematics, Technion-Israel Institute of Technology, 32000 Haifa, Israel.*

Professor Uwe Kaiser, *Department of Mathematics, Brandeis University, Waltham, MA 02254, U.S.A.*

Professor O. Kakimizu, *Department of Education, Igarashi-Nino Machi, Migata-shi 950-21, Japan.*

Professor E. Kalfagianni, *Department of Mathematics, Columbia University, New York, NY 10027, U.S.A.*

Professor C. Kearton, *Department of Mathematics, Durham University, Durham, DH1 3LE, England.*

Professor M. Kidwell, *U.S. Naval Academy, Chauvenet Hall, Annapolis, MD 21402-5002, U.S.A.*

Professor U. Koschorke, *Mathematics Institute, University of Siegen, FB6, Holderlinst 3, 59 Siegen 21, Germany.*

Professor C. Lescop, *Ecole Normale Superieure de Lyon, 46 Allee d'Italie, 69364 Lyon cedex 07, France.*

Professor K. Jones, *Department of Mathematics, University of Texas, Austin, TX 78712, U.S.A.*

Professor A. Lubotzky, *Department of Mathematics, Hebrew University of Jerusalem, 91904 Jerusalem, Israel.*

Professor D. Markushevich, *Department of Mathematics, Technion-Israel Institute of Technology, 32000 Haifa, Israel.*

Professor Yoshihiko Marumoto, *Department of Mathematics, Osaka Sangyo University, Nakagaito 3-1-1, Daito, Osaka 574, Japan.*

Professor Y. Mathieu, *Department of Mathematics, University of Provence, Case H, F13331 Marseille 3, France.*

Professor W. Menasco, *Department of Mathematics, SUNY at Buffalo, Buffalo, NY 14214, U.S.A.*

Professor K. Miyazaki, *Department of Mathematics, Tsuda College, Tsuda-Machi Kodaira-shi, Tokyo 187, Japan.*

Professor B. Moishezon, *Department of Mathematics, Hebrew University of Jerusalem, 91904 Jerusalem, Israel.*

Dr. Yoav Moriah, *Department of Mathematics, Technion-Israel Institute of Technology, 32000 Haifa, Israel.*

Professor J. Levine, *Department of Mathematics, Brandeis University, Waltham, MA 02254, U.S.A.*

Professor V. Lin, *Department of Mathematics, Technion-Israel Institute of Technology, 32000 Haifa, Israel.*

Professor F. Paulin, *Department of Mathematics, Ecole Normale Super de Lyon, 46 Allec d'Italie, 69364 Lyon 07 cedex, France.*

Dr. V. Polterovitch, *Department of Mathematics, Tel Aviv University, Ramat Aviv, 69978 Tel Aviv, Israel.*

Professor Michael Polyak,, *School of Mathematics, Tel Aviv University, Ramat Aviv, 69978 Tel Aviv, Israel.*

Dr. Yoav Rieck, *Department of Mathematics, University of Texas at Austin, Austin, TX 78712, U.S.A.*

Professor D. Rolfsen, *Department of Mathematics, University of British Columbia, 121-1984 Mathematics Road, Vancouver, BC V6T 1Y4, Canada.*

Professor G. Rosenberger, *Fach Math., University of Dortmund, Postfach 50 05 00, D-4600, Germany.*

Dr. Emil Saucan, *Department of Mathematics, Technion-Israel Institute of Technology, 32000 Haifa, Israel.*

Professor K. Morimoto, *Department of Mathematics, Takushoku University, Tatemochi, Hachigi, Tokyo 193, Japan.*

Dr. Ramin Naimi, *Department of Mathematics, Technion-Israel Institute of Technology, 32000 Haifa, Israel.*

Professor P. Shalen, *Department of Mathematics, University of Illinois at Chicago, Chicago, IL 60680, U.S.A.*

Dr. Yafim Schwartzman, *Department of Mathematics, Technion-Israel Institute of Technology, 32000 Haifa, Israel.*

Professor K. Taniyama, *Department of Mathematics, School of Education, Waseda University, Shinjuku, Tokyo, Japan.*

Professor A. Thompson, *Department of Mathematics, University of California, Davis, CA 95616, U.S.A.*

Professor V. Touraev, *Department of Mathematics, University Louis Pasteur, 7 rue Rene Descartes, 67084 Strasbourg, France.*

Professor J. Viro, *Dept. of Mathematics & Computer Science, University of California, Riverside, CA 95521, U.S.A.*

Professor O. Viro, *Dept. of Mathematics & Computer Science, University of California, Riverside, CA 95521, U.S.A.*

Professor M. Scharlemann, *Department of Mathematics, University of California, Santa Barbara, CA 93106, U.S.A.*

Professor P. Scott, *Department of Mathematics, University of Michigan, Ann Arbor, MI 48109, U.S.A.*

Dr. Zlil Sela, *Department of Mathematics, Princeton University, Princeton, NJ 08544, U.S.A.*

Professor P. Vogel, *Department of Mathematics, University of Paris VII, 2 Place Jussiev, Paris F75251, France.*

Professor B. Wajnryb, *Department of Mathematics, Technion-Israel Institute of Technology, 32000 Haifa, Israel.*

Professor S. Wang, *Department of Mathematics, Beijing University, Beijing China.*

Dr. E. Weinstein, *Department of Mathematics, Hebrew University of Jerusalem, 91904 Jerusalem, Israel.*

Dr. D. Yavin, *Max-Planck-Institut fur Mathematik, Gottfried-Claren Str. 26, 5300 Bonn 3, Germany.*

Contemporary Mathematics
Volume **164**, 1994

THE 3–TORUS IS KERVAIRE

Andrew J. Duncan and James Howie

ABSTRACT. The 3-dimensional torus $S^1 \times S^1 \times S^1$ has a canonical cell-structure, with one cell in dimensions 0 and 3, and three cells in dimensions 1 and 2. Its 2-skeleton is thus the geometric realization of the group presentation

$$< X, Y, Z, |\ XYX^{-1}Y^{-1}, YZY^{-1}Z^{-1}, ZXZ^{-1}X^{-1} > .$$

We show that this 2-complex K is *Kervaire* , in other words that any system of equations over a group G, modelled on K, has a solution in some larger group.

INTRODUCTION

In this note we consider a complex associated to the 3-Torus $S^1 \times S^1 \times S^1$ and show that any system of equations over a group G, modelled on this complex has a solution in some larger group. This means that the conclusion of the *Kervaire-Laudenbach conjecture* holds for such a system of equations. In the remainder of this section we outline the Kervaire-Laudenbach conjecture and in the following section prove our result.

An early and general statement of the "Kervaire-Laudenbach Conjecture" is the *Adjunction Problem* of B.H. Neumann [N]. This asks when the following property holds.

Property A. *Let G be a group and F the free group on generators $x_1, \cdots, x_n$. If $w_1, \cdots, w_m$ are elements of $G * F$ then the canonical map*

$$G \longrightarrow \frac{G * F}{N}$$

is injective, where N denotes the normal closure of $< w_1, \cdots, w_m >$ in $G * F$.

Given words $w_1, \cdots, w_m$ in the free product $G * F$ we may consider the expressions $w_1 = 1, \cdots, w_m = 1$ as a system of equations in unknowns x_i with coefficients from G. This system of equations is said to have a *solution over G* if there exists a group H and a map $\phi : G * F \longrightarrow H$ such that $\phi(w_i) = 1$, for $i = 1, \cdots, m$, and $\phi|_G$ is an injection. Since every such map ϕ factors through $(G * F)/N$, Property A is equivalent to the statement that every system of equations has a solution over G.

1980 *Mathematics Subject Classification* (1985 *Revision*). Primary 20F05; Secondary 20F32.
Research supported by SERC grant GR/E 88998 .

This paper is in final form and no version of it will be submitted for publication elsewhere.

Neumann shows that Property A holds for words w of the form gx^k, where $g \in G$ and $k \neq 0$, and hence that any group G can be embedded in a group H in which every element is a k-th power, for all $k > 0$. Levin [L] generalizes Neumann's results to show that Property A holds for words w of the form $g_1xg_2x\ldots g_tx$ and proves a corresponding result on embeddings.

Clearly Property A does not hold in general (e.g. $w_1 = x, w_2 = x^{-1}g, 1 \neq g \in G$) and the problem is to find necessary and sufficient conditions under which it holds. Conditions can be imposed on the group G or on the words w_i. One such is that $\{w_1, \cdots, w_m\}$ be *non-singular*: that is that the n-tuples $(d_{i,1}, \cdots, d_{i,n})$, $i = 1, \cdots, m$, are linearly independent, where $d_{i,j}$ is the exponent sum of x_j in w_i. The *Kervaire-Laudenbach Conjecture* is

Kervaire-Laudenbach Conjecture. If $\{w_1, \cdots, w_m\}$ is non-singular then Property A holds.

(The attribution to Kervaire originates from a special case of the problem which arises from consideration of higher dimensional knot groups [K], whilst Serre has attributed the same special case to Laudenbach [B, Problem Section, p. 734].)

The set $\{w_1, \cdots, w_m\} \subseteq G * F$ is called *Kervaire* if Property A holds. The following result of Gerstenhaber and Rothaus gives strong support to the Kervaire-Laudenbach conjecture.

Theorem 1 [GR,R]. *Let G be a locally residually finite group. Then $\{w_1, \cdots, w_m\}$ is Kervaire if it is non-singular.*

(A group H is *residually finite* if, given any $h \in H$ there exists a finite group K and a homomorphism $\phi : H \longrightarrow K$ with $\phi(h) \neq 1$. A group is *locally* **P** if every non-trivial finitely generated subgroup is **P** (where **P** is some property.)) Moreover Howie [H1] has shown

Theorem 2. *If G is locally indicable then $\{w_1, \cdots, w_m\}$ is Kervaire if it is non-singular.*

(A group H is *indicable* if there exists an epimorphism from H to Z.)
If $w \in G*F(x_1, \cdots, x_n)$ then the *x-length* of w is $\sum_{j=1}^n \epsilon_j$ where ϵ_j is the number of occurrences of $x_j^{\pm 1}$ in w. Gersten shows [G1] that to prove the Kervaire-Laudenbach conjecture it is sufficient to prove it under the additional assumption that the x-length of w is at most 3, for $i = 1, \cdots, m$. Furthermore he shows (*loc. cit.*) that any set $\{w_1, \cdots, w_m\}$ of non-singular words of x-length at most 2 is Kervaire. In addition Howie [H2] has shown that any non-singular set of two words of x-length at most 3 is Kervaire.

In addition to the results of Neumann and Levin mentioned above there are several special cases of Kervaire words with $n = m = 1$. For instance Edjvet [EM1,EM2] shows that ax^kdx^{-l} is Kervaire, provided that $k \neq l$ and $(|a|, |d|) \neq (2, 3)$. Edjvet and Howie [EH] show that if w is a non-singular word in $G * F(x)$ of x-length at most 4 then w is Kervaire.

The Kervaire conjecture remains unsettled. However even if it proves to be true it doesn't tell the whole story. For example let $H = <G, x : x^{-1}ax = b>$, where a and b are elements of G of equal order. Then G embeds in H in spite of the word $x^{-1}axb^{-1}$ being non-singular. (*cf.* Edjvet's results above.)

The problem can be recast geometrically ([H1]). Let K be a CW-complex with $\pi_1(K) = G$ and form a new complex as follows. Corresponding to each generator x_i of F attach a 1-cell to (the base point of) K, to give a complex K' with $\pi_1(K') = G * F$. Now attach a 2-cell α_j to K' with attaching map in the class of w_j, for $j = 1, \cdots, m$. This gives a complex L with $\pi_1(L) = (G*F)/N$. Moreover $\{w_1, \cdots, w_m\}$ is Kervaire if and only if the canonical map from $\pi_1(K)$ to $\pi_1(L)$ is an injection. The chain map $C_2(L, K) \longrightarrow C_1(L, K)$, is given, with respect to the obvious bases, by $(d_{i,j})$, so $\{w_1, \cdots, w_m\}$ is non-singular if and only if $H_2(L, K) = 0$.

Conversely given a relatively 2-dimensional pair of 2-complexes (K, L) then $(L/K)^{(1)}$ generates a free group F and the 2-cells of $L\backslash K$ determine a set of words $\{w_1, \cdots, w_m\}$ in $\pi_1(K) * F$. In this case we say that the system of equations $w_1 = 1, \cdots, w_m = 1$ is *realized* by (K, L). If X is a CW-complex and (K,L) is a relatively 2-dimensional pair of 2-complexes with $L/K = X$ then we say that the system of equations realized by (K, L) is *modelled* on X. A 2-complex X is said to be *Kervaire* if Property A holds for every system of equations modelled on X.

The 3-Torus

The 3-dimensional torus $S^1 \times S^1 \times S^1$ has a canonical cell-structure, with one cell in dimensions 0 and 3, and three cells in dimensions 1 and 2. Its 2-skeleton is thus the geometric realization of the group presentation

$$< X, Y, Z \,|\, XYX^{-1}Y^{-1}, YZY^{-1}Z^{-1}, ZXZ^{-1}X^{-1} > .$$

We will show that this 2-complex K is Kervaire. This was proved by Gersten [G2] in the case where G is torsion-free.

Fix a group G, and let $a, b, c, d, e, f, g, h, i, j, k, l$ be elements of G. We must find a group H, containing G as a subgroup, and elements $x, y, z \in H$, such that

$$axbycx^{-1}dy^{-1} = eyfzgy^{-1}hz^{-1} = izjxkz^{-1}lx^{-1} = 1 \text{ in } H.$$

We distinguish several cases, according to the values in G of the eight elements $\alpha = eai$, $\beta = lbe^{-1}$, $\gamma = f^{-1}cl^{-1}$, $\delta = i^{-1}df$, $\alpha' = h^{-1}aj^{-1}$, $\beta' = k^{-1}bh$, $\gamma' = gck$, and $\delta' = jdg^{-1}$, which we will call the *corner labels*. The *star graph* of the system (see [BP,EH,G2]) is the 1-skeleton of the octahedron (the dual of the cube by which the 3-cell of the torus is attached), and these eight elements are the labels of the eight triangular faces (or, equivalently, the corner labels of the eight corners of the cube). We associate them with the corners of the cube according to the scheme depicted in Figure 1. In general, some or all of these elements may be trivial in G. The different cases we consider reflect the numbers of different nontrivial corner labels.

Theorem 1. *Suppose there is a pair of opposite faces of the cube of Figure 1, such that two or more of the corner labels on each of these faces are nontrivial elements of G. Then the system of equations has a solution in an overgroup of G.*

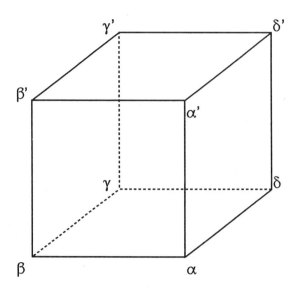

FIGURE 1.

Corollary. *If five or more of the eight corner labels are nontrivial elements of G, then the system of equations has a solution in an overgroup of G.*

Proof. If five corners of a cube are coloured, then it is possible to choose a pair of opposite faces, each of which has at least two coloured corners (try it!).

Proof of Theorem 1. We construct the solution group H of Theorem 1 in a series of HNN extensions beginning from G. Before beginning let us fix once and for all a pair of opposite faces of the cube. Without loss of generality, we choose the top and bottom face (as in Figure 1). Thus we are assuming that at least two of $\alpha, \beta, \gamma, \delta$ are nontrivial elements of G, and similarly for $\alpha', \beta', \gamma', \delta'$. Let $G_0 = G * <x>$, the HNN extension of G with trivial associated subgroup. Note that each of the elements $axb, cx^{-1}d$ has infinite order in G_0, so we may form the HNN extension

$$G_1 = <G_0, y \mid (axb)y(cx^{-1}d)y^{-1}>$$

of G_0.

Lemma 2. *If at least two of $\alpha, \beta, \gamma, \delta$ are nontrivial elements of G, and similarly for $\alpha', \beta', \gamma', \delta'$, then the elements $eyf, lx^{-1}i$ form a basis for a free subgroup of rank 2 in G_1, and similarly for $gy^{-1}h, jxk$.*

We then form the HNN extension

$$H = <G_1, z \mid (eyf)z(gy^{-1}h)z^{-1}, (lx^{-1}i)z(jxk)z^{-1}>$$

of G_1, using Lemma 2 to verify that the associated subgroups are each free of rank 2. It is clear that G is a subgroup of G_0, hence of G_1 and hence of H, and that the three required equations hold in H.

Proof of Lemma 2. Set $s = lx^{-1}i \in G_0$, and $t = eyf \in G_1$. Then we may regard G_1 as the HNN extension

$$< G_0, t \mid (\alpha s^{-1}\beta)t(\gamma s\delta)t^{-1} >$$

with stable letter t. An easy argument, involving the normal form for words in an HNN extension, shows that s, t generate a free subgroup of G_1 of rank 2 provided the subgroup $< s >$ has trivial intersection with each of the associated subgroups $< \alpha s^{-1}\beta >$, $< \gamma s\delta >$, and from the free product structure of G_0 as $G * < s >$ it is clear that this holds provided at least one of α, β is nontrivial in G, and at least one of γ, δ is nontrivial in G. By hypothesis at least two of $\alpha, \beta, \gamma, \delta$ are nontrivial, so we must consider the case where either $\alpha = 1 = \beta$, $\gamma \neq 1 \neq \delta$, or $\alpha \neq 1 \neq \beta$, $\gamma = 1 = \delta$. But then it again follows that s, t generate a free subgroup of rank 2 by a similar argument, this time regarding G_1 as an HNN extension

$$< (G * < t >), s \mid (\delta t^{-1}\alpha)s^{-1}(\beta t\gamma)s > .$$

The argument to show that $jxk, gy^{-1}h$ generate a free rank 2 subgroup of G_1 is entirely analogous, and we omit it.

Theorem 3. *If some face of the cube in Figure 1 has all four corner labels trivial in G, then the system of equations has a solution in an overgroup of G.*

Proof. Without loss of generality we assume that the hypothesis of the Theorem applies to the top face of the cube, in other words that $\alpha' = \beta' = \gamma' = \delta' = 1$ in G. Define
$$G_2 = < G, x, y \mid [jxk, gy^{-1}h] = [lx^{-1}i, eyf] = 1 > .$$

Then G_2 contains G as a subgroup, because the natural map $G \to G_2$ is a split injection. A splitting can be defined, for example, by $x \mapsto il$, $y \mapsto hg$.

If N is the normal closure of G in G_2, then $G_2/N \cong \mathbf{Z}^2$, with basis $\{xN, yN\}$. It follows easily that the subgroups of G_2 generated by $\{jxk, gy^{-1}h\}$ and $\{lx^{-1}i, eyf\}$ respectively are each free abelian on the given sets of generators. We can thus form the HNN extension

$$H = < G_2, z \mid z(jxk)z^{-1}(lx^{-1}i), z(gy^{-1}h)z^{-1}(eyf) >,$$

which contains G_2, and hence G, as a subgroup.

Note that we have the equation

$$axbycx^{-1}dy^{-1} = (hj)x(kh^{-1})y(g^{-1}k^{-1})x^{-1}(j^{-1}g)y^{-1}$$
$$= 1$$

in G_2, and hence in H, using the identities $\alpha' = \beta' = \gamma' = \delta' = 1$ and the first defining relation of G_2. The other two desired equations also hold in H by the defining relations of H as an HNN extension.

Theorems 1 and 3 take care of most possible combinations of trivial and nontrivial corner labels, but unfortunately not all. To complete the proof we need to introduce a technique of changing variables. We can do this whenever there are two adjacent trivial corner labels on the cube, and the result is to change the system to a new one in which the corner labels are the same elements of G, but differently distributed around the cube.

There are twelve different ways in which we can have two adjacent trivial corner labels (one for each edge of the cube). For each of these there are two possible changes of variable (one for each face incident at that edge). There are thus twenty-four potential ways in which this change of variable can take place. Although the effects of these twenty-four moves are different, the idea in each case is the same, so we shall describe only one of them, by way of illustration. We shall then invoke cubical symmetry to show that all twenty-four possible changes of variable are permissible.

Let us suppose that $\gamma = \delta = 1$ in G, in other words that $cl^{-1} = f = d^{-1}i$. In this case we shall replace z by a new variable $w = yfz$, noting that the equation $eyfzgy^{-1}hz^{-1} = 1$ becomes $ewgy^{-1}hw^{-1}yf = 1$ in terms of this new variable. Similarly, in the presence of the equation $axbycx^{-1}dy^{-1} = 1$, the equation $izjxkz^{-1}lx^{-1} = 1$ is equivalent to $axbycl^{-1}zk^{-1}x^{-1}j^{-1}(yd^{-1}iz)^{-1} = 1$, or (using the identities $ycl^{-1}z = w = yd^{-1}iz$) to $axbwk^{-1}x^{-1}j^{-1}w^{-1} = 1$.

Hence we may replace our original set of three equations in x, y, z by a new set of equations

$$axbycx^{-1}dy^{-1} = yfewgy^{-1}hw^{-1} = a^{-1}wjxkw^{-1}b^{-1}x^{-1} = 1,$$

(also modelled on K) in x, y, w, with corner labels cyclic conjugates of

$$\{1, 1, \beta, \alpha, \alpha', \beta', \gamma', \delta'\}$$

in place of the labels

$$\{\alpha, \beta, 1, 1, \alpha', \beta', \gamma', \delta'\}$$

of Figure 1. In other words, we have essentially interchanged the edge that had two trivial corner labels with one of its nearest parallels.

Now let F be any face of the cube and let E_1 and E_2 be parallel edges incident at F. Suppose that E_1 has trivial corner labels and that we wish to change variables so as to interchange E_1 and E_2. We choose a symmetry σ of the cube taking E_1 to $\gamma\delta$ and E_2 to $\beta\alpha$. The above argument (adapted to the new labels) at α, β, γ and δ is used to interchange $\beta\alpha$ and $\gamma\delta$. The symmetry σ^{-1} restores the original labelling with E_1 and E_2 interchanged.

Theorem 4. *The standard 2-skeleton K of the 3-torus is Kervaire.*

Proof. By the Corollary to Theorem 1, we may assume that we have at most four nontrivial corner labels. Moreover, if there are precisely four, then by Theorem 1 the result still follows unless there is no pair of parallel faces, each containing two of the nontrivial corner labels. Given that precisely four corner labels are non-trivial assume, for the sake of argument that α' is non-trivial. Then if $\beta' = \delta' = \alpha = 1$, either $\beta = 1$, in which case the top and bottom faces each have two non-trivial labels, or $\beta \neq 1$ in which case the front and back faces each have two non-trivial labels. Hence we may assume the non-trivial corner labels occur in pairs connected by edges. It then follows from Theorem 1, again, that we may assume that the four nontrivial labels are distributed at one vertex and its three neighbours, say $\alpha, \alpha', \beta', \delta'$. But then a change of variable move as described above changes our system to one where the four nontrivial labels are distributed as $\alpha', \beta', \delta', \delta$, and the result again follows from Theorem 1.

Suppose next that precisely three corner labels are nontrivial. If they all belong to a common face of the cube, we may apply Theorem 3, so suppose that this does not happen. If two of the nontrivial labels are at adjacent corners, then we may assume without loss of generality that the three nontrivial labels are α, β, δ'. Then the change of variable move gives us a system to which we may apply Theorem 3. Finally, if no two of the three labels lie at adjacent corners, we may assume they are α, β', δ'. The change of variable move reduces us to the previous case, where two nontrivial labels are adjacent.

Finally, suppose at most two labels are nontrivial. Then Theorem 3 applies except in the case where these labels are at opposite corners, say α, γ'. Apply the change of variable move once again to produce a system with two nontrivial labels on a common face, and then apply Theorem 3.

Remarks. In the case of no nontrivial corner labels, the solution to the system is trivial. As pointed out by Gersten [G2], if one of the corner labels is nontrivial, then so is at least one more. If precisely two labels are nontrivial, then standard arguments reduce us to the case where G is cyclic. In this case, an alternative argument can be given: construct a central extension of G by $\mathbf{Z}^3$ in which the equations hold, namely that corresponding to the 2-cocycle $(abcd, efgh, ijkl) \in G^3 \cong H^2(\mathbf{Z}^3, G)$.

REFERENCES

[BP] W. Bogley and S. J. Pride, *Aspherical relative presentations*, Proc. Edinburgh Math. Soc. **35** (1992), 1–39.

[B] A. Dold and B. Eckmann (eds), *Proceedings of the Second International Conference on the Theory of Groups, Australian National University, Canberra, 1973*, Lecture Notes in Mathematics, 372, Springer, Berlin; New York, 1974.

[E1] M. Edjvet, *The solution of certain sets of equations over groups*, Groups-St. Andrews, 1989 (C.M. Campbell and E.F. Robinson, eds.), London Math. Soc. Lecture Notes Series, vol. 159, Cambridge Univ. Press, 1991, pp. 105–123.

[E2] M.Edjvet, *Equations over Groups and a Theorem of Higman, Neumann and Neumann*, Proc. London Math. Soc. (3) **62** (1991), 563–589.

[EH] M. Edjvet and J. Howie, *The solution of length four equations over groups*, Trans. Amer. Math. Soc. **326** (1991), 345–369.

[G1] S.M. Gersten, *Nonsingular Equations of small weight over groups*, Combinatorial Group Theory and Topology (S.M. Gersten and J.R. Stallings, eds.), Ann. of Math. Studies, vol. 111, Princeton Univ. Press, 1987, pp. 121–144.

[G2] S. M. Gersten, *On certain equations over torsion free groups*, Preprint.

[GR] M. Gerstenhaber and O.S. Rothaus, *The solution of Sets of Equations in Groups*, Proc. Nat. Acad. Sci. U.S.A. **48** (1962), 1531–1533.

[H1] J. Howie, *On pairs of 2-Complexes and Systems of Equations over Groups*, J. Reine. Angew. Math. **324** (1981), 165-174.

[H2] J. Howie, *Nonsingular systems of two length three equations over a group*, Math. Proc. Camb. Phil. Soc. **110** (1991), 11-24.

[K] M.A. Kervaire, *On higher dimensional knots*, Differential and Combinatorial Topology (S.S. Cairns, ed.), A symposium in Honor of Marston Morse, Princeton Mathematical Series, vol. 27, Princeton Univ. Press, 1965, pp. 105–119.

[L] F. Levin, *Solutions of Equations over Groups*, Bull. Amer. Math. Soc **68** (1962), 603–604.

[N] B.H.Neumann, *Adjunction of elements to groups*, Jour. London Math. Soc. **18** (1943), 4–11.

[R] O.S. Rothaus, *On the non-triviality of some group extensions given by generators and relations*, Ann. of Math. (2) **106** (1977), 599–612.

DEPARTMENT OF MATHEMATICS, HERIOT-WATT UNIVERSITY, RICCARTON, EDINBURGH EH14 4AS

Current address: DEPARTMENT OF MATHEMATICS AND STATISTICS, THE UNIVERSITY, NEWCASTLE UPON TYNE NE1 7RU

E-mail: A.Duncan@ncl.ac.uk

DEPARTMENT OF MATHEMATICS, HERIOT-WATT UNIVERSITY, RICCARTON, EDINBURGH EH14 4AS

E-mail: jim@cara.ma.hw.ac.uk

Contemporary Mathematics
Volume **164**, 1994

A Topological Interpretation of the
Atiyah-Patodi-Singer Invariant

M. FARBER AND J. LEVINE

0. Let M be a closed, connected, oriented manifold of dimension $n = 2l - 1$ and let $\pi = \pi_1(M)$. For any unitary representation $\alpha : \pi \to U_k$, Atiyah, Patodi and Singer in [APS, **II**] define a numerical invariant $\rho_\alpha(M) \in \mathbb{R}$ as follows. Choose a Riemannian structure for M and then consider the self-adjoint elliptic differential operator B_α on the space of all differential forms of even degree with values in the flat bundle defined by α by the formula $B_\alpha = i^l(-1)^{p+1}(*d_\alpha - d_\alpha*)$ on forms of degree $2p$, where d_α is the covariant derivative of the flat bundle defined by α and $*$ is the duality operator defined by the Riemannian structure. They then consider the eta-function $\eta_\alpha(s) = \sum_{\lambda \neq 0}(\text{sign}\,\lambda)|\lambda|^{-s}$, where λ runs over all nonzero eigenvalues of B_α counting multiplicities. Atiyah, Patodi and Singer show, in [**APS, I, II**], that $\eta_\alpha(s)$ defines an analytic function for $\Re(s)$ large, which can be analytically continued to have a finite value at $s = 0$. They then define $\rho_\alpha(M) = \eta_\alpha(0) - k\eta(0)$, where $\eta(s)$ is the eta-function of the trivial representation. It is an immediate consequence of their Index Theorem that $\rho_\alpha(M)$ is independent of the choice of metric and that the reduction of $\rho_\alpha(M)$ to $\mathbb{R}/\mathbb{Z}$ depends only on the oriented bordism class of M.

INDEX THEOREM ([**APS,II**]). *If M is the oriented boundary of an oriented compact Riemannian manifold V, such that the Riemannian structure on V is a product near M, and the representation α extends to a unitary representation β of $\pi_1(V)$, then*

$$\text{sign}\,_\alpha(V) = k \int_V L(p) - \eta_\alpha(0).$$

In this formula, $L(p)$ is the Hirzebruch polynomial in the Pontriagin forms of V and $\text{sign}\,_\alpha(V)$ is the signature of the intersection form on V over the twisted coefficient system defined by α.

1991 *Mathematics Subject Classification.* Primary 58G11; Secondary 57N65.

The research was partially supported by the USA - Israel Binational Science Foundation

The second author was supported in part by a NSF Grant

This paper is in final form and no version of it will be submitted for publication elsewhere.

COROLLARY. Given $M = \partial V$, with $\alpha = \beta|_M$, we have $\rho_\alpha(M) = k\mathrm{sign}\,(V) - \mathrm{sign}\,_\alpha(V)$, where $\mathrm{sign}\,(V)$ is the ordinary signature of V.

We address two problems in this work.

(1) Give an intrinsic topological definition of $\rho_\alpha(M)$.
(2) To what extent is $\rho_\alpha(M)$ an invariant of homotopy type?

Viewing $\rho_\alpha(M) = \rho(M) \cdot \alpha$ as a real-valued function on the variety of unitary representations of $\pi = \pi_1(M)$, our results consist of:

(a) a formula for the jumps in $\rho(M)$ at discontinuities, and
(b) a formula for the "differential" of $\rho(M)$, reduced mod $\mathbb{Z}$.

Both of these formulae depend only on the homotopy type of M and give an intrinsic homotopy invariant definition of $\rho(M)$, up to a locally constant function (which vanishes on the component of the trivial representation). It is known that homotopy invariance fails if π is finite [**W**] and in many cases when π has torsion [**We**].

Earlier work of Neumann [**N**] and Weinberger [**We**] showed that $\rho(M)$ is a homotopy invariant for a large class of $\pi_1(M)$.

1. For any group π we can consider the set of k-dimensional unitary representations of π, denoted $R_k(\pi)$. If π is finitely generated this set is, in a natural way, a real algebraic variety – any representation of π (with m generators) leads to an obvious manifestation of $R_k(\pi)$ as a subvariety of $U(k) \times \cdots U(k)$ (m times). It is not hard to see that this algebraic structure is independent of the presentation of π. In [**L**], [**L1**] the Atiyah-Patodi-Singer invariant $\rho_\alpha(M)$ is considered as a function $\rho(M) : R_k(\pi) \to \mathbb{R}$, where $\rho(M) \cdot \alpha = \rho_\alpha(M)$ and $\pi = \pi_1(M)$. It is shown in [**L1**] that there is a stratification of $R_k(\pi)$ by subvarieties $R_k(\pi) = \Sigma_0 \supseteq \Sigma_1 \supseteq \cdots \supseteq \Sigma_i \supseteq \ldots$ such that $\rho(M)|(\Sigma_i - \Sigma_{i+1})$ is continuous, for $i \geq 0$. Specifically, Σ_i is defined as follows. Define $d_\alpha = \sum_i \dim H_i(M; \alpha)$, where $H_i(M; \alpha)$ is homology with the twisted coefficient system defined by α. If $d = \min\{d_\alpha : \alpha \in R_k(\pi)\}$, then $\Sigma_i = \{\alpha : d_\alpha \geq d + i\}$. Note that this stratification depends only on the homotopy type of M.

We propose to study the discontinuities of $\rho(M)$. In [**APS-III**] it is shown that the reduction $\bar{\rho}_\alpha(M)$ of $\rho_\alpha(M)$ to $\mathbb{R}/\mathbb{Z}$ depends continuously on α. (In fact they give a (K-theoretic) formula for this reduction). This shows that the "jump" in $\rho(M)$ at a discontinuity is integral.

Set γ be an analytic curve in $R_k(\pi)$. Analyticity means that γ lies in some Σ_i so that it intersects Σ_{i+1} in a discrete set of points. We may assume that, for some $\epsilon > 0$ and $|t| < \epsilon$, the inclusion $\gamma(t) \in \Sigma_{i+1}$ holds if and only if $t = 0$. Then $\rho(M) \circ \gamma$ is continuous except for some (integer) jump at $t = 0$. More precisely, $\rho(M) \circ \gamma$ is continuous at $t \neq 0$ and, since $\bar{\rho}(M) \circ \gamma$ ($\bar{\rho}(M)$ is the reduction of $\rho(M)$ to $\mathbb{R}/\mathbb{Z}$) is continuous at $t = 0$, there is a well-defined limit of $\rho(M) \circ \gamma(t)$ as $t \to +0$, which agrees with $\rho(M) \circ \gamma(0) \bmod \mathbb{Z}$. We propose to find a formula for the difference.

We first interpret γ as a representation of π over P, the ring of power series with a positive radius of convergence, since the entries of the matrix $\gamma(t)$ are elements of P. We can then use γ to define, for example, a right action of π on the free P-module P^k of rank k, by regarding P^k as row vectors over P and using right multiplication by $\gamma(t)$. That each $\gamma(t)$ is unitary means that this action preserves the canonical P-valued Hermitian form on P^k. If P^k_γ denotes the right $\mathbb{C}\pi$-module defined by γ, then the conjugate left $\mathbb{C}\pi$-module $\overline{P}^k_\gamma$ (recall $\overline{P}^k_\gamma = P^k_\gamma$ with π-action defined by $g \cdot \alpha = \alpha \cdot g^{-1}$, for any $g \in \pi$, $\alpha \in P^k_\gamma$) is isomorphic to the module defined by regarding P^k as column vectors and using left multiplication by $\gamma(t)$.

We may use P^k_γ as a local coefficient system over $M(\pi = \pi_1(M))$ and define $H_*(M;\gamma) = H_*(P^k_\gamma \otimes_\pi C(\tilde{M}))$ and $H^*(M;\gamma) = H_*(\mathrm{Hom}\,_\pi(C(\tilde{M}), \overline{P}^k_\gamma))$. Now Poincaré duality applied to M shows that the intersection pairing over $\mathbb{Z}\pi$ on $\tilde{M}$ induces an isomorphism:

$$\overline{H_i(M;A)} \approx H^{n-i}(M;\bar{A}), \qquad 0 \le i \le n,$$

where A is any $(R, \mathbb{Z}\pi)$-bimodule (R any ring with involution) and, generally, $A \to \bar{A}$ denotes the usual passage from $(R, \mathbb{Z}\pi)$-bimodules to $(\mathbb{Z}\pi, R)$-bimodules. The duality isomorphism is one of right R-modules. We apply duality with $A = \bar{P}^k_\gamma$, $R = P$ with involution defined by complex conjugation.

Since P is a discrete valuation ring with fundamental ideal generated by t, we see that $H_i(M;\gamma)$ is determined by its rank r_i over P and its torsion submodule $T_i(M;\gamma)$. The universal coefficient theorem shows that $H^j(M;\gamma)$ has rank r_j and its torsion-module is the "dual module" of $T_{j-1}(M;\gamma)$ – if T is a (left) torsion P-module then the dual module is $T^* = \mathrm{Hom}\,_P(T; \hat{P}/P)$, where $\hat{P}$ is the quotient field of P, a (right) P-module. Now Poincaré duality tells us that $r_i = r_{n-i}$, $\bar{T}_i(M;\gamma) \approx T_{n-i-1}(M;\gamma)^*$ and, furthermore, there is a *non-singular*, sesquilinear, $\pm$Hermitian pairing $\langle\ ,\ \rangle$:

$$T_q(M;\gamma) \times T_q(M;\gamma) \to \hat{P}/P \quad (n = 2q+1) \quad \text{with}$$
$$\langle \lambda\alpha, \beta \rangle = \lambda\langle \alpha, \beta \rangle, \ \langle \alpha, \beta \rangle = \pm\overline{\langle \beta, \alpha \rangle}.$$

Now, non-singular, sesquilinear, $\pm$Hermitian pairings of a torsion P-module T can be classified by a collection of signatures. Specifically, let $\langle\ ,\ \rangle : T \times T \to \hat{P}/P$ be such a pairing and define $\Delta_i(T)$ to be quotient $\ker t^i/(t \ker t^{i+1} + \ker t^{i-1})$. It is easy to check that $\langle\ ,\ \rangle$ induces a non-singular, bilinear, $\pm$-Hermitian pairing (over $\mathbb{C}$) $\langle\ ,\ \rangle_i \colon \Delta_i(T) \times \Delta_i(T) \to \mathbb{C}$ by the formula:

$$t^i \langle \alpha, \beta \rangle_i \equiv \langle [\alpha], [\beta] \rangle \qquad \mathrm{mod}\, tP.$$

Note $\langle \alpha, \beta \rangle \in t^{-i}P/P$. Now we can define $\sigma_i(\langle\ ,\ \rangle) = \mathrm{sign}\,(\langle\ ,\ \rangle_i)$ and it is not hard to prove:

PROPOSITION 1. *Suppose that* $\langle \, , \, \rangle$ *and* $\langle \, , \, \rangle'$ *are two non-singular, sesquilinear, ϵ-Hermitian pairings ($\epsilon = \pm 1$) defined on the same torsion P-module T as above. Then* $\langle \, , \, \rangle$ *and* $\langle \, , \, \rangle'$ *are congruent if and only if* $\sigma_i(\langle \, , \, \rangle) = \sigma_i(\langle \, , \, \rangle')$ *for all i.*

Now, we go back to our analytic curve γ in $R_k(\pi)$ and define $\sigma_i(M;\gamma) = \sigma_i(\langle \, , \, \rangle)$, where $\langle \, , \, \rangle$ is the pairing defined above on $T_q(M;\gamma)$.

Our main result is:

THEOREM 1.

$$\lim_{t\to+0} \rho(M) \cdot \gamma(t) - \rho(M) \cdot \gamma(0) = \sum_{i=1}^{\infty} \sigma_i(M;\gamma).$$

The first order part of the linking pairing described above was considered in [KK]; cf. also [KK1], where the first order part of Theorem 1 is proven in the case of manifolds with boundary. In [KK1] there is a discussion of an alternative version of the higher order part of Theorem 1 which would use cup products and higher Massey products.

We remark that it is a consequence of the *curve selection lemma* that the discontinuities of $\rho(M)$ along analytic curves determine $\rho(M)$ up to a continuous function on $R_k(\pi)$. More explicitly, suppose we define:

$$\sigma(M;\gamma) = \lim_{t\to+0} \rho(M) \cdot \gamma(t) - \rho(M) \cdot \gamma(0)$$

where γ is an analytic curve in $R_k(\pi)$ defined on a neighborhood of 0. In fact, if ρ is any real-valued function on $R_k(\pi)$ whose reduction mod $\mathbb{Z}$ is continuous and which is "piecewise-continuous" in the sense that, for some stratification of $R_k(\rho)$ by subvarieties $R_k(\pi) = \Sigma_0 \supseteq \Sigma_1 \supseteq \cdots \supseteq \Sigma_i \supseteq \ldots$, the function $\rho(M)|(\Sigma_i - \Sigma_{i+1})$ is continuous for all $i \geq 0$, then we can define $\sigma(\rho,\gamma)$ by the above formula. Then we have:

PROPOSITION 2. *If ρ_1, ρ_2 are two piecewise-continuous real-valued functions on $R_k(\pi)$ which are continuous $\mathrm{mod}\,\mathbb{Z}$, then $\rho_1 - \rho_2$ is continuous if and only if $\sigma(\rho_1,\gamma) = \sigma(\rho_2,\gamma)$ for every analytic curve γ in $R_k(\pi)$.*

The proof of Theorem 1 begins with a general formula for the spectral jump at $t = 0$ of a path A_t of elliptic differential operators in terms of the associated signatures of a "linking" pairing on a torsion P-module, with values in $\hat{P}/P$, defined directly from A_t. Using a parametrized Hodge decomposition [K: Ch. VII, Th. 3.9] it is then shown that these signatures, for $A_t = B_{\gamma(t)}$, coincide with the signatures of the "topological" linking pairing $\langle \, , \, \rangle$. The details will appear in a future paper [FL].

Originally we proved the formula of Theorem 1 under the rather stringent hypothesis that $\gamma(t)$ extends to an analytic path in $R_k(\pi_1(V))$ for some compact oriented manifold V bounded by M. We then use the Index Theorem to give a purely topological proof as follows. We put ourselves in a more general context.

Let (V, M) be an algebraic Poincaré (AP) pair of finite type over the ring P, in the sense of Miščenko [**M**], with dimension $V = 2q + 2$. Then $H_q(M)$ supports a non-singular torsion pairing, on its torsion submodule, with value in $\hat{P}/P$, and we can define $\sigma_i(M)$ to be the signatures of the associated $\pm$Hermitian forms, as above. For any $\epsilon \geq 0$, let P_ϵ be the subring of P consisting of all power series with radius of convergence $> \epsilon$. We may assume that (V, M) comes from an AP-pair over P_ϵ, for some $\epsilon > 0$. Then we can define $(V_c, M_c) = (V, M) \otimes_c \mathbb{C}$, for $0 \leq c \leq \epsilon$, where $\mathbb{C}$ is regarded as a P_ϵ-module via the ring homomorphism $P_\epsilon \to \mathbb{C}$ defined by $f \mapsto f(c)$, and let $\sigma_c(V)$ be the signature of the intersection pairing on V_c. In fact, it is not hard to see that $\sigma_c(V)$ is constant for c in the interval $(0, \delta)$, for some $0 < \delta \leq \epsilon$ – let us denote this constant value by $\sigma_+(V)$. Then, our general result is $\sigma_+(V) - \sigma_0(V) = \sum_{i \geq 1} \sigma_i(M)$.

To prove this, first consider the special case in which $H_i(V) = 0$ for $i \leq q$. Then $H_{q+1}(V)$ is a free module and the intersection pairing is represented by a matrix B which can be decomposed into a block sum of matrices of the form $t^i B_i(t)$, where $B_i(0)$ is non-singular over $\mathbb{C}$, plus a 0 matrix. Thus for $0 < c \leq \epsilon$, $\sigma_c(V) = \sum_{i \geq 1} \operatorname{sign} B_i(c)$ and $\sigma_0(V) = \operatorname{sign} B_0(0)$. If $B_i(c)$ is non-singular in an interval $0 \leq c < \delta$, then, in this interval, $\sigma_c(V) = \sum_{i \geq 1} \operatorname{sign} B_i(0)$.

We now turn to the torsion pairing on $H_q(M)$. It is a standard fact, in this situation, that a matrix representative B of the intersection pairing on $H_{q+1}(V)$ is also a presentation matrix for $H_q(M)$ and the inverse of the non-degenerate part of B represents the torsion-pairing. Thus we see that $t^i B(t)$ presents the free P/t^i summand of $H_q(M)$ and, in addition, the associated Hermitian pairing on $\Delta_i(TH_q(M))$ is represented by the matrix $B_i(0)^{-1}$, for $i \geq 1$. Putting all these observations together gives the desired result.

For the general case we reduce to the special case by doing algebraic surgery on V as in [**M**]. It is not hard to see that we may kill all the homology of V of dimension $< q$ without changing M, but it is necessary to check that $\sigma_+(V)$ and $\sigma_0(V)$ are not changed by these surgeries. In fact, the only surgeries which change $H_{q+1}(V)$ are those to kill $H_q(V)$. When a class $\alpha \in H_q(V)$ is killed by a surgery then the effect on $H_{q+1}(V)$ and the intersection pairing are as follows:

Case 1. *α is a torsion class of $H_q(V)$ – say $t^r \alpha = 0$; then a rank 2 orthogonal summand is added to $H_q(V)$ with the intersection pairing on this new summand represented by the matrix $\begin{bmatrix} 0 & t^r \\ \pm t^r & 0 \end{bmatrix}$.*

Case 2. *α has "infinite order" in $H_q(V)$, but its image in $H_q(V, M)$ is torsion: then the rank of $H_{q+1}(V)$ is increased by one but the new element is totally isotropic, i.e. its intersection with all elements of $H_{q+1}(V)$ is zero.*

Case 3. *The image of α in $H_q(V, M)$ has infinite order: then $H_{q+1}(V)$ is unchanged.*

But in all three cases neither $\sigma_+(V)$ nor $\sigma_0(V)$ is changed and so the theorem follows.

2. We now study the reduction to $\mathbb{R}/\mathbb{Z}$ of $\rho(M)$. Denote by $\bar{\rho}(M): R_k(\pi) \to \mathbb{R}/\mathbb{Z}$ the function $\bar{\rho}(M) \cdot \alpha = \bar{\rho}_\alpha(M)$. As remarked above, it is proved in [**APS, III**] that $\bar{\rho}(M)$ is continuous. Moreover, they obtain a formula for $\bar{\rho}(M)$ as follows. For any unitary representation α of a discrete group Γ, there is associated an element $\beta(\alpha) \in K^{-1}(B_\Gamma; \mathbb{R}/\mathbb{Z})$, where B_Γ is the classifying space of Γ. Then, if $\alpha \in R_k(\pi)$ and $\phi: M \to B_\pi$ is the classifying map for $\pi \approx \pi_1(M)$, we have $\phi^* \beta(\alpha) \in K^{-1}(M; \mathbb{R}/\mathbb{Z})$. Now let $\sigma \in K^1(\tau M)$ be the "self-adjoint symbol" of the signature operator on M (τM is the Thom space of the cotangent bundle). The "index theorem for flat bundles" of [**APS, III**] then asserts that $\bar{\rho}_\alpha(M) = \phi^* \beta(\alpha) \cdot \sigma$, using the product

$$K^{-1}(M; \mathbb{R}/\mathbb{Z}) \times K^1(\tau M) \to K^0(\tau M; \mathbb{R}/\mathbb{Z}) \overset{\text{ind}}{\to} \mathbb{R}/\mathbb{Z}.$$

Using the ideas of this general result we can obtain a more explicit, though less definitive, determination of $\bar{\rho}(M)$. If $\alpha \in R_k(\pi)$ then let $\det \alpha \in R_1(\pi)$ be the obvious representation $(\det \alpha)(g) = \det \alpha(g)$. We can then define $\arg \det \alpha \in H^1(M; \mathbb{R}/\mathbb{Z})$ by the formula $\det \alpha(g) = \exp(2\pi i (\arg \det \alpha)(g))$, for any $g \in \pi$. We now define $\tilde{\rho}(M): R_k(\pi) \to \mathbb{R}/\mathbb{Z}$ by the cohomological formula:

$$\tilde{\rho}(M) \cdot \alpha = -(2(\arg \det \alpha) \cup L(M))[M]$$

where $L(M)$ is the Hirzebruch L-polynomial in the Pontriagin classes of M, defined by the generating series $x/\tanh(x)$. This makes sense, since the component of $L(M)$ in H^{2l-2} lifts to an integral class [**No**], but depends on the particular lift (here we assume that $\dim M = 2l - 1$). However, we note that changing the lift will only change $\tilde{\rho}(M)$ by a locally constant $\mathbb{Q}/\mathbb{Z}$-valued function on $R_k(\pi)$. An alternative definition of $\tilde{\rho}(M)$ goes as follows. Chooose a basis $z'_1, \ldots, z'_m$ of $H^1(M; \mathbb{Z})$ and $z_1, \ldots, z_m \in H_1(M; \mathbb{Z})$ such that $z'_i \cdot z_j = \delta_{ij}$. Let τ_i be the signature of an oriented closed submanifold of M representing the Poincaré dual of z'_i in $H_{2l-2}(M; \mathbb{Z})$. Then, up to a locally constant function on $R_k(\pi)$:

$$\tilde{\rho}(M) \cdot \alpha = -2 \sum_{i=1}^{m} \tau_i \arg \det \alpha(z_i)$$

See [**L1**] for a special case.

THEOREM 2. $\bar{\rho}(M) - \tilde{\rho}(M)$ *is constant on connected components of* $R_k(\pi)$. *For example, we have* $\bar{\rho}(M) = \tilde{\rho}(M)$ *on the component of the trivial representation.*

SKETCH OF THE PROOF. Suppose α_t ($0 \le t \le 1$) is a path in $R_k(\pi)$. Then we can associate to α_t a "Hermitian" bundle ξ (i.e. unitary bundle with a connexion) over $I \times M$, such that on $t \times M$ the induced connexion is flat and has monodromy α_t. The curvature form ξ has the form $\Omega = dt \wedge \omega$, where ω is a 1-form on $I \times M$ with coefficients in $\text{Hom}(\xi, \xi)$. Now the Index Theorem of [**APS, I**] applied to the generalized signature operator D_ξ on $I \times M$ with coefficient in ξ gives the formula:

$$\text{Index } D_\xi = \int_{I \times M} 2^l \text{ch}\, \xi \cdot \mathcal{L}(I \times M) - (\eta_{\alpha_1}(M) - \eta_{\alpha_0}(M))$$

where $\mathcal{L}$ is the Hirzebruch form with generating series: $\frac{x/2}{\tanh(x/2)}$, $\dim(I \times M) = 2l$ and $\text{ch}\, \xi$ is the Chern character form of ξ. But this form reduces to $k + \frac{1}{2\pi i}\text{Trace}\,(\Omega)$, since $\Omega = dt \wedge \omega$, and so we can derive from the Index Theorem the equation:

$$\rho_{\alpha_1}(M) - \rho_{\alpha_0}(M) \equiv \frac{2^{l-1}}{\pi i} \int_{I \times M} dt \wedge \text{Trace}\,(\omega) \wedge \mathcal{L}(M) \qquad \text{mod } \mathbb{Z}$$

Now the 1-form $\text{Trace}(\omega)$ on $I \times M$ defines a 1-parameter family of cohomology classes $(\text{Tr}\,\omega)_t \in H^1(M; \mathbb{R})$ and so we have

$$\rho_{\alpha_1}(M) - \rho_{\alpha_0}(M) \equiv \frac{-2^{l-1}}{\pi i} \left[\left[\int_0^1 (\text{Tr}\,\omega)_t dt \right] \cup \mathcal{L}(M) \right] \cdot [M] \quad \text{mod } \mathbb{Z}$$

where we use $\mathcal{L}(M)$ as above to denote the Hirzebruch polynomial in the Pontriagin classes of M. We now use the fact that

$$(\text{Tr}\,\omega)_t(g) = \text{Trace}\,\left(\frac{d\alpha_t}{dt}(g) \circ \alpha_t^{-1}(g) \right) = \frac{d}{dt}\left(\log \det \alpha_t(g) \right)$$

to obtain our final formula:

$$\rho_{\alpha_1}(M) - \rho_{\alpha_0}(M) \equiv -2((\arg \det \alpha_1(g) - \arg \det \alpha_0(g)) \cup L(M)) \cdot [M]$$

which implies the Theorem.

A formula for $\rho(M) \mod \mathbb{Q}$, in terms of Cheeger-Chern-Simons classes is given in [**CS**], Corollary 9.3.

3. We now discuss the implication of these Theorems for the question of the homotopy invariance of the ρ-invariant. Suppose M, M' are homotopy equivalent manifolds (odd-dimensional closed, oriented) and $\pi_1(M) \approx \pi_1(M')$ identified by the homotopy equivalence. Let $\Delta(M, M') = \rho(M) - \rho(M') : R_k(\pi) \to \mathbb{R}$. Now $\tilde{\rho}(M) = \tilde{\rho}(M')$ (by Novikov [**No**]) and so, by Theorem 2, $\Delta(M, M')$ reduced mod $\mathbb{Z}$ is constant on each component of $R_k(\pi)$. Furthermore, it is clear that $\sigma_i(M, \gamma) = \sigma_i(M', \gamma)$, for any analytic curve γ in $R_k(\pi)$, and so it follows from Theorem 1 and Proposition 2 that $\Delta(M, M')$ is continuous. Now putting these together we conclude that $\Delta(M, M')$ is constant on each component of $R_k(G)$ (0 at the trivial representation). (Weinberger [**We**]) proves this under the assumption that M and M' are rationally cobordant over G or, alternatively, whenever G satisfies the Novikov conjecture).

The analysis of homotopy lens spaces in Wall [**W**] gives many examples where $\Delta(M, M')$ takes non-zero rational values when π is finite cyclic. Weinberger shows in [**We**] that $\Delta(M, M')$ is rational if π satisfies the Novikov conjecture, and has now announced a proof that $\Delta(M, M')$ is always rational, using the results of this paper.

REFERENCES

[APS, I,II,III] M. Atiyah, V. Patodi, I. Singer, *Spectral asymmetry and Riemannian geometry*, Math. Proc. Camb. Phil. Soc. (I) **77** (1975a), pp. 43–69; (II) **78** (1975b), pp. 405–432; (III) **79** (1976), pp. 71–99.

[CS]   J. Cheeger, J. Simons, *Differential characters and geometric invariants*, Geometry and Topology, Proceedings of the Special year , 1983-84, University of Maryland, Lecture Notes in Mathematics, vol. 1167, Springer-Verlag, 1985, pp. 50-80.

[FL]   M. Farber, J. Levine, *Deformations of the Atiyah-Patodi- Singer eta-invariant*, Preprint.

[K]   T. Kato, *Perturbation Theory for Linear Operators*, Grund. der Math. Wissenschaften, Bd. 132, Springer-Verlag, Heidelberg, 1966.

[KK]   P. Kirk, E. Klassen, *Computing spectral flow via cap products*, J. of Differential Geometry, to appear.

[KK1]   P. Kirk, E. Klassen, *The spectral flow of the odd signature operator on a manifold with boundary*, Preprint.

[L]   J. Levine, *Signature invariants of homology bordism with applications to links*, Knots 90: Proceedings of the Osaka knot theory conference.

[L1]   J. Levine, *Link invariants via the eta invariant*, Commentarii Math. Helv., to appear.

[M]   A. Miščenko, *Homotopy invariants of simply-connected manifolds I. Rational invariants*, Izv. Akad. Nauk SSSR Ser. Mat. **34** (1970), 501–514; English transl. in Math. USSR, Izv. **4** (1970), 506–519.

[N]   W. Neumann, *Signature-related invariants of manifolds – I. Monodromy and γ-invariants*, Topology **18** (1979), 147–172.

[No]   S. Novikov, *Manifolds with free abelian fundamental groups and their applications*, Izv. Akad. Nauk SSSR Ser. Mat. **30** (1966), 207–246; English transl. in Amer. Math. Soc. Transl. **(2) 71** (1968), 1–42.

[W]   C.T.C. Wall, *Surgery on Compact Manifolds*, Academic Press, New York, 1971.

[We]   S. Weinberger, *Homotopy invariance of η-invariants*, Proc. Nat. Acad. Sci. USA **85** (1988), 5362–5363.

DEPARTMENT OF MATHEMATICS, TEL AVIV UNIVERSITY, TEL AVIV, RAMAT AVIV, ISRAEL 69978

DEPARTMENT OF MATHEMATICS, BRANDEIS UNIVERSITY, WALTHAM, 02254, MA, USA

E-mail address: farber@math.tau.ac.il,   levine@binah.cc.brandeis.edu

Contemporary Mathematics
Volume **164**, 1994

ON GROUPS OF
SPECIAL NEC TYPE

A FREIHEITSSATZ AND RELATED

RESULTS FOR A CLASS OF MULTI-RELATOR GROUPS

BENJAMIN FINE AND GERHARD ROSENBERGER

1. Introduction. The **Freiheitssatz** (FHS) or **independence theorem** for one-relator groups says that, if $G = < x_1, ..., x_n; R >$ and R is a cyclically reduced word involving all the generators, then the free group $< x_1, ..., x_{n-1}; >$ injects into G [19]. Much recent work has been concerned with extensions of this classical result of Magnus. J. Howie [13],[14] has proven a Freiheitssatz for one-relator products where the relator is a proper power of order 4 or greater. Fine, Howie and Rosenberger [7], using representation techniques, proved a Freiheitssatz for one-relator products of cyclics where the relator has order 2 or greater. As a consequence of both this theorem and the representation techniques it was shown that the class of one-relator products of cyclics share many properties with linear groups, such as the Tits Alternative, virtual torsion-freeness and SQ-universality even though they themselves may not be linear,[8],[10],[11],[16]. These results were further extended to

---

Mathematics 1991 Subject Classification Code: Primary 20F05,20F10, Secondary 20F32,20H15 : Keywords : Fuchsian Group, NEC Group, SQ-Universal

This is a final version of this paper.

groups of F-type which are abstract algebraic generalizations of Fuchsian groups [9].

More recently I. Anshel proved a FHS for a class of two-relator groups where the generators and relators satisfy an "independence condition" [1]. Her results were generalized by B.Bogley and S. Pride to groups with semi-staggered presentations [4],[5],[20]. H. Short in a private communication has indicated a further generalization of the Anshel result [27].

In the present paper we consider a class of groups which arise in the study of NEC groups. We call the groups in this class **groups of SN-type** for **Special NEC-type**. A precise formulation will be given in the next section but, concisely, they can be described as iterated amalgams of one-relator products of cyclics. We show that a natural FHS exists for this class and that these groups admit essential representations into $PSL_2(\mathbb{C})$. As a consequence of these representations we obtain results on linearity properties for groups in this class analagous to those on one-relator products of cyclics. In particular we show that any group in this class is either SQ-universal or infinite solvable.

The class of groups of SN-type contains as a subclass a collection of special NEC groups studied by Zieschang and Kaufman [30]. Our techniques allow us to generalize a result in this latter paper on the ranks of these groups.

Finally our techniques allow us to handle a related class of groups, {termed **generalized tetrahedron groups** by Vinberg, who studied them independently [28]} which also arose in our study of one-relator products of cyclics. We will present our results on these groups in a subsequent paper.

2. Preliminaries and Notation. If G is a group and $\{A\}$ a subset of G then $< A >_G$ denotes the subgroup of G generated by A. If $H = < A; rel.A >$ and $G = < A, B; rel.A, rel.(A, B) >$ then we say that G satisfies a **Freiheitssatz** relative to H if $< A >_G = H$.

If $\{G_\mu\}$ is a class of groups then a **one-relator product** of this class is a group $G = (\star G_\mu)/N(R)$ where R is a non-trivial, cyclically reduced word in the free product $(\star G_\mu)$ of syllable length ≥ 2 and $N(R)$ is its normal closure. The G_μ are called the **factors** and R is the **relator**. A Freiheitssatz for a one-relator product G means that each factor G_μ injects into G.

If each G_μ is cyclic we obtain a **one-relator product of**

cyclics. Such groups have presentations of the form

(2.1) $G =< a_1, ..., a_n; a_1^{e_1} = ... = a_n^{e_n} = 1, R(a_1, .., a_n) = 1 >$

where $e_i = 0$ or $e_i \geq 2$ for $i = 1, .., n$ and R is a non-trivial, cyclically reduced word in the free product on the $\{a_i\}$. We assume throughout that R involves all the $a_i's$ and that R is not a proper power. In [7] a FHS for groups of form (2.1) was proved when $R = S^m$ with $m \geq 2$. The proof technique involved representing such groups in $PSL_2(\mathbb{C})$. A representation ρ of G in a Linear Group is **essential** if $\rho(a_i)$ has infinite order if $e_i = 0$ or order e_i if $e_i \geq 2$ for $i = 1, .., n$, and $\rho(S)$ has exact order m. The results in [7] followed from showing that when $R = S^m, m \geq 2$ then there exists an essential representation into $PSL_2(\mathbb{C})$ which is faithful on the free product $< a_1, ..., a_{n-1}; a_1^{e_1} = ... = a_{n-1}^{e_{n-1}} = 1 >$. From this representation it was shown that these groups satisfy certain properties in common with linear groups. In most cases they satisfy the **Tits Alternative** - that is they either contain free subgroups of rank two or are virtually solvable; under certain conditions they are virtually torsion-free and finally in most cases they are either SQ-universal or virtually solvable. Recall a group G is SQ-universal if every countable group is embeddable in a quotient of G.

The groups we consider in the present paper have the form

(2.2) $G =< a_1, ..., a_n, b_1, ..., b_k; a_1^{e_1} = ... = a_n^{e_n} =$

$$= b_1^{f_1} = ... = b_k^{f_k} = R_1^{m_1} = ... = R_k^{m_k} = 1 >$$

where $n \geq 1, k \geq 2, e_j = 0$ or $e_j \geq 2$ for $i = 1, .., n, f_i = 0$ or $f_i \geq 2$ for $i = 1, .., k, m_i \geq 2$ for $i = 1, .., k$ and for each $i = 1, ..., k, R_i = R_i(a_1, .., a_n, b_i)$ is a cyclically reduced word in the free product on $\{a_1, .., a_n, b_i\}$ which involves b_i and at least one of the $a_i's$ and which is not a proper power. In this case we call the $\{R_i\}$ the **relators**.

Groups of this form arise in the study of NEC and other planar discontinuous groups and hence we call them **groups of SN-type** for **Special NEC-type**.

Extending the previous definition we say that a representation ρ of a group of SN-type is **essential** if $\rho(a_i)$ has infinite order if $e_i = 0$ or exact order e_i if $e_i \geq 2$ for $i = 1, .., n, \rho(b_i)$ has infinite order if $f_i = 0$ or exact order f_i if $f_i \geq 2$ for $i = 1, .., k$ and $\rho(R_i)$ has exact order m_i for $i = 1, .., k$. In the next section we show that each group of SN-type admits an essential representation in $PSL_2(\mathbb{C})$ and satisfies a FHS. From this follows many of the linearity properties of one-relator products of cyclics.

3. The FHS and Essential Representations. We first establish the FHS and the existence of essential representations for groups of SN type.

Theorem 1. *Let G be a group of SN-type so that G has a presentation of the form*

$$(2.2) \qquad G = < a_1, ..., a_n, b_1, ..., b_k; a_1^{e_1} = ... = a_n^{e_n} =$$

$$= b_1^{f_1} = ... = b_k^{f_k} = R_1^{m_1} = ... = R_k^{m_k} = 1 >$$

where $n \geq 1, k \geq 2, e_j = 0$ or $e_j \geq 2$ for $i = 1, .., n, f_i = 0$ or $f_i \geq 2$ for $i = 1, .., k, m_i \geq 2$ for $i = 1, .., k$ and for each $i = 1, ..., k, R_i = R_i(a_1, .., a_n, b_i)$ is a cyclically reduced word in the free product on $\{a_1, .., a_n, b_i\}$ which involves b_i and at least one of the $a_i's$ and which is not a proper power. Then

(1) $< a_1, ..., a_n >_G = < a_1; a_1^{e_1} > \star ... \star < a_n; a_n^{e_n} >$
(2) *For any subset $\{b_{i_1}, ..., b_{i_t}\}$ we have*

$$< a_1, .., a_n, b_{i_1}, ..., b_{i_t} >_G =$$

$$< a_1, .., a_n, b_{i_1}, .., b_{i_t}; a_1^{e_1} = .. = a_n^{e_n} = b_{i_1}^{f_{i_1}} = .. = b_{i_t}^{f_{i_t}} =$$

$$= R_{i_1}^{m_{i_1}} = .. = R_{i_t}^{m_{i_t}} = 1 >$$

(3) *If each R_i involves all $\{a_1, ..., a_n\}$ then for any proper subset $\{a_{i_1}, ..., a_{i_t}\}$*

$$< a_{i_1}, .., a_{i_t}, b_j >_G = < a_{i_1}; a_{i_1}^{e_{i_1}} > \star ... \star < a_{i_t}; a_{i_t}^{e_{i_t}} > \star < b_j; b_j^{f_j} >$$

(4) *There exists an essential representation of G in $PSL_2(\mathbb{C})$ which is faithful on $< a_1, ..., a_n >_G$*

Proof. We suppose first that $k = 2$. Therefore the group G has the form

$$G = < a_1, .., a_n, b_1, b_2; a_1^{e_1} = .. = a_n^{e_n} = b_1^{f_1} = b_2^{f_2} = R_1^{m_1} = R_2^{m_2} = 1 >$$

where e_i, f_i are as before, $m_1, m_2 \geq 2$ and $R_i = R_i(a_1, .., a_n, b_i), i = 1, 2$. Let

$$G_1 = < a_1, .., a_n, b_1; a_1^{e_1} = .. = a_n^{e_n} = b_1^{f_1} = R_1^{m_1} = 1 > \quad \text{and}$$

$$G_2 = < a_1, .., a_n, b_2; a_1^{e_1} = .. = a_n^{e_n} = b_2^{f_2} = R_2^{m_2} = 1 > .$$

G_1 and G_2 are one-relator products of cyclics. Since both m_1 and m_2 are at least 2 we can apply the FHS of [7] to say that $< a_1, ..., a_n >_{G_1} = < a_1, ..., a_n >_{G_2} = < a_1, .., a_n; a_1^{e_1} = .. = a_n^{e_n} = 1 >$. It follows that identifying $\{a_1, ..., a_n\}$ in G_1 with $\{a_1, ..., a_n\}$ in G_2 gives a subgroup isomorphism and therefore G is the free product of G_1 with G_2 amalgamated over $< a_1, ..., a_n >$. From the structure theorems on free products with amalgamation, G_1, G_2 and $< a_1, ..., a_n >$ inject into G. If $\{a_{i_1}, .., a_{i_r}\}$ is any proper subset of $\{a_1, ..., a_n\}$ then $\{a_{i_1}, .., a_{i_r}, b_i\}$ generate a subgroup of G_i. Again from the FHS in [7] this must be the free product. This establishes (1),(2) and (3) for the case $k = 2$.

If $k = 3$ then G has the form

$$G = < a_1, .., a_n, b_1, b_2; a_1^{e_1} = .. = a_n^{e_n} = b_1^{f_1} = b_2^{f_2} = b_3^{f_3}$$

$$= R_1^{m_1} = R_2^{m_2} = R_3^{m_3} = 1 > .$$

Using the result for $k = 2$ and the FHS of [7] G must now be the amalgamated free product of

$$G_1 = < a_1, .., a_n, b_1, b_2; a_1^{e_1} = .. = a_n^{e_n} = b_1^{f_1} = b_2^{f_2} = R_1^{m_1} = R_2^{m_2} = 1 >$$

$$\text{and } G_2 = < a_1, .., a_n, b_3; a_1^{e_1} = .. = a_n^{e_n} = b_3^{f_3} = R_3^{m_3} = 1 > .$$

amalgamated over $< a_1, ..., a_n >$. Parts (1),(2) and (3) of the theorem now follow as before.

For general k the first three parts now follow easily by induction.

We now establish the existence of an essential representation. Again first suppose that $k = 2$. Let G, G_1 and G_2 be as before with $k = 2$. From [7] there exists an essential representation ρ_1 of G_1 in $PSL_2(\mathbb{C})$ which is faithful on $< a_1, ..., a_n >$. Let $A_1, .., A_n, B_1$ be the images of $a_1, .., a_n, b_1$ under ρ_1. Let

$$B_2 = \pm \begin{pmatrix} t & \star \\ \star & 2cos(\pi/f_2) - t \end{pmatrix}$$

with t a variable to be determined. Since $tr(B_2) = 2cos(\pi/f_2)$ we have $B_2^{f_2} = 1$ for every choice of t. We show there exists a choice of t so that the image of R_2 has order m_2 giving the essential representation.

Suppose $R_2 = w_1 b_2^{k_1} w_2 ..... b_2^{k_t}$ where $w_1, w_2, ...$ are non-trivial words in $\{a_1, ..., a_n\}$. Let $\overline{R_2} = W_1 B_2^{k_1} W_2 ... B_2^{k_t}$ where W_i are the corresponding images of $w_1, w_2, ..., w_n$ under the representation ρ_1.

Since ρ_1 is faithful on $< a_1, ..., a_n >$ it follows that $W_1, W_2, ...$ are non-identity projective matrices. Therefore the same arguments as in [7] can be used to assert that there exists a choice {perhaps by conjugating the original $A_1, ..., A_n$ if necessary} so that

$$W_1 B_2^{k_1} W_2 ... B_2^{k_t} = \pm \begin{pmatrix} f_1 & f_2 \\ f_3 & f_4 \end{pmatrix}$$

where $f_1 + f_4$ is a <u>non-constant</u> polynomial in t. From the fundamental theorem of algebra there is a choice t_1 so that $(f_1 + f_4)(t_1) = 2cos(\pi/m_2)$. With this t_1 in B_2 we then have $tr(\overline{R_2}) = 2cos(\pi/m_2)$ and hence $\overline{R_2}^{m_2} = 1$

Let ρ be the representation which takes $a_1, ..., a_n, b_1, b_2$ to $A_1, ..., A_n, B_1, B_2(t_1)$ {or whichever conjugates of $A_1, ..., A_n, B_1$ were necessary so that the trace polynomial was non-constant} respectively. This representation ρ is then faithful on $< a_1, ..., a_n >$ and from the trace conditions $\rho(b_1)\{= B_1\}, \rho(b_2)\{= B_2\}, \rho(R_1)$ and $\rho(R_2)$ have exactly the correct orders - that is $\rho(b_i)$ has infinite order if $f_i = 0$ or order f_i if $f_i \geq 2$ and $\rho(R_i)$ has exact order m_i for $i = 1, 2$. Therefore the representation ρ is essential.

Once the essential representation for the case $k = 2$ is established the same argument can be iterated for the general case. This completes the theorem.

In [7] it was further established using a modification of the above argument that if A and B are groups which admit faithful representations in $PSL_2(\mathbb{C})$ and R is a non-trivial word of syllable length at least 2 in $A \star B$ then the one-relator product $G = (A \star B)/N(R^m)$ with $m \geq 2$ admits a representation ρ into $PSL_2(\mathbb{C})$ such that $\rho_{|A}$ is faithful, $\rho_{|B}$ is faithful and $\rho(R)$ has order m. Using this result and an analagous proof to that of Theorem 1 we obtain:

Theorem 2. *Let $G_0 =< A;\ rel\ A >$ and, for i=1,..,k, let $G_i =< B_i; rel\ B_i >$. Suppose that*

$$G =< A, B_1, .., B_k;\ rel\ A,\ rel\ B_1, ..,\ rel\ B_k, R_1^{m_1} = ... = R_k^{m_k} = 1 >$$

where $k \geq 2$ and, for each $i = 1,..,k$, $R_i = R_i(A, B_i)$ is a non-trivial cyclically reduced word in the free product $A \star B_i$ of syllable length at least two and each $m_i \geq 2$. Assume further that each G_i for $i = 0,1,..,k$ admits a faithful representation into $PSL_2(\mathbb{C})$. Then

(1) *Each G_i for $i = 0,1,...,k$ injects in G*
(2) *For each subset $\{i_1, .., i_t\}$ of $1,2,....,k$ we have*

$$< A, B_{i_1}, .., B_{i_t} >_G=$$

$$=< A, B_{i_1}, .., B_{i_t}; \text{ rel } A, .., \text{ rel } B_{i_t}, R^m_{i_1 i_1} = .. = R^m_{i_t i_t} = 1 >$$

(3) There exists a representation ρ of G into $PSL_2(\mathbb{C})$ such that $\rho_{|G_i}$ is faithful for $i = 0,1..,k$ and $\rho(R_i)$ has order m_i for $i = 1,2,..,k$.

We note that the conclusions of Theorem 1 are also valid for certain modifications of the presentation type (2.2). For example the same proof can be used for groups with presentations of the form
(2.3)
$$G =< a_1, .., a_{n+k}; a_1^{e_1} = .. = a_{n+k}^{e_{n+k}} = R_1^{m_1} = ... = R_k^{m_k} = 1 >$$
where $n \geq 1, k \geq 2, e_j = 0$ or $e_j \geq 2$ for $i = 1, ..., n + k$, $m_i \geq 2$ for $i = 1, 2, .., k$ and for each $i = 1, .., k$, $R_i = R_i(a_i, ..., a_{i+n})$ is a non-trivial, cyclically reduced word in the free product on $\{a_i, ..., a_{i+n}\}$.

In general it is not clear what happens relative to essential representations for general multi-relator groups. Below we present an example of a group G very similar in form to a group of SN-type but which has no essential representation in $PSL_2(\mathbb{C})$.

Let $G =< x, y, z; x^4 = y^3 = z^7 = (xy[x,y])^2 = (x^3z(x[x,y])^2z^{-1})^5 = 1 >$. Suppose X, Y, Z are the images of x, y, z under a supposedly essential representation in $PSL_2(\mathbb{C})$. Then without loss of generality let $\text{tr}(X) = a = \sqrt{2}$, since $x^4 = 1$, and $\text{tr}(Y) = b = 1$ since $y^3 = 1$. Suppose $\text{tr}(XY) = c$ then from a direct computation we have $\text{tr}X[X,Y] = a(a^2 + b^2 + c^2 - abc - 3)$ and $\text{tr}XY[X,Y] = c(a^2 + b^2 + c^2 - abc - 3)$. Since $(XY[X,Y])^2 = 1$ we have, if the representation is essential, that $\text{tr}(XY[X,Y]) = 0$. Therefore either $c = 0$ or $a^2 + b^2 + c^2 - abc - 3 = 0$. However if $c = 0$ since $a = \sqrt{2}$ and $b = 1$ we have that $a^2 + b^2 + c^2 - abc - 3 = 0$ also. Therefore $\text{tr}(X[X,Y]) = 0$ and so $(X[X,Y])^2 = 1$. However then from the second relation it would follow that $X^{15} = 1$ which coupled with $X^4 = 1$ gives $X = 1$ and hence the representation is not essential.

4. Linearity Results. In this section we show that groups of SN-type satisfy much the same linearity properties as one-relator products of cyclics. Specifically we examine the Tits Alternative, SQ-universality and virtual torsion-freeness.

Theorem 3. *Let G be a group of SN-type. Then either G is SQ-universal or G is infinite solvable.*

Proof. For a group of SN-type having form (2.2) let $N = n + k$. We first prove the result for $k = 2$ so that $N = n + 2$. The result for general k then follows in a straightforward manner. Thus we suppose that our group G has the form

$$G = < a_1, .., a_n, b_1, b_2; a_1^{e_1} = .. = a_n^{e_n} = b_1^{f_1} = b_2^{f_2} = R^m = S^t = 1 >$$

where $n \geq 1, e_j = 0$ or $e_j \geq 2$ for $i = 1, ..., n, f_i = 0$ or $f_i \geq 2$ for $i = 1, 2, m \geq 2, t \geq 2$ and $R = R(a_1, .., a_n, b_1)$ is a cyclically reduced word in the free product on $\{a_1, .., a_n, b_1\}$ and $S = S(a_1, .., a_n, b_2)$ is a cyclically reduced word in the free product on $\{a_1, .., a_n, b_2\}$.

Since, if a quotient of a group is SQ-universal, the group is itself SQ-universal we can, without loss of generality, assume that $e_j \geq 2$ for $i = 1, ..., n$ and $f_i \geq 2$ for $i = 1, 2$, passing to a quotient if necessary.

From Theorem 1, G admits an essential representation ρ into $PSL_2(\mathbb{C})$ which is faithful on the free product on $\{a_1, .., a_n\}$. Therefore from Selberg's theorem on finitely generated subgroups of linear groups [26] there exists a normal subgroup H of finite index in $\rho(G)$ such that $\rho(a_i)$ has order e_i modulo H, $\rho(b_i)$ has order f_i modulo H, $\rho(R)$ has order m modulo H and $\rho(S)$ has order t modulo H. Thus the composition of maps {where π is the canonical map}

$$G \xrightarrow{\rho} \rho(G) \xrightarrow{\pi} \rho(G)/H$$

gives an essential representation ϕ of G onto a finite group.

Let $X = < a_1, ..., a_n, b_1, b_2; a_1^{e_1} = .. = a_n^{e_n} = b_1^{f_1} = b_2^{f_2} = 1 >$ be the free product on $\{a_1, .., a_n, b_1, b_2\}$. There is a canonical epimorphism $\beta : X \to G$. We therefore have the sequence

$$X \xrightarrow{\beta} G \xrightarrow{\phi} \rho(G)/H$$

Let $Y = \ker(\phi \circ \beta)$. Then Y is a normal subgroup of finite index in X and Y is torsion-free. Since X is a free product of cyclics and Y is torsion-free, Y is a free group of finite rank r.

Suppose $|X : Y| = j$. Regard X as a Fuchsian group with finite hyperbolic area $\mu(X)$. {Every fintely generated free product of cyclics can be faithfully represented as such a Fuchsian group}. From the Riemann- Hurwitz formula :

$$j\mu(X) = \mu(Y)$$

where

$$\mu(Y) = r - 1$$

and

$$\mu(X) = N - 1 - (\frac{1}{e_1} + .. + \frac{1}{e_n} + \frac{1}{f_1} + \frac{1}{f_2}),$$

we obtain

$$r = 1 - j(\frac{1}{e_1} + .. + \frac{1}{e_n} + \frac{1}{f_1} + \frac{1}{f_2} - N + 1)$$

G is obtained from X by adjoining the relations R^m and S^t and so $G = X/K$ where K is the normal closure of R^m and S^t. Since K is contained in Y the quotient Y/K can be considered as a subgroup of finite index in G. Applying the Reidemeister- Schreier process or repeated applications of Corollary 3 in [3] Y/K can be defined on r generators subject to $(j/m) + (j/t)$ relations. The deficiency d of this presentation for Y/K is then

$$d = r - \frac{j}{m} - \frac{j}{t} = 1 - j(\frac{1}{e_1} + .. + \frac{1}{e_n} + \frac{1}{f_1} + \frac{1}{f_2} + \frac{1}{m} + \frac{1}{t} - N + 1)$$

If $N \geq 5$, or $N = 4$ and $(e_1, e_2, f_1, f_2, m, t) \neq (2,2,2,2,2,2)$, then $d \geq 2$. From the work of B. Baumslag and S. Pride [2] Y/K and hence G has a subgroup of finite index mapping onto a free group of rank 2. Therefore Y/K is SQ-universal and since this has finite index in G, G is also SQ-universal. Therefore if $N \geq 5$, or $N = 4$ and $(e_1, e_2, f_1, f_2, m, t) \neq (2,2,2,2,2,2)$ then G is SQ-universal.

Next suppose that $N = 4$ and $(e_1, e_2, f_1, f_2, m, t) = (2,2,2,2,2,2)$, so now G has the presentation

$$G = < a_1, a_2, b_1, b_2; a_1^2 = a_2^2 = b_1^2 = b_2^2 = R^2 = S^2 = 1 >$$

where $R = R(a_1, a_2, b_1)$ and $S = S(a_1, a_2, b_2)$. We may assume that R and S are not proper powers since otherwise it reverts back to the previous arguments where one of the exponents is not 2. Without loss of generality we must consider the following four cases:

(1) $R = a_1 b_1$ and $S = a_1 b_2$
(2) $R = a_1 b_1$ and $S = a_2 b_2$
(3) $R = a_1 b_1$ and S involves both a_1 and a_2
(4) Both R and S involve a_1 and a_2.

In case (1) if $R = a_1 b_1$ and $S = a_1 b_2$ then

$$G = < a_1, a_2, b_1, b_2; a_1^2 = a_2^2 = b_1^2 = b_2^2 = (a_1 b_1)^2 = (a_1 b_2)^2 = 1 > .$$

Setting $a_1 = 1$, G can be mapped onto $< a_2, b_1, b_2; a_2^2 = b_1^2 = b_2^2 = 1 > = \mathbb{Z}_2 \star \mathbb{Z}_2 \star \mathbb{Z}_2$ which is SQ-universal and therefore G is. {Recall that any non-trivial free product except the infinite dihedral group $\mathbb{Z}_2 \star \mathbb{Z}_2$, is SQ-universal}.

In case (2), where $R = a_1 b_1$ and $S = a_2 b_2$ then by setting $b_1 = 1$, G can be mapped onto $< a_1, a_2, b_2; a_1^2 = a_2^2 = b_2^2 = (a_2 b_2)^2 = 1 > = \mathbb{Z}_2 \star D_2$ {where D_2 is the Klein 4-group }. Since this is a non-trivial free product and not infinite dihedral, G is SQ-universal.

In case (3) where $R = a_1 b_1$ and S involves both a_1 and a_2 set $a_1 = 1$, to obtain as an image of G, $< a_2, b_1, b_2; a_2^2 = b_1^2 = b_2^2 = (a_2 b_2)^{2w} = 1 >$. This is a non-trivial free product $\mathbb{Z}_2 \star T$ where T has the presentation $< a_2, b_2; a_2^2 = b_2^2 = (a_2 b_2)^{2w} = 1 >$. For no value of w is this cyclic of order 2 and therefore $\mathbb{Z}_2 \star T$ is SQ-universal, and hence G is.

Finally suppose case (4) where both R and S involve a_1 and a_2. Let H be the subgroup of G generated by $a = a_1 a_2, b = a_1 b_1$ and $c = a_1 b_2$. Then H has index 2 in G and there are three possibilities for R and S relative to H.

(1) Both R and S are elements of H
(2) R is an element of H but S is not {or vice versa}.
(3) Both R and S are not elements of H.

If both R and S are elements of H then $R = T(a, b)$ where $T(a, b)$ is a freely reduced word in the free group on $\{a, b\}$ and $S = U(a, c)$ where $U(a, c)$ is a freely reduced word in the free group on $\{a, c\}$. Then H has a presentation

$$H = < a, b, c; T^2(a, b) = T^2(a^{-1}, b^{-1}) = U^2(a, c) = U^2(a^{-1}, c^{-1}) = 1 > .$$

Using the arguments in [18 pg. 293] we may assume that one of the generators a, b has exponent sum zero in $T(a, b)$. If b has exponent sum zero, let $a = 1$ to obtain the quotient $< b, c; c^{2w} = 1 > = \mathbb{Z} \star \mathbb{Z}_{2w}$ if $w \geq 1$ or $\mathbb{Z} \star \mathbb{Z}$ if $w = 0$. Either is SQ-universal and therefore G is. If a has exponent sum zero let $b = 1$ to obtain the quotient $< a, c; U^2(a, c) = U^2(a^{-1}, c^{-1}) = 1 >$. From a theorem of M. Edjvet [6] on groups which have balanced presentations, this quotient has a subgroup of finite index mapping onto a free group of rank 2, and is therefore SQ-universal. Thus G is SQ-universal. This completes the situation where both R and S are elements of H.

Now suppose that R is an element of H but S is not. Suppose $R = T(a, b)$ as above then $S = a_1 U(a, c)$ where $U(a, c)$ is a freely reduced word in the free group on $\{a, c\}$. The subgroup H now has the presentation

$$H = < a, b, c; T^2(a, b) = T^2(a^{-1}, b^{-1}) = U(a, c)U(a^{-1}, c^{-1}) = 1 > .$$

Let $a = 1$ to obtain the quotient $< b, c; b^{2w} = 1 > = \mathbb{Z} \star \mathbb{Z}_{2w}$ which as above is SQ-universal and therefore G is.

Finally suppose both R and S are not in H. Then $R = a_1 T(a, c), S = a_1 U(a, c)$ and H has the presentation

$$H = < a, b, c; T(a, b)T(a^{-1}, b^{-1}) = U(a, c)U(a^{-1}, c^{-1}) = 1 > .$$

Letting $a = 1$ we obtain the quotient $< b, c; > = \mathbb{Z} \star \mathbb{Z}$ which is SQ-universal, and therefore G is. This completes the situation when $N = 4$ and all exponents are 2.

We must now consider the case when $N = 3$. Thus G has the presentation

$$(2.4) \qquad G = < a_1, b_1, b_2; a_1^{e_1} = b_1^{f_1} = b_2^{f_2} = R^m = S^t = 1 > .$$

From the proof given above for $N \geq 4$ a group with a presentation (2.4) is SQ-universal if the deficiency d of the subgroup Y/K is 2 or greater, or equivalently, if $(1/e_1) + (1/f_1) + (1/f_2) + (1/m) + (1/t) < 2$. By a lengthy argument on the possible values of (e_1, f_1, f_2, m, t) the SQ-universality or not of groups with presentations of form (2.4) can be established directly. However this can be simplified using a result of Lossov [17]. He proved that if $G = A \star_H B$ with $|H| < \infty, |A : H| \geq 2$ and $|B : H| \geq 3$ then G is SQ-universal. In (2.4) we have $G = G_1 \star_H G_2$ where $G_1 = < a_1, b_1; a_1^{e_1} = b_1^{f_1} = R^m = 1 >, G_2 = < a_1, b_2; a_1^{e_1} = b_2^{f_2} = S^t = 1 >$ and $H = < a_1; a_1^{e_1} = 1 >$. Since we can assume that $e_1 \geq 2$, H is finite cyclic and Lossov's theorem applies unless both G_1 and G_2 are finite with $|G_1 : H| = 2$ and $|G_2 : H| = 2$. This would imply that both G_1 and G_2 are finite dihedral and G would have the presentation

$$(2.5) \quad G = < a_1, b_1, b_2; a_1^{e_1} = b_1^2 = b_2^2 = (a_1 b_1)^2 = (a_1 b_2)^2 = 1 > .$$

This has the structure of a free product of two isomorphic finite dihedral groups amalgamated over their cyclic subgroup of index 2. This group is infinite solvable. Therefore if $N = 3$ either G is SQ-universal or G has a presentation of form (2.4) and G is infinite solvable.

This completes the proof for $k = 2$.

If $k > 2$ using the same argument as before, employing an essential representation of G onto a finite group we obtain a subgroup Y/K of finite index in G having a presentation with deficiency d given by

$$d = 1 - j(\frac{1}{e_1} + .. + \frac{1}{e_n} + \frac{1}{f_1} + .. + \frac{1}{f_k} + \frac{1}{m_1} + .. + \frac{1}{m_k} - (n+k) + 1)$$

$$= j[(n+k) - 1] - j(\frac{1}{e_1} + .. + \frac{1}{e_n} + \frac{1}{f_1} + .. + \frac{1}{f_k} + \frac{1}{m_1} + .. + \frac{1}{m_k}).$$

As in the case when $k = 2$ the deficiency is 2 or greater unless $n = 1$ or $n = 2$ together with all exponents 2. Thus G is SQ-universal except possibly in these latter two cases. If $n = 1$, Lossov's theorem applies to show that G is SQ-universal if $k > 2$. If $n = 2$ and all exponents are 2 then G has the presentation

$$G = < a_1, a_2, b_1, b_2, ..., b_k; a_1^2 = a_2^2 = b_1^2 = ... = b_k^2 = R_1^2 = ... = R_k^2 = 1 >.$$

Setting $a_1 = a_2 = 1$ we get the quotient $< b_1, b_2, ..., b_k; b_1^2 = ... = b_k^2 = 1 >$. This is a free product of cyclic groups of order 2 and since $k > 2$ it has more than two factors and is thus SQ-universal. Therefore G is SQ-universal. This completes the proof.

Notice that, if a group is SQ-universal, it must contain a free subgroup of rank 2. Recall that a group G satisfies the Tits Alternative if either G contains a free subgroup of rank 2 or G is virtually solvable. This is a "linear" property since Tits proved that all finitely generated linear groups satisfy this.

Corollary 1. Let G be a group of SN-type then G satisfies the Tits Alternative.

Corollary 2. Let G be a group of SN-type then G is infinite solvable if and only if G has a presentation $< a_1, b_1, b_2; a_1^{e_1} = b_1^2 = b_2^2 = (a_1b_1)^2 = (a_1b_2)^2 = 1 >$. Otherwise G is SQ-universal.

Selberg's theorem [26] says that any f.g. linear group is virtually torsion-free, that is it contains a torsion-free subgroup of finite index. For groups of SN-type, because of the existence of essential representations, this occurs under certain conditions on the relators.

Theorem 4. *Let G be a group of SN-type. Assume each relator R_i is not a proper power and satisfies one of the following conditions:*

(1) *$R_i = U_i V_i$ where $U_i = U_i(a_1, .., a_k)$ and $V_i = V_i(a_{k+1}, .., a_n, b_i), 1 \leq k \leq n$ are non-trivial words in the free product on $\{a_1, .., a_k\}$ and $\{a_{k+1}, .., a_n, b_i\}$ respectively.*

(2) *R_i is not conjugate in the free product on $\{a_1, .., a_n\}$ to a word of the form XY where X, Y are elements of orders $p \geq 2, q \geq 2$ respectively with $(1/m_i) + (1/p) + (1/q) > 1$ and $m_i \geq 4$.*

(3) *R_i is arbitrary but $e_i = 0$ for $i = 1,...,n$, $f_i = 0$ for $i = 1,...,k$.*

Then G is virtually torsion-free.

Proof. Let H be a one-relator product of cyclics with presentation

$$H = < a_1, .., a_n, b_1, b_2; a_1^{e_1} = .. = a_n^{e_n} = R^m = 1 > .$$

If $m \geq 2$ and anyone of the three conditions in the statement of the theorem are satisfied then any element of finite order in H is conjugate to a power of one of the generators a_i or a power of the relator R [11]. Now let G be a group of SN-type. From the proof of Theorem 1, G is an amalgamated free product of one-relator products of cyclics. Hence, from the torsion theorem for such amalgams any element of finite order in G is conjugate to an element of finite order in one of the factors. Combining this with the statement above on one-relator products of cyclics we have that if each relator R_i satisfies one of the stated conditions then any element of finite order in G is conjugate to a power of a generator $a_1, ...a_n, b_1, ..., b_k$ or a power of a relator $R_1, .., R_k$.

Let $\rho : G \to PSL_2(\mathbb{C})$ be an essential representation. From Selberg's theorem $\rho(G)$ contains a torsion-free subgroup $N^\star$ of finite index. Let N be the pull-back of $N^\star$ in G; then N has finite index in G. Let g be an element of finite order in N. Then g is conjugate to a power of a generator or a power of a relator. Since $N^\star$ is torsion-free $\rho(g) = 1$. However ρ is essential, so $\rho(a_i)$ has infinite order if $e_i = 0$ or exact order e_i if $e_i \geq 2$ for $i = 1, .., n$, $\rho(b)$ has infinite order if $f_i = 0$ or exact order f_i if $f_i \geq 2$ and $\rho(R_i)$ has exact order m_i for $i = 1, .., k$. Therefore no power of a generator or of a relator (except the identity) is in ker ρ. Therefore $g = 1$ and N is torsion-free.

Corollary 3. *Let G be a group of SN-type satisfying the conditions of theorem 4 then the conjugacy classes of torsion elements* are precisely given by the powers of the generators and the powers of the relators.

5. Rank Conditions for Certain NEC Groups.
Zieschang and Kaufman [30] have recently considered the rank { minimal number of generators } of groups with presentations of the form

$$(5.1) \quad < a_1, a_2, a_3; a_1^2 = a_2^2 = a_3^2 = (a_1 a_2)^h = (a_1 a_3)^k = 1 > .$$

These groups are contained in the class of groups of SN-type and are NEC groups with reflections. Zieschang and Kaufman [30] proved that if $2 \leq h \leq k$, then G has rank 3 if $h > 2$ or $h = 2$ and k is even. Otherwise G has rank 2. We now give some extensions of this.

Theorem 5. *Let* $G =< a_1, a_2, a_3; a_1^{e_1} = a_2^{e_2} = a_3^{e_3} = (a_1 a_2)^h = (a_1 a_3)^k = 1 >$ *with* $e_i = 0$ *or* $e_i \geq 2$ *for* $i = 1,2,3$, $2 \leq h \leq k$. *Then* $2 \leq rankG \leq 3$ *and rank* $G = 2$ *if and only if* G *has the presentation* $< a, b, c; a^p = b^2 = c^2 = (ab)^2 = (ac)^k = 1 >$ *with* $p = 0$ *or* $p \geq 2$ *and* $k \geq 3$, k *odd.*

{ *In the case where* G *has rank 2,* G, *as above, is generated by* $x = a$ *and* $y = cb$ *and* G *is an epimorphic image of the Fuchsian group* $F =< s_1, s_2, s_3, s_4; s_1^2 = s_2^2 = s_3^2 = s_4^k = s_1 s_2 s_3 s_4 = 1 >$ *with* $k \geq 3$ *and* k *odd.*}

Proof. Let G be as above. If $e_1 = e_2 = e_3 = 2$ then the result follows from Zieschang-Kaufman. Assume that $e_1 \neq 2$. As before $G = G_1 \star_H G_2$ where $G_1 =< a_1, a_2; a_1^{e_1} = a_2^{e_2} = (a_1 a_2)^h = 1 >$,$G_2 =< a_1, a_3; a_1^{e_1} = a_3^{e_3} = (a_1 a_3)^k = 1 >$ and $H =< a_1; a_1^{e_1} = 1 >$. Clearly $2 \leq$ rank $G \leq 3$. Assume $G =< x, y >$ so that rank $G = 2$. We use Nielsen reduction on $\{x, y\}$. From [9] we have the following result.

Lemma. {*Lemma A of [9]*} *Let* $G = G_1 \star_A G_2$ *and assume that a length* L *and an order are introduced in* G *as in [29] and [24]. If* $\{x_1, ..., x_m\}$ *is a finite system of elements in* G *then there is a Nielsen transformation from* $\{x_1, ..., x_m\}$ *to a system* $\{y_1, ..., y_m\}$ *for which one of the following cases holds.*

  (1) *Each element* $w \in< y_1, ..., y_m >$ *can be written as* $\prod_{i=1}^{q} y_{v_i}^{\epsilon_i}$, *with* $\epsilon_i = \pm 1, \epsilon_i = \epsilon_{i+1}$ *if* $v_i = v_{i+1}$ *with* $L(y_{v_i}) \leq L(w)$ *for* $i = 1,..,q$.
  (2) *There is a product* $a = \prod_{i=1}^{q} y_{v_i}^{\epsilon_i}, a \neq 1$ *with* $y_{v_i} \in A$ (i $= 1,...,q$) *and in one of the factors* G_j *there is an element* $x \in A$ *with* $x^{-1} a x \in A$

(3) *Of the y_i there are $p \geq 1$, contained in a subgroup of G conjugate to G_1 or G_2 and a certain product of them is conjugate to a non-trivial element of A.*

(4) *There is a $g \in G$ such that for some $i \in \{1, ..., m\}$ we have $y_i \notin gAg^{-1}$, but for a suitable natural number k we have $y_i^k \in gAg^{-1}$.*

The Nielsen transformation can be chosen so that $\{y_1, ..., y_m\}$ is smaller than $\{x_1, ..., x_m\}$ or the lengths of the elements of $\{x_1, ..., x_m\}$ are preserved. Further if $\{x_1, ..., x_m\}$ is a generating system of G, then in the case (iii), we find that $p \geq 2$ because then conjugations determine a Nielsen transformation. If we are interested in the combinatorial description of $< x_1, ..., x_m >$ in terms of generators and relations we find again that $p \geq 2$ in case (iii) possibly after suitable conjugations.

Case (i) of the lemma cannot occur for $\{x, y\}$ in G for otherwise G would be a free product of cyclics. Each G_i is an ordinary triangle group, so there is no $h \in G_i - H$ with $h^t \in H$. Therefore case (iv) of the lemma cannot occur. If case (iii) occurs we may assume that x,y are both in G_1 or both in G_2, which is a contradiction since $G_1 \neq H \neq G_2$.

Therefore case (ii) of the lemma occurs. Without loss of generality assume that $x = a_1^t$ for some positive integer t with $1 \leq t \leq e_1$ and that there is an element v in $G_i - H$ with $v^{-1}a_1^u v = a_1^w$ for some $u, w \neq 0$ and $a_1^u \in < x > = < a_1^t >$. Regard G_1 as a subgroup of $PSL_2(\mathbb{C})$. Then a direct computation shows that $u = -w$ and $v^2 = 1$. Therefore $v^{-1}a_1 v = a_1^{-1}$ so $(va_1)^2 = 1$. Hence $e_2 = m = 2$ and we may assume that $x = a_1$ because $G = < x, y > = < a_1^r, y > \subset < a_1, y >$. Assume that either $e_3 = 0$ and k even or both e_3 and k are positive even. Adjoining the relations $a_3^2 = (a_1 a_3)^2 = a_1 = 1$ we obtain the quotient $< a_2, a_3; a_2^2 = a_3^2 = 1 >$ which is not cyclic. It follows that at least one of e_3 or k is odd and therefore there is no relation $w^{-1}a_l^t w = a_1^s$ for some non-zero integers s, t and some $w \in G_2 - H$. Analagously as above if $e_3 \geq 2$ then the gcd of e_2 and k is 1. Now we come back to our generating system $\{x, y\}$ with $x = a_l$.

Using the above arguments and Nielsen reduction we may assume that $y = h_1 a_2 h_2 a_2 ... h_m a_2$ with $m \geq 1$ and the h_i coset representatives of H in G_2. Because there is no relation $w^{-1}a_l^t w = a_1^s$ we must have $m = 1$ for otherwise x and y cannot generate G. Hence $y = ha_2$ for some $h \in G_2 - H$. Further x and y can generate G only if $yxy^{-1} = ha_1^{-1}h^{-1}$ or, equivalently, a_l and $ha_1^{-1}h^{-1}$

generate G_2. From the classification of generating pairs of triangle groups $\{G_2$ is also an ordinary triangle group$\}$ [15] and [22] we get that one of e_3 or k is 2. Therefore without loss of generality we have that $e_3 = 2$ and $k \geq 3$ and hence G has a presentation $< a, b, c; a^p = b^2 = c^2 = (ab)^2 = (ac)^k = 1 >$ with p = 0 or $p \geq 2$ and $k \geq 3, k$ odd. This completes the proof when $e_1 \neq 2$.

Note that, if G has a presentation as above, let $u = ab$ so that G also has the presentation $< u, b, c; u^2 = b^2 = c2 = (ub)^p = (ubc)^k = 1 >$. Let $x = a = ub$ and $y = cu = cba^{-1}$. Then $[x, y] = (ubc)^2$ and therefore ubc, c, u, b are in $< x, y >$ since k is odd. Therefore $\{x, y\}$ generate G. From the second presentation it is clear that G is an image of the Fuchsian group F with presentation $< s_1, s_2, s_3, s_4; s_1^2 = s_2^2 = s_3^2 = s_4^k = s_1 s_2 s_3 s_4 = 1 >$ with k odd and $k \geq 3$.

Now suppose that $e_1 = 2$ and $e_2 \leq m$ and $e_3 \leq k$. Let $\{x, y\}$ be a generating pair for G. Again we use Lemma A from [9] and case (ii) of that lemma must hold. Therefore we may assume that $x = a_1$ is in A and without loss of generality there is a $v \in G_1 - H$ with $v a_1 v^{-1} = a_1$. Regarding G_1 as a subgroup of $PSL_2(\mathbb{C})$, and using direct calculations we get that $e_2 = 2$ and $m = 2r \geq 2$ and $(a_1 a_2)^r$ commutes with a_1. As for the case when $e_1 \neq 2$ we use the classification of generating pairs for triangle groups to obtain that at least one of e_3 or k is odd and the gcd of e_3 and k is 1. Hence we get that $e_3 = 2$ and k odd. From this we must have $m = 2$ for, if $m \geq 4$ then G has rank 3 from the Zieschang-Kaufman result. This completes the proof.

Using modifications of the same type of cancellation arguments as in the proof above we can give the following generalization which we just state.

Theorem 6. Let $G = < a_1, a_2, a_3, ..., a_n, b_1, b_2; a_1^{e_1} = a_2^{e_2} = ... = a_n^{e_n} = b_1^{f_1} = b_2^{f_2} = (a_1 a_2 .... a_n b_1)^h = (a_1 a_2 .... a_n b_2)^k = 1 >$ with $e_j = 0$ or $e_j \geq 2$ for i = 1,2,...n , $f_i = 0$ or $f_i \geq 2$ for i = 1,2 and $h \geq 2, k \geq 2$. Then $n + 1 \leq$ rank $G \leq n + 2$.

REFERENCES

[1] I. Anshel -"A Freiheitssatz for a Class of Two-Relator Groups" - in Topology and Combinatorial Group Theory Springer Lecture Notes in Math. No. 1440 1-22

[2] B.Baumslag and S.J.Pride - "Groups with two more generators than relators" - J. London Math. Soc. (2) 17 (1987) 425-426

[3] G. Baumslag, J. Morgan and P. Shalen - "Generalized triangle groups " - Math. Proc. Camb. Phil. Soc. 102 (1987) 25-31

[4] W. Bogley - "An Identity Theorem for Multi-Relator Groups" - to appear Math. Proc. Camb. Phil. Soc.

[5] W.Bogley and S.J. Pride - "Aspeherical Relative Presentations" - Proc. Edin. Math. Soc. 35 (1992) 1-39

[6] M. Edjvet - "On a Certain Class of Group Presentations" - Math. Proc. Camb. Phil. Soc. 105 (1989) 25-35

[7] B. Fine, J. Howie and G. Rosenberger - "One-relator quotients and free products of cyclics" - Proc. Amer. Math. Soc. 102 (1988) 1-6

[8] B. Fine and G. Rosenberger - "Complex Representations and One-Relator Products of Cyclics " - in Geometry of Group Representations - Contemporary Math. 74 (1987) 131-149

[9] B. Fine and G. Rosenberger - "Generalizing Algebraic Properties of Fuchsian Groups" - in Proceedings of Groups St. Andrews 1989 - London Mathematical Soc. Lecture Notes Series 159, (1989) Vol. 1, 124-148

[10] B. Fine, F. Levin and G. Rosenberger - "Free Subgroups and Decompositions of One-Relator Products of Cyclics: Part 1: The Tits Alternative " - Arch. Math. 50 (1988) 97-109

[11] B. Fine, F. Levin and G. Rosenberger - "Free subgroups and decompositions of one-relator products of cyclics; Part 2 : Normal Torsion-Free Subgroups and FPA Decompositions" - J. Indian Math. Soc. 49 (1985) 237-247

[12] B. Fine and G. Rosenberger - "Conjugacy Separability of Fuchsian Groups" - Contemporary Math 109 (1990) 11-19

[13] J. Howie - "The Quotient of a Free Product of Groups by a Single High-Powered Relation I. Pictures. Fifth and Higher Powers" - Proc. London Math. Soc. (3) 59 (1989) 507-540

[14] J. Howie - "The Quotient of a Free Product of Groups by a Single High-Powered Relation II. Fourth Powers" - Proc. London Math. Soc. (3) 61 (1990) 33-62

[15] A.W. Knapp - "Doubly Generated Fuchsian Groups" - Mich. Math. J. 15 (1968) 289-304

[16] F. Levin and G. Rosenberger - "Free Subgroups of Generalized Triangle Groups: Parts I,II" - Part I Alg. i. Logika 28 (1989) 227-240, Part II to appear

[17] K.I. Lossov - "SQ-Universality of Free Products with Amalgamated Finite Subgroups" - Sib. Math. Z. 27 (1986) 128-139

[18] R.Lyndon and P. Schupp - Combinatorial Group Theory - Springer-Verlag 1977

[19] W. Magnus, A. Karass and D. Solitar - Combinatorial Group Theory - Wiley Interscience, 1968

[20] S.J. Pride - "Groups with Presentations in Which Each Defining Relator Involves Exactly Two Generators" - J. London Math. Soc. (2) 36 (1987) 245-256

[21] G. Rosenberger - "Faithful Linear Representations and Residual Finiteness of Certain One-Relator Products of Cyclics" - J. Siberian Math. Soc. 32 (1991) 204-206

[22] G. Rosenberger - "All Generating Pairs of All Two-Generator Fuchsian Groups" - Archiv. Math. 46 (1986) 198-204

[23] G. Rosenberger-"The SQ-universality of one-relator products of cyclics" - Result. in Math. to appear

[24] G. Rosenberger - "Zum Rang und Isomorphieproblem fur freie Produkte mit Amalgam" - Habilitationsschrift, Hamburg (1974)

[25] G.Rosenberger - "On One-Relator Groups that are Free Products of Two Free Groups with Cyclic Amalgamation" Groups St Andrews 1981, Cambridge University Press, 328-344

[26] A. Selberg - "On Discontinuous Groups in Higher Dimensional Symmetric Spaces" - Int. Colloq. Function Theory, Tata Institute, Bombay 1960, 147-164

[27] H. Short - private communication

[28] E.B. Vinberg - "On Groups with Periodic Relations" - to appear

[29] H. Zieschang-"Uber die Nielsensche Kurzungsmethode in freien Produckten mit Amalgam" - Invent. Math. 10 (1970) 4-37

[30] H. Zieschang and R. Kaufman - "On the Rank of NEC Groups" - *Discrete Groups and Geometry* - London Math. Soc. Lecture Notes 173 (1992) 137-148

BENJAMIN FINE, DEPARTMENT OF MATHEMATICS, FAIRFIELD UNIVERSITY, FAIRFIELD, CONNECTICUT 06430, UNITED STATES

GERHARD ROSENBERGER, FACHBEREICH MATHEMATIK, UNIVERSITAT DORTMUND, POSTFACH 50 05 00, 4600 DORTMUND 50, FEDERAL REPUBLIC OF GERMANY

Contemporary Mathematics
Volume **164**, 1994

Collapsing and reconstruction of manifolds

DAVID GILLMAN, SERGEI MATVEEV AND DALE ROLFSEN

Dedicated to the memory of Professor Israel Berstein.

Introduction

We discuss some consequences of a 3-manifold reconstruction theorem which recently appeared in [GR2], and announce the following generalization to n-dimensions. We would like to thank the organizers for a most stimulating and well-planned conference.

THEOREM OF RECONSTRUCTION. *Let M^n be a compact, connected, orientable PL n-dimensional manifold. Then M collapses to an $(n-1)$-dimensional polyhedron K with the property that $K \times I$ collapses to a (PL) homeomorphic copy of M.*

In other words, M can be "reconstructed" from the product of an appropriate spine with the unit interval $I = [0, 1]$. We'll state it more precisely below; the proof will appear in a subsequent paper. The polyhedra K needed for this theorem are the *special* polyhedra introduced in higher dimensions by Matveev [Mat], with an additional "orientability" condition which will also be discussed below. As a consequence we show that all special spines K^{n-1} of the n-ball obey the Zeeman hypothesis (K contractible implies $K \times I$ collapsible) and answer some related questions regarding collapsibility of other contractible spaces. Another application of the reconstruction theorem is that any manifold M, as above, admits a codimension zero embedding in a cartesian product of trees.

All discussion in this paper should be understood to be within the category of compact polyhedra (or finite simplicial complexes) and piecewise-linear maps. We assume the reader is familiar with standard PL terminology, as defined for

1991 *Mathematics Subject Classification.* Primary 57N35; Secondary 57M20, 57N15.

An expanded version of this paper, with proofs and generalizations, is being submitted for publication elsewhere.

example in [R-S] or [Zee1]. In particular $X \searrow Y$ means X collapses to Y, using some triangulation, and we call Y a *spine* of X; if $Y = *$, a point, X is said to be *collapsible*. Non-collapsible, yet contractible, 2-complexes exist, but E. C. Zeeman [Zee2] conjectured that any contractible K^2 has the property that $K \times I$ is collapsible. Zeeman pointed out that his conjecture would imply the famous conjecture of Poincaré: every simply-connected 3-manifold is homeomorphic with the sphere S^3. Both conjectures are unsolved at the time of this writing.

M. Cohen defined a space X to be *q-collapsible* if $X \times I^q$ is collapsible, and showed in [Coh1] that there exist contractible polyhedra, of each dimension greater than 2, which fail to be 1-collapsible; in other words, Zeeman's hypothesis fails in higher dimensions. He showed, with Berstein and Connelly, [BCC], that for each positive integer q there exist contractible complexes which fail to be q-collapsible. On the positive side, Cohen showed in [Coh2] that any contractible k-dimensional complex K^k is q-collapsible for some q; in fact one can take $q = 2k$ for $k \geq 3$ and $q = 6$ for $k = 2$.

Special polyhedra and orientations

For a given positive integer k, the class of *special* k-dimensional polyhedra may be defined as the (compact) polyhedra which are locally modelled on the k-skeleton of the $k + 2$-simplex, Δ_k^{k+2}. For the case $k = 2$, this class has been well-studied, under various names such as fake surfaces and standard complexes, and reflects the singularities of a generic soap film. By PL invariance of links, there is a well-defined stratification of a special K^k by its intrinsic i-skeleta , $0 \leq i \leq k$, consisting of points matched with Δ_i^{k+2}. We also assume, with Matveev, that the intrinsic i-skeleta are unions of open i-cells. In other words, a special polyhedron is a CW complex with respect to its intrinsic skeleta. If K^{n-1} is a special complex embedded in a PL manifold M^n, we will also assume the embedding is nice, in that the pair (M, K) is assumed to be locally homeomorphic to the pair $(\Delta_n^{n+1}, \Delta_{n-1}^{n+1})$, where $\Delta_n^{n+1} = \partial \Delta^{n+1}$ is the standard PL n-sphere. This assumption is equivalent to assuming the embedding does not contain a counterexample to the Schönflies conjecture.

Note that each component of the $k-1$-skeleton of a special k-dimensional complex K^k, is locally incident with three sheets of the k-skeleton. A "fundamental class" for K consists of orientations of the components of the k-skeleton in such a way that at each $k - 1$ stratum, there is partial cancellation in the boundaries of the 3 incident k-strata. More formally, consider the chain groups $C_*(K^k)$. The components of the k-skeleton and the $k - 1$ skeleton define $c_1^k, \ldots, c_{r_k}^k \in C_k(K^k)$ and $c_1^{k-1}, \ldots, c_{r_{k-1}}^{k-1} \in C_{k-1}(K^k)$ respectively. A fundamental class is a k-chain of the form

$$c^k = \sum_1^{r_k} \epsilon_i c_i^k,$$

for some choice of $\epsilon_i = \pm 1$, with the property that the boundary has the form

$$\partial c^k = \sum_1^{r_{k-1}} \delta_i c_i^{k-1},$$

with all $\delta_i = \pm 1$ (and not ± 3, as is possible). If K^k possesses such a fundamental class, we say that K is *orientable*.

Examples: (1) The famous "house with two rooms" is a special spine of the 3-ball with two vertices. It is known to be 1-collapsible. The complex associated with the group presentation $\{x, y; x = 1, xyx^{-1}y^{-2} = 1\}$ is a special spine of B^3 with a single vertex. They are both 1-collapsible, indeed we showed in [GR1] that all special spines of the 3-ball are 1-collapsible.

(2) The Poincaré homology sphere can be constructed from a solid regular dodecahedron, by identifying opposite faces, after a $\pi/5$ twist. If one removes the interior of the dodecahedron, taking the boundary with identifications, the resulting complex K is a special spine of the punctured Poincaré manifold. It is orientable, but of course is not contractible. The suspension $\Sigma(K)$ is an example of a contractible 3-complex which fails to be 1-collapsible (see [BCC]).

(3) The complex K associated to the group presentation $\{x; x^3 = 1\}$ is a spine of the punctured lens space $L(3, 1)_0$. It fails to be a special complex only because it fails to have a 0-skeleton. The reconstruction theorem fails for this example – Gillman has shown that $L(3, 1)_0$ cannot essentially embed in $K \times I$.

(4) The punctured lens space $L(5, 2)_0$ has a special spine obtained as follows. On the torus boundary T of a solid torus, consider a meridian curve J and a curve $L \subset T$ of homology class 5 meridians + 2 longitudes. $J \cup L$ separates T into a number of rectangles – let R be one of them. Let K be the union of $T - int(R)$, a meridian disk sewn to J and another disk sewn to L. Then K is a spine of $L(5, 2)_0$ with a single vertex. It is *not* orientable.

The following were proved by Casler [Cas], for n=3, and for higher dimensions by Matveev [Mat] – except for the orientability condition in the existence theorem, which is not difficult to arrange.

THEOREM: EXISTENCE. *Every compact, connected PL manifold M^n, with nonvoid boundary, collapses to an orientable, $(n - 1)$-dimensional special spine K.*

THEOREM: UNIQUENESS. *Suppose $n \geq 3$ and M_i^n are compact manifolds with special spines K_i^{n-1}, $i = 1, 2$. If K_1 is homeomorphic to K_2, then M_1 is homeomorphic to M_2. In fact any homeomorphism $K_1 \to K_2$ extends to a homeomorphism $M_1 \to M_2$*

Example: The surface RP^2 is a spine of both $RP^2 \times I$ and punctured RP^3, so the uniqueness result would fail if we weaken the definition of special complex to include manifolds, not insisting on an intrinsic CW structure. The same holds for the following, our main result.

THEOREM: RECONSTRUCTION. *If K^{n-1} is an orientable special spine of the PL manifold M^n, then $K \times I$ collapses to a subset PL homeomorphic with M.*

Consequences of the reconstruction theorem

First of all, Zeeman's conjecture holds for certain high-dimensional polyhedra.

COROLLARY 1.. *If K is an orientable special $(n-1)$-dimensional spine of a PL n-ball, then K is 1-collapsible.*

PROOF. $K \times I \searrow B^n \searrow *$.

Recall that Cohen showed that contractible 2-complexes are 6-collapsible, while Zeeman conjectured 1-collapsibilty. This broad gap in our knowledge can be narrowed somewhat in the case of special spines.

COROLLARY 2.. *Let K^2 be a contractible special complex which embeds in a 3-manifold. Then K is 4-collapsible. If the 4-dimensional PL Poincaré conjecture holds, K is 3-collapsible. If the 3-dimensional Poincaré conjecture holds, K is 1-collapsible.*

PROOF. If $M^3 \searrow K^2$, then by reconstruction, and abuse of notation, $K \times I \searrow M$. Now we know that $M \times I^3$ is a 6-ball, since it is contractible and its boundary is simply-connected, and the PL Poincaré conjecture is true in dimensions 5 and 6. Therefore $K \times I^4 \searrow M \times I^3 = B^6 \searrow *$. The other parts follow similarly.

As another application of the reconstruction theorem, we will show that compact PL manifolds with boundary can be embedded in "small" collapsible complexes. If X and Y are complexes, let $X * Y$ denote their (unreduced) join, and let $c(X)$ denote the cone on X; $c(X) = X * \{pt\}$. For any positive integer r, let r also denote the discrete space consisting of r points, so that $c(r)$ is a tree.

THEOREM: EMBEDDING. *Let M^n be a compact, connected PL manifold with nonempty boundary. Then there exist positive integers $r_1, \ldots, r_n$ and an embedding of M^n in $c(r_1) \times \cdots \times c(r_n)$.*

PROOF. By reconstruction, M embeds in $K^{n-1} \times I$ and hence it also embeds in the cone $c(K)$. For each $i = 1, \ldots, n$, let r_i be a finite set in one-to-one correspondence with the (simplicial) $i-1$-skeleton of K. By an elementary argument, one can embed K in $r_1 * \cdots * r_n$. Therefore, M embeds in $c(r_1 * \cdots * r_n)$. The theorem follows from the PL identity $c(X * Y) \cong c(X) \times c(X)$, which implies $c(r_1 * \cdots * r_n) \cong c(r_1) \times \cdots \times c(r_n)$.

An easy construction shows that one can embed any connected, compact 2-manifold with nonempty boundary in the fixed 2-complex: $c(2) \times c(3)$. It has been recently announced by Zhongmou Li that all compact 3-manifolds with nonvoid boundary embed in $c(2) \times c(3) \times c(3)$. We don't know if, for general n in the embedding theorem, one can conclude that the r_i depend only on n and not on M.

References

BCC. I. Berstein, M. Cohen, R. Connelly, *Contractible, non-collapsible products with cubes*, Topology **17** (1978), 183–187.

Cas. B. Casler,, *An embedding theorem for connected 3-manifolds with boundary*, Proc. Amer. Math. Soc. **16** (1965), 559–566.

Coh. M. Cohen, *Dimension estimates in collapsing $X \times I^q$*, Topology **14** (1975), 253–256.

GR1. D. Gillman, D. Rolfsen, *The Zeeman conjecture for standard spines is equivalent to the Poincaré conjecture*, Topology **22** (1983), 315–323.

GR2. D. Gillman, D. Rolfsen, *Three-manifolds embed in small 3-complexes*, International J. Math. **3** (1992), 179–183.

Mat. S. Matveev, *Special spines of piecewise-linear manifolds*, Math USSR Sbornik **21** (1973), 279–291.

RS. C. Rourke, B. Sanderson, *Introduction to piecewise-linear topology*, vol. 69, Springer, 1972.

Zee1. E. C. Zeeman, *Seminar on combinatorial topology*, IHES Notes, 1963.

Zee2. E. C. Zeeman, *On the dunce hat*, Topology **2** (1964), 341–358.

DEPARTMENT OF MATHEMATICS, UNIVERSITY OF CALIFORNIA, LOS ANGELES, USA

DEPARTMENT OF MATHEMATICS, UNIVERSITY OF CHELYABINSK, RUSSIA

DEPARTMENT OF MATHEMATICS, UNIVERSITY OF BRITISH COLUMBIA, VANCOUVER, CANADA

Contemporary Mathematics
Volume **164**, 1994

Metrics on Manifolds with Convex or Concave Boundary

JOEL HASS

ABSTRACT. This paper begins an investigation of the restrictions on the topology of a manifold imposed by assuming the existence of a metric of positive or negative curvature with convex or concave boundary.

1. Introduction

In this paper we investigate what restrictions on the topology of a smooth manifold with boundary are imposed by the assumption that the manifold has a metric of positive or negative curvature and convex or concave boundary. The boundary conditions put restrictions on the possible metrics that the manifold carries. For example a 3-manifold with convex boundary and positive sectional (or even scalar) curvature is homeomorphic to a 3-ball [6]. In contrast, a metric can be constructed on any 3-manifold with non-empty boundary in which the manifold has negative sectional curvature and the boundary is concave [1].

We obtain here a complete solution for the two-dimensional case. In three dimensions we obtain a complete answer in three out of four cases, and partial results in the other case.

Understanding the topology of bounded manifolds with restricted metrics is a natural question from several points of view. For example, in the general theory of

1991 Mathematics Subject Classification Primary 53A10; Secondary 57M35.

Supported by NSF Grant DMS90-24796.

This paper is in final form and no version of it will be submitted for publication elsewhere.

relativity, one encounters four-dimensional space-times which contain black holes. A space-like slice in such a space-time is a 3-manifold with concave boundary.

The existence results for certain types of metrics are not what one might expect from Thurston's geometrization conjecture for closed 3-manifolds. Thurston's conjecture implies that closed 3-manifolds with negative sectional curvature metrics admit hyperbolic metrics. In contrast, there are 3-manifolds with negative sectional curvature and concave boundary which do not have hyperbolic metrics with concave boundary.

We will take concave and convex to mean strictly concave and strictly convex. Thus the boundary of a manifold is concave if its second fundamental form is positive definite with respect to the outward pointing normal.

2. Dimension two

We begin in dimension two, where a complete answer can be given to the questions of existence of metrics with prescribed conditions.

2.1 Convex boundary and positive curvature. Only the disk admits this type of metric among surfaces with boundary, since a boundary curve of any other surface can be homotoped to a non-trivial shortest geodesic in its homotopy class. The second variation formula for the length of curves [4] shows that shortest geodesics in a homotopy class cannot occur in positive curvature. Alternately the Gauss-Bonnet theorem implies that the Euler characteristic of such a surface must be positive, and so it must be a disk.

2.2 Concave boundary and positive curvature. This condition is less restrictive than the previous one.

THEOREM 1. *Any bounded surface admits a metric with positive sectional curvature and concave boundary.*

Proof: The proof proceeds by explicit construction of the metric. First consider an orientable punctured surface. Immerse the punctured surface on the 2-sphere so that the boundary is concave, as indicated in Figure 1.2.2 for a punctured torus, and pull back the metric.

Figure 1

By using 2g handles as in Figure 1, we obtain a metric of positive curvature and concave boundary on a once punctured orientable surface of genus g. Removing additional small disks allows us to increase the number of punctures.

Non-orientable surfaces which are the connect sum of an odd number of projective planes, with some punctures removed, can be obtained in a similar fashion by starting with a projective plane with a small disk removed and adding handles just as above. The result is an immersion of the surface into RP^2 with concave boundary. The pulled back metric satisfies the desired conditions.

It remains to construct a metric on a punctured Klein bottle with some number of handles attached. To do so we remove a disk from a projective plane leaving a boundary which is concave except along a geodesic segment. The boundary of this segment makes two right angles with the rest of the curve, as shown in Figure 2.

Figure 2

By gluing two of these punctured projective planes together along their respective geodesic segments, we obtain a punctured Klein bottle with concave boundary and positive sectional curvature. Small disks can be removed from this surface and handles can be added as before to get the remaining non-orientable surfaces.

2.3 Convex boundary and negative curvature. Any bounded surface admits a metric with convex boundary and negative curvature.

A bounded surface admits a complete hyperbolic metric with each end having infinite volume. Cutting off each end near infinity gives a hyperbolic metric on the surface with convex boundary.

2.4 Concave boundary and negative curvature. Any bounded surface admits a metric with concave boundary and negative curvature, except for the disk, Mobius band, and annulus.

A bounded surface admits a hyperbolic metric with finite area, unless it is homeomorphic to the disk, Mobius band or annulus. Cutting off the cusps of such a surface along a horocircle gives a hyperbolic metric with concave boundary. The Gauss-Bonnet theorem rules out the existence of such a metric on the disk, Mobius band, and annulus.

3. Dimension three

In dimension three, the question of which manifolds admit metrics of positive or negative curvature and convex or concave boundary is solved except for one case - positive curvature and concave boundary. The known results are not direct extensions of the 2-dimensional case.

3.1 Positive curvature and convex boundary. Only the 3-ball admits such a metric [3][6]. The boundary consists of a union of 2-spheres, by the Gauss-Bonnet Theorem. The idea of the proof of this case is that one can isotop the boundary 2-sphere to a least area 2-sphere. Such a 2-sphere cannot exist in a manifold of positive scalar curvature, by an application of the formula for the second variation of area. So the boundary is isotopic to a point and the manifold is the 3-ball.

3.2 Positive curvature and concave boundary. As in the previous case, the Gauss-Bonnet Theorem implies that a 3-manifold with positive sectional curvature and concave boundary has a boundary which consists of a union of 2-spheres. Such metrics can be constructed on a punctured spherical space forms by removing a collection of small balls from the manifold. We do not know if this gives rise to all manifolds of this type.

3.3 Negative curvature and concave boundary. The two-dimensional case suggests that few manifolds should have metrics with negative curvature and concave boundary. It turns out however that all bounded 3-manifolds admit such a metric [1]. The construction of such a metric starts by removing a knotted graph from the interior of the manifold to produce a manifold which admits a hyperbolic metric with totally geodesic boundary. The boundary is then pushed in slightly to obtain a concave boundary. One finally adds 2-handles to recover the topology of the original 3-manifold. If the attaching curves of the 2-handles are sufficiently long, then the properties of negative curvature and concave boundary can be retained.

3.4 Negative curvature and convex boundary. We get a complete characterization for this case by applying Thurston's work on hyperbolic Haken manifolds [5][7]. A 3-manifold has a metric with negative curvature and convex boundary if and only if it is irreducible and contains no π_1-injective tori.

In a 3-manifold with convex boundary one can minimize area in the homotopy or isotopy class of a surface without worrying that the surface might try to bump against the boundary [2]. Thus if there are essential 2-spheres or tori in M, one could find a least area essential torus or 2-sphere in their homotopy or isotopy class. For our purposes an essential 2-sphere means an embedded 2-sphere not bounding a ball and an essential torus means a π_1-injective torus, possibly boundary parallel. The Gauss normal curvature formula and the Gauss-Bonnet theorem imply that least area immersed surfaces of non-negative Euler characteristic do not exist in a negatively curved manifold, so that M is irreducible and contains no essential tori. It follows that many punctured 3-manifolds do not admit such a metric.

If M has no essential 2-spheres or tori, Thurston's hyperbolization theorem for Haken manifolds implies the existence of a complete, geometrically finite hyperbolic metric. See Theorem A' of [5]. There are no cusps since there are no boundary tori on M. The convex core gives a compact manifold homeomorphic to M. Enlarging the convex core by a small value $\varepsilon>0$ gives a hyperbolic manifold homeomorphic to M and having smooth convex boundary.

4. Some open problems

The following are among many open problems concerning metrics on bounded manifolds.

QUESTION 1. *Are these the only 3-manifolds with positive sectional curvature and concave boundary?*

QUESTION 2. *Characterize manifolds which admit negative sectional curvature metrics with concave boundary in dimension larger than three.*

QUESTION 3. *Does the n-dimensional ball admit a negative sectional curvature metric with concave boundary for all $n\geq3$?*

QUESTION 4. *Characterize n-manifolds which admit positive curvature metrics with concave boundary, n≥3.*

REFERENCES

1. J. Hass, Bounded 3-manifolds admit negatively curved metrics with concave boundary, preprint.

2. J. Hass and G. P. Scott, The existence of least area surfaces in 3-manifolds, Trans. Amer. Math. Soc. **310**, (1988) 87-114.

3. W. Meeks, L. Simon and S.T. Yau, Embedded minimal surfaces, exotic spheres and manifolds of positive Ricci curvature, Annals of Math. **116** (1982) 621-659.

4. J. Milnor, *Morse Theory*, Princeton University Press (1969) Princeton, New Jersey.

5. J. Morgan, On Thurston's Uniformization Theorem for Three-Dimensional Manifolds, *Proc. Conf. on the Smith Conjecture*, J. Morgan and H. Bass (ed), Academic Press, (1984).

6. J.H. Rubinstein, Embedded minimal surfaces in 3-manifolds with positive scalar curvature, University of Melbourne Research Report 8 (1984).

7. W. Thurston, *The geometry and topology of 3-manifolds*, Princeton University Lecture Notes (1978).

Department of Mathematics, University of California, Davis, CA 95616
E-mail address: hass@math.ucdavis.edu

Contemporary Mathematics
Volume **164**, 1994

Geometric Structures on Branched Covers over Universal Links

KERRY N. JONES

ABSTRACT. A survey of recent results is presented which bear on the question of what geometric information can be gleaned from the representation of a three-manifold as a branched cover over a fixed universal link. Results concerning Seifert-fibered manifolds, graph manifolds and hyperbolic manifolds are discussed.

0. Introduction

Closed, orientable three-manifolds admit a variety of universal constructions, that is, constructions by which all manifolds of that class are obtainable, e.g., Heegaard diagrams, surgery diagrams, etc. One of the difficulties faced in three-manifold topology is the decision as to which of the universal constructions is most likely to yield a solution to a particular problem. In this paper, we present several recent results which arise in the use of the universal link construction to work toward the Thurston Geometrization Conjecture.

More specifically, we present a structure theorem for nonpositively curved Euclidean cone manifolds without vertices, which allows us to deduce which geometries are possible for the pieces of the torus decomposition of certain branched covers, and, in fact, to construct the characteristic submanifold for such covers (in this context, "curvature" refers to a combinatorial condition on the branching indices of a branched cover). We also obtain results concerning hyperbolic structures on negatively curved hyperbolic cone manifolds. Cone manifold structures (cone metrics) are the natural geometric structures to consider in connection with branched covers, since a cone metric on the base space of a branched covering map may be lifted to the branched cover.

1991 *Mathematics Subject Classification.* Primary 57M12; Secondary 57R15, 57M25.
Key words and phrases. branched cover, universal link, cone manifold.
This paper is in final form and no version of it will be submitted for publication elsewhere

We also present the negative result that, at least for some universal links, a geometric structure on the cover is not always reflected in the branched covering map itself. More specifically, there exist hyperbolic manifolds, all of whose cone manifold structures arising from branched covering maps over a fixed universal link (the Borromean rings) have some positive curvature.

As this paper is, for the most part, a survey of existing results, only sketches of proofs (with references to full proofs) will be presented. Only Theorem 3.4 (and its corollaries) are results that have not appeared elsewhere.

1. Universal Links

We begin by defining branched covers as well as fixing the notation we will use subsequently

DEFINITION. *A branched covering map is a continuous map of pairs* $\rho :$ $(\hat{M}, \hat{L}) \to (M, L)$ *where* $\hat{M}, M$ *are n-manifolds and* $\hat{L}, L$ *are* $(n-2)$-*subcomplexes of* $\hat{M}, M$, *respectively, such that* $\rho^{-1}(L) = \hat{L}$ *and* ρ *is a covering map when restricted to both* $\hat{L}$ *and* $\hat{M} - \hat{L}$. *We say that* $\hat{M}$ *is a branched cover of* M, *branched over* L.

For the purposes of this paper, we are interested in the case in which $n = 3$ and $\hat{L}, L$ are links. In this case, we may associate to each component $\hat{L}_0$ of $\hat{L}$ its *branching index*, which is the ratio of the degree of ρ, restricted to the boundary of a regular neighborhood of $\hat{L}_0$ to the degree of ρ, restricted to $\hat{L}_0$ itself. Equivalently, this is the degree of ρ restricted to a disk in the regular neighborhood of $\hat{L}_0$ transverse to $\hat{L}_0$. We also note that the branched cover is completely determined by the *associated covering map*, the restriction of ρ to $\hat{M} - \hat{L}$. We will also use the fact that covering maps of degree d over $M - L$, and thus branched covering maps over (M,L) are in 1-1 correspondence with conjugacy classes of transitive representations of $\pi_1(M - L)$ into S_d (that is, representations whose image acts transitively on the set $\{0, 1, \ldots, d - 1\}$). This *monodromy representation* associated to a branched covering map is a convenient tool for working with a branched cover. In particular, the branching indices of preimages of a component L_0 of L are the cycle lengths of the monodromy evaluated on a meridian of L_0.

Concerning universal links, we make the following formal

DEFINITION. *A universal link is a link in* S^3 *with the property that all closed, orientable 3-manifolds are representable as branched covers over that link.*

It is, of course, by no means obvious that universal links exist at all, but the following theorem tells us that they in fact exist in abundance.

THEOREM 1.1 [6] AND [7]. *Any hyperbolic 2-bridge link is universal, as are the Borromean rings, the Whitehead link and the knot* 9_{46}.

In fact, at the present time, no hyperbolic links are known to be *not* universal. The only known obstruction to universality is that a universal link cannot be the connected sum of iterated torus knots or links (see [8]). The problem here is that any branched covering of such a link either contains an essential 2-sphere or is Seifert-fibered.

Y. Uchida has recently shown (see [18]) that hyperbolicity implies universality for another class of links called *chains* (a chain is a $(\epsilon, \epsilon, \ldots, \epsilon, 2, 2, \ldots, 2)$ pretzel link where $\epsilon = \pm 1$).

2. Cone Manifolds and Branched Covers

DEFINITION. *A Euclidean cone manifold is a metric space obtained as the quotient space of a disjoint union of a collection of geodesic n-simplices in $\mathbb{E}^n$ by an isometric pairing of codimension-one faces in such a combinatorial fashion that the underlying topological space is a manifold. Hyperbolic and spherical cone manifolds are defined similarly.*

One may define cone manifolds in more generality than this (see [9], for example), but this definition will suffice for our purposes. Such a space possesses a Riemannian metric of constant sectional curvature on the union of the top-dimensional cells and the codimension-1 cells. On each codimension-2 cell, the structure is completely described by an angle, which is the sum of the dihedral angles around all of the codimension-2 simplicial faces which are identified to give the cell. The *cone locus* of a cone manifold is the closure of all the codimension-2 cells for which this angle is not 2π (the Riemannian metric may be extended smoothly over all cells whose angle is 2π). In this paper, we are principally interested in the case where $n = 3$ and the cone locus is a link (which must have constant cone angle on each component). We will refer to such cone manifolds as *cone manifolds without vertices*, meaning no vertices in the cone locus graph with valence other than 2. The structure at points of lower-dimension cells is not quite so easily described, but we will content ourselves with the observation that some neighborhood of any point in any cone manifold of dimension n is a metric cone on a spherical cone manifold of dimension $n-1$. Thus, each point in a cone manifold determines a spherical cone metric on S^{n-1}, after normalization so that the smooth portions of this sphere have curvature $+1$. We will refer to this metric on S^{n-1} as the *normalized link metric* of a point. In the "no vertices" case, this metric is always either the usual smooth metric on S^2 or a cone metric with two diametrically opposed cone points with equal cone angles.

We should note here that there is a strong connection between cone manifolds and orbifolds (see [16]), namely that orbifolds are cone manifolds with all cone angles of the form $2\pi/k$ for some integer k (k is the order of the isotropy of a non-vertex cone point). Orbifolds are (generally) the quotient of a simply-connected manifold by a properly discontinuous, but not necessarily free group action. Note, however, that there need not be such a group action lurking in the background for cone manifolds in general.

One of the principal reasons we are interested in cone manifold structures in connection with universal links is that cone metrics may be lifted to branched covers. More particularly, if $(\hat{M}, \hat{L})$ is a branched cover over (M, L) and M admits a cone metric in which the cone locus is contained in L, then we may lift this cone metric to a cone metric on $\hat{M}$ in which the cone locus is contained in $\hat{L}$ (providing L is geodesic in M). The cone angles on the components of $\hat{L}$ are the cone angles on the corresponding components of L, multiplied by the corresponding branching indices of the branched covering.

We now have the machinery to make the following

DEFINITION. *Let M be a 3-dimensional orbifold with singular set Σ a link in M and let L be a geodesic link containing Σ. Then, a branched cover $(\hat{M}, \hat{L})$ over (M, L) is said to be sufficiently branched if all of the branching indices over each component of L are greater than or equal to the order of the isotropy group of that component of L. Similarly, a branched cover is said to be totally insufficiently branched if all of the branching indices over each component of L are less than or equal to the order of the isotropy group of that component.*

Note that if we lift the orbifold structure to a sufficiently branched cover, all cone angles are $> 2\pi$ and if we lift to a totally insufficiently branched cover, all cone angles are $< 2\pi$.

Geodesics in a cone manifold are piecewise geodesics (with respect to the underlying constant curvature model) which join at points of the cone locus in such a way as to have an angle of at least π between them. Angles between geodesics at a cone point are measured by considering the points of intersection between the two geodesics and the spherical link mentioned earlier. The angle is the distance in the normalized link metric between the two points of intersection. The upshot of this in the "no vertices" case is that if a geodesic encounters a component of the cone locus with cone angle $< 2\pi$, there is no way to continue this geodesic and if a geodesic encounters a component of the cone locus with cone angle $> 2\pi$, there are an infinite number of distinct ways to continue that geodesic. In the "no vertices" case, there is also an alternate way of measuring the angle between geodesics, namely by projecting to a totally geodesic disk perpendicular to the component of cone locus at the intersection and measuring the angle in this 2-dimensional cone manifold. As an intuition-building exercise, show that these methods do in fact differ and that in the "no vertices" case, the former method always yields an angle of at most π, so that geodesic continuation can be recognized by continuation along an angle of exactly π.

As might be conjectured by considering the Gauss-Bonnet theorem, there is a very strong analogy between cone angle and curvature, with cone angles greater than 2π behaving like negative curvature and cone angles less than 2π behaving like positive curvature. As an example of this behavior, consider two parallel geodesics in a Euclidean cone manifold that pass on either side of a component of cone locus perpendicular to their common (local) plane. If the cone angle is less than 2π, the geodesics intersect after passing by the cone geodesic and if the cone angle is greater than 2π, they diverge after passing by the cone geodesic. This analogy is made precise by the following

THEOREM 2.1. *Let M be a 3-dimensional cone manifold with no vertices. Then,*

(1) *if M is spherical and all cone angles are less than 2π, M admits a Riemannian metric of positive sectional curvature.*

(2) *if M is Euclidean and all cone angles are less than 2π, M admits a Riemannian metric of nonnegative sectional curvature.*

(3) *if M is Euclidean and all cone angles are greater than 2π, M admits a Riemannian metric of nonpositive sectional curvature.*

(4) *if M is hyperbolic and all cone angles are greater than 2π, M admits a Riemannian metric of negative sectional curvature.*

IDEA OF PROOF. In each case, one constructs explicitly a smooth metric by altering the cone metric in a tubular neighborhood of the cone locus and verifies the sectional curvature bounds. See [10] for details. Note that this immediately translates into a statement about sufficiently branched and totally insufficiently branched covers over various flavors of orbifolds.

This smoothing technique will be very useful for us in using the machinery of differential geometry to assist in some of our proofs (although direct cone manifold proofs could probably be constructed). Gromov and Thurston in [4] use this technique in higher dimensions to deduce the existence of negatively curved manifolds which are not hyperbolic. The theorem is probably true even allowing vertices, but it is not clear how to go about explicitly constructing a metric for which the curvature bounds can be verified.

We conclude this section by mentioning the work of Aitchison and Rubinstein (see [1]) on polyhedral metrics (which are essentially Euclidean cone metrics). The focus of their work is on cone manifolds that *do* have vertices, but which are built up from pieces of very restricted shapes such as Euclidean cubes and "flying saucers" (generalized cubes which have a cone geodesic with cone angle $2\pi k/3$ between diagonally opposed vertices and have $2k$ square faces). They deduce some very strong results about such cone manifolds when they are nonpositively curved (i.e., when there are no geodesics of length $< 2\pi$ in the normalized metric of any vertex). However, as we will see, it is likely that the manifolds to which their methods and ours both apply are generally Haken manifolds, so there does not appear to be a great deal of useful interplay between the two approaches.

3. Surfaces in Branched Covers

In this section, we investigate the π_1-injectively immersed surfaces (especially tori) in sufficiently branched covers over Euclidean orbifolds (or, equivalently, Euclidean cone manifolds with all cone angles $> 2\pi$).

For the smoothing results of the last section to be truly useful, we must know that the geometry of the singular metric is in fact closely related to the geometry of the smoothed metric. The next few results tell us that, at least for torus submanifolds, this is indeed the case. Details of these are found in [11].

Most of these results are probably valid for surfaces in general, but the proofs in
[11] need tori, and we won't need the additional generality here.

PROPOSITION 3.1. *If M is a Euclidean cone manifold with all cone angles
$> 2\pi$ and S is a totally geodesic (cone metric) immersed torus, then there is a
smooth metric (as in Theorem 2.1) in which S is homotopic to a totally geodesic
torus which has the same intersection pattern with the smoothing neighborhood
that S does with the cone locus.*

PROOF. See [11], Lemma 1.3.

PROPOSITION 3.2. *If M is a Euclidean cone manifold with all cone angles
$> 2\pi$ and S is a π_1-injectively immersed torus in M, then S is homotopic to a
totally geodesic (cone metric) immersed torus.*

IDEA OF PROOF. See [11], Lemma 1.4. Here we are using the fact that
π_1-injectively immersed surfaces in Riemannian 3-manifolds are homotopic to
minimal surfaces [14] as well as the fact that minimal tori in a 3-manifold of
nonpositive sectional curvature must be totally geodesic (a straightforward Gauss
equation argument).

PROPOSITION 3.3. *If M is a Euclidean cone manifold with all cone angles
$> 2\pi$ and S is a totally geodesic surface in M, then S is a π_1-injective immersed
surface in M.*

IDEA OF PROOF. One simply uses the fact that geodesics in the universal
cover of M diverge at least linearly to show that there can be no homotopically
trivial closed geodesics in M.

We will use these results in the next section, but first we will use Proposition
3.3 to deduce the following rather unfortunate

THEOREM 3.4. *Let M be a branched cover over S^3, branched over the Borromean rings with no branching indices equal to one. Then, M is Haken.*

PROOF. S^3 admits a Euclidean orbifold structure with singular locus the
Borromean rings and all cone angles π that arises from "folding" up the faces of
a cube as in Figure 3.1. Consider the horizontal plane that cuts through the lines
between the faces labelled A and A' (and D and D') one quarter of the distance up
the cube. Developing this plane one finds that in all directions the "next" plane
is either this same plane or a parallel plane three-quarters of the distance up the
cube. These two planes close up to yield an imbedded totally geodesic sphere in
S^3 that only intersects one component of the singular locus (four times). Lifting
this sphere to M, we find an imbedded totally geodesic surface, which, by Prop.
3.3, must be π_1-injective. It is clearly 2-sided, and hence incompressible. Since
M admits a metric of nonpositive sectional curvature, it must be irreducible by
the Cartan-Hadamard Theorem, and thus, M is Haken.

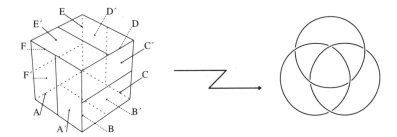

FIGURE 3.1

It may seem a bit curious to have labelled this theorem "unfortunate," since, normally, proving that manifolds are Haken is a cause for rejoicing. However, in this case we have shattered any illusion that geometric structures on manifolds might be always "realizable" by some branched cover representation over any given universal link. More specifically, we have

COROLLARY 3.5. *There exist infinitely many hyperbolic 3-manifolds M for which every representation as a branched cover over the Borromean rings has some branching index equal to 1 (and thus positive curvature in every lifted cone manifold structure).*

PROOF. Let M be any non-Haken hyperbolic 3-manifold and use Theorem 3.4.

We do, however, also obtain the following two more agreeable corollaries:

COROLLARY 3.6. *Let M be a closed, orientable 3-manifold. Then M is double-branch-covered by a Haken manifold.*

PROOF. Represent M as a branched cover over the Borromean rings (possible since they are universal) and let L be the index-1 branching locus of M relative to this branched covering. Then, the 2-fold cyclic branched cover of M, branched over L is Haken, by Theorem 3.4.

This is not a new result (see, for example, [**3**]), but is a much simpler proof than any of the earlier proofs of this result.

COROLLARY 3.7. *Let M be a branched cover of S^3, branched over the Borromean rings, with no branching indices equal to 1 or 2. Then, M is hyperbolic.*

PROOF. By Theorem 2.1 and the fact that there is a hyperbolic orbifold structure on the Borromean rings with all cone angles $2\pi/3$, M admits a metric of negative sectional curvature (since all branching indices are 3 or greater and thus all lifted cone angles are 2π or greater, but any cone angles equal to 2π are nonsingular and are not considered part of the cone locus of M). By Theorem 3.4 and Thurston's Hyperbolization Theorem for Haken manifolds, M is hyperbolic.

4. Torus Decomposition of Branched Covers

In this section, we will consider the torus decomposition of a sufficiently branched cover over a Euclidean orbifold and investigate the possible geometries that may arise. The reader is referred to [15] for a thorough and beautiful discussion of the eight three-dimensional homogeneous geometries. It is quite easy to see that if there is a proper decomposition, the Seifert-fibered components must have $\mathbb{H}^2 \times \mathbb{R}$ or $\mathbb{E}^3$ geometry, since a Seifert-fibered manifold with boundary must have one of the three product geometries and the only $S^2 \times \mathbb{R}$ manifold with boundary is a solid torus, in which case the splitting torus would be compressible. It is also fairly easy to distinguish these two cases geometrically in the following manner: first, homotope the splitting tori to totally geodesic tori (or possibly Klein bottles) using standard minimal surface techniques and the theorems of the previous section. Then, if a Seifert-fibered component of the (proper) torus decomposition contains a component of the cone locus, it must have $\mathbb{H}^2 \times \mathbb{R}$ geometry, and if it does *not* contain a component of the cone locus, it must have $\mathbb{E}^3$ geometry (see [10], essentially this is a corollary of a fact, proven there, that a closed Euclidean 3-manifold cannot admit a Euclidean cone manifold structure with all cone angles greater than 2π). For manifolds with empty torus decomposition (Seifert-fibered closed manifolds), we have the result (also in [10], proved by a straightforward "growth of groups" argument) that a *closed* Seifert-fibered Euclidean cone manifold with cone angles $> 2\pi$ must have $\mathbb{H}^2 \times \mathbb{R}$, Nil, or $\widetilde{SL_2\mathbb{R}}$ geometry.

Some more recent results (which we will now discuss) give improvements on this by eliminating Nil from the list and by providing an easy way to construct the torus decomposition of a Euclidean cone manifold with all cone angles $> 2\pi$. Details for these results may be found in [12].

First, a structure theorem for such manifolds:

THEOREM 4.1. *Let M be a closed, orientable 3-dimensional Euclidean cone manifold with no vertices and all cone angles $> 2\pi$. Then there is a canonical compact 2-complex C in M such that*

(1) *The components of the complement of C (denoted by $M_1, \ldots, M_n$) are each the interior of a compact Seifert-fibered manifold (possibly with boundary).*

(2) *Each M_i may be given a convex Euclidean cone metric.*

(3) *M is homotopically atoroidal if and only if each M_i is an open solid torus.*

This canonical 2-complex is easily constructed from the cone manifold by considering product neighborhoods of cone geodesics and looking at the intersections of maximal such neighborhoods. The Seifert-fibration on each of the M_i is such that any cone geodesics that intersect the M_i are in fact fibers of the Seifert-fibration. This is another example of just how "canonical" the notion of a Seifert-fibration seems to be: there are a number of results now of the general form "If a Seifert-fibered manifold admits structure X, then it admits a structure X which is nice with respect to the fibration." Some of the known replacements for "structure X" are "incompressible surface," "essential lamination" and "Euclidean cone metric with cone angles $> 2\pi$" (see [**2**] for the essential lamination result).

One of the most useful aspects to this structure theorem, however, is that the boundary tori of the M_i which are *not* solid tori (after pulling them in slightly along a collar neighborhood) form a collection of tori which contains the characteristic tori of Jaco-Shalen/Johannson. Furthermore, the only case in which "extra" tori arise is where some of the Seifert-fibered pieces (M_i) may be fibered in more than one way and thus the fiberings that are constructed in the structure theorem might be incompatible across a given torus but there might be another way to fiber some M_i so that there *would* be compatibility across a torus, and so the torus could be removed. However, the only spaces that actually arise in the decomposition that may be fibered in multiple ways are I-bundles over the torus and Klein bottle (see [**15**] again) so this situation is easily handled to compute the actual torus decomposition. Furthermore, all the Seifert-fibered components of the complement of this collection of tori either have negatively curved base orbifold or are actually Euclidean, so we have the elimination of Nil as claimed above.

Another interesting corollary of this theorem is the following: suppose M is a Euclidean cone manifold with all cone angles $> 2\pi$. Suppose further that M admits a π_1-injective immersed torus but no incompressible tori. We use the structure theorem to assert that M must be Seifert-fibered, since if C is nonempty, any component of its complement that isn't a solid torus gives rise to a nontrivial torus decomposition as above, but *all* of the components being solid tori forces M to be homotopically atoroidal. Thus, the only possibility is that C is empty, forcing M to be Seifert-fibered. Thus, we recover the result proven recentyl by Casson and Jungreis and independently by Gabai (see [**5**]) that irreducible non-Haken 3-manifolds containing $Z \oplus Z$ subgroups in their fundamental groups must be Seifert-fibered.

For convenience, we collect these in the following

COROLLARY 4.2. *If M is a closed, orientable Euclidean cone manifold with no vertices and all cone angles $> 2\pi$, then*

(1) *If M admits a π_1-injective torus but no incompressible torus, M must be Seifert-fibered.*

(2) *The collection of boundary-parallel tori in each non-solid torus component of M_i forms a collection of tori containing the Jaco-Shalen/ Johannson characteristic tori.*

(3) *If M is Seifert-fibered, it must have $\mathbb{H}^2 \times \mathbb{R}$ or $\widetilde{SL_2\mathbb{R}}$ geometry.*

5. Hyperbolic Structures on Branched Covers

Thus far, we have discussed primarily Euclidean cone manifolds. We will conclude by turning briefly to hyperbolic cone manifolds. Here, one of the key questions is that of *deformability*. That is, how far can one deform a hyperbolic cone manifold without running into some degeneracy or other. Hyperbolic cone manifolds can always be deformed locally, but there are several kinds of degeneracy that may eventually arise (see [9] for a more detailed discussion). The possible kinds of degeneracy are

(1) Volume tending to zero.
(2) Developing a cusp along a submanifold which is becoming Euclidean.
(3) The cone locus "bumping into itself" (that is, changing its combinatorial type).

It has long been conjectured that cone angles may always be decreased. One reason for this conjecture (other than the fact that no counterexamples are known) is that when the cone angles are being decreased, the first kind of degeneracy cannot happen – volume always increases when all cone angles are decreasing. Furthermore, as the cone angles are decreased, the existence of the submanifolds that are "becoming Euclidean" can often be ruled out on other grounds. In particular, if we have a hyperbolic cone manifold with all cone angles $> 2\pi$, neither of the first two kinds of degeneracy can take place as we deform all cone angles toward 2π since the manifold is known to be atoroidal and aspherical (all of the submanifolds along which a cusp would develop must be essential spheres and tori). The third kind of degeneracy has proven to be quite difficult to rule out *a priori*, and very little progress in this direction has been made (although some recent progress has been announced by Hodgson and Kerckhoff [unpublished]).

However, if one is seeking to show that hyperbolic cone manifolds with cone angles $> 2\pi$ are in fact hyperbolic, there are other means besides simply deforming the cone metric. One may also try using the smoothing results above to obtain a metric of pinched negative sectional curvature and attempt a deformation of this metric toward constant sectional curvature. Of course, this is an extremely difficult problem as well, and is in fact known to be impossible in dimensions higher than 3 (see [4]). However, there is a result of Tian [17] on deformation of negatively curved metrics to Einstein metrics (which in dimension 3 have constant sectional curvature) that does enable us to at least deduce a sort of asymptotic result. This theorem (details of which are found in [13]) essentially says that for a fixed hyperbolic orbifold, we can find a constant such that if the maximum and minimum branching indices of a branched cover have a ratio less than this constant, the branched cover must be hyperbolic. This is the essence of the theorem, although there are some unfortunate details that complicate its statement – the "constant" is not quite constant, but depends on the number of components of the branching locus that don't have the minimum branching index.

The actual statement of the theorem is

THEOREM 5.1. *Let (M, L) be a 3-manifold and a link such that $M - L$ admits a hyperbolic metric of finite volume. Let $L_1, \ldots, L_q$ be the components of L. Let $(\hat{M}, \hat{L})$ be a branched cover over (M, L) with minimum branching index n_i over L_i and maximum branching index N_i over L_i. Denote by Q_i the number of components of branching locus over L_i with branching index not equal to n_i (counted with the appropriate multiplicity in the case of longitudinal wrapping). Then, there exist integers $(m_1, \ldots, m_q)$ and functions $(K_1, \ldots, K_q)$ with $K_j : \mathbb{Z} \to \mathbb{R}$ and $K_j(i) > 1$, only depending on (M, L), such that $(\hat{M}, \hat{L})$ admits a hyperbolic metric if $n_i \geq m_i$ and $N_i \leq n_i K_i(Q_i)$ for all $i = 1, \ldots, q$.*

IDEA OF PROOF. One takes a particular smoothing of the lifted cone metric on $\hat{M}$ for which explicit bounds on the sectional curvature may be computed, and then applies Tian's theorem to deduce the existence of a nearby Einstein metric (again, see [13] and [17] for details).

Although this result is far from ideal (we have no idea what order of magnitude the m's and K's have, so we cannot use the result to show that any given cover is hyperbolic), it is the first result giving combinatorial conditions on the branched covering map under which irregular branched covers of hyperbolic links must be hyperbolic (if all branching indices are equal, one may use the Hyperbolic Dehn Surgery Theorem to get hyperbolic structures).

REFERENCES

1. I.R. Aitchison and J.H. Rubinstein, *An introduction to polyhedral metrics of non-positive curvature on 3-manifolds*, LMS Lecture Note Series **151** (1990), 127–162.

2. M. Brittenham, *Essential Laminations in Seifert-Fibered Spaces*, thesis, Cornell University, 1990.

3. R. Brooks, *On branched coverings of 3-manifolds which fiber over the circle*, Crelles J. **362** (1985), 87–101.

4. M. Gromov and W. Thurston, *Pinching constants for hyperbolic manifolds*, Invent. Math. **89** (1987), 1–12.

5. David Gabai, *Convergence groups are Fuchsian groups*, preprint.

6. H.M. Hilden, M.T. Lozano, and J.M. Montesinos, *The Whitehead link, the Borromean rings and the Knot 9_{46} are universal*, Collec. Math. **34** (1983), 19–28.

7. _____, *On knots that are universal*, Topology **24** (1985), 499–504.

8. _____, *Non-simple universal knots*, Math. Proc. Camb. Phil. Soc. **102** (1987), 87–95.

9. Craig Hodgson, *Geometric structures on 3-dimensional orbifolds: notes on Thurston's proof*, preprint.

10. Kerry N. Jones, *Cone manifolds in 3-dimensional topology: applications to branched covers*, thesis, Rice University (1990).

11. _____, *Injectively immersed tori in branched covers over the figure eight knot*, Proc. Edinburgh Math. Soc. **36** (1993), 231–259.

12. _____, *The structure of closed nonpositively curved Euclidean cone 3-manifolds*, Pac. J. Math. (to appear).

13. _____, *Hyperbolic structures on branched covers over hyperbolic links*, preprint, 1992.

14. R. Schoen and Shing-Tung Yau, *Existence of incompressible minimal surfaces and the topology of three dimensional manifolds with non-negative scalar curvature*, Ann. of Math. **110** (1979), 127–142.

15. Peter Scott, *The geometries of 3-manifolds*, Bull. LMS **15** (1983), 401–487.
16. William P. Thurston, *The geometry and topology of three-manifolds*, Princeton University Course Notes, 1979.
17. Gang Tian, *A pinching theorem on manifolds with negative curvature*, preprint.
18. Y. Uchida, *Universal chains*, Kobe J. Math. **8** (1991), 55–65.

THE UNIVERSITY OF TEXAS, AUSTIN, TX 78712

Current address: Ball State University, Muncie, IN 47306

E-mail address: 01KNJONES@bsuvc.bsu.edu

Contemporary Mathematics
Volume **164**, 1994

Link homotopy and skein modules of 3-manifolds

UWE KAISER

ABSTRACT. For each oriented 3-manifold we define sequences of modules from link homotopy and the Conway relation. Algebraic properties, relations between these modules and to the Hoste-Przytycki homotopy skein module are proved. An interpretation of the modules based on an obstruction theory for link homotopy is developed, which gives an interpretation of torsion in the module.

Introduction

In 1954 J. Milnor proposed to study 3-manifolds via the classification of links up to link homotopy [**7**]. In contrast to isotopy, in a link homotopy each component of a link is allowed to intersect itself. But link homotopy *sets* are still hard to compute, in particular because of a lack of algebraic structures. So only recently the classification problem for the 3-sphere [**2**] and simply connected 3-manifolds [**1**] has been finished.

In 1987 J. Hoste and J. Przytycki associated to each oriented 3-manifold a *module* as follows (for details compare section 1): Consider the free $\mathbb{Z}[z]$-module on the set of link homotopy classes of oriented links K in M. Then the homotopy skein module $\mathfrak{S}(M)$ is the quotient by the submodule, which is generated by all elements $K_+ - K_- - zK_0$ (Conway relations for all crossings of distinct components). It is proved in [**3**] that $\mathfrak{S}(F \times I)$ is free for each oriented surface F.

We note that homotopy skein modules have two aspects: (i) they define a functor $M \to \mathfrak{S}(M)$ (compare [**9**] and section 2), and (ii) the image of a link in the module is a link homotopy invariant.

The plan of this paper is the following:
(1) We study algebraic properties of the modules $\mathfrak{S}(M)$.

1991 *Mathematics Subject Classification*. Primary 57M25, 57N10.
This paper is in final form and no version of it will be submitted for publication elsewhere.

(2) We define modules $\mathfrak{T}(M)$ and prove that $\mathfrak{T}(M)$ and $\mathfrak{S}(M)$ are isomorphic.

(3) We generalize both (1) and (2) to sequences of modules.

(1): In (3.2) and (4.6) we prove the following theorem.

THEOREM I.

(a) *Homotopy skein modules are never finitely generated.*

(b) *Free and torsion free are equivalent properties of homotopy skein modules.*

(c) $\mathfrak{S}(S^2 \times S^1)$ *contains torsion.*

The following result is an application.

THEOREM II. *Let M and M' be two oriented 3-manifolds with non-empty boundary. Let $M \natural M'$ be a connected sum along disks in their boundaries. Then $\mathfrak{S}(M \natural M')$ free implies that $\mathfrak{S}(M)$ and $\mathfrak{S}(M')$ are free, and the natural homomorphism $\mathfrak{S}(M) \to \mathfrak{S}(M \natural M')$ is injective.*

(2): In order to understand torsion in $\mathfrak{S}(M)$ we develop an obstruction theory for link homotopy in 3-manifolds based on the Conway relation (this idea is implicit in [2], [8] and [11], and all computations of skein modules): Choose a set σ of links (up to link homotopy), such that each link in M is regularly homotopic to precisely one element of σ. For each regular homotopy between a link in M and its *partner* in σ, consider the smoothings K_0 at crossings $K_\pm \to K_\mp$ of distinct components as the first order obstruction. Choose regular homotopies from the smoothings to links in σ. Repeat the process. Since each smoothing strictly decreases the number of components the *unlinking procedure* will stop after a finite number of steps. From the set of final links in σ an element of the free $\mathbb{Z}[z]$-module with basis σ is defined (the power of z measures the order of the obstructions). Different choices of regular homotopies for the same link can define distinct elements of the free module. This determines an indeterminacy submodule $\mathfrak{t}(M, \sigma)$. Let $\mathfrak{T}(M, \sigma) := \mathbb{Z}[z](\sigma)/\mathfrak{t}(M, \sigma)$. In (5.10) and (6.7) we prove the following (not surprising) result.

THEOREM III. *The modules $\mathfrak{S}(M)$ and $\mathfrak{T}(M, \sigma)$ are isomorphic. Moreover, $\mathfrak{S}(M)$ contains torsion if and only if $\mathfrak{t}(M, \sigma) \neq 0$ (this does not depend on σ).*

(3): For the definition of link homotopy invariants, sequences of *restricted* modules $\mathfrak{S}_p(M)$ and $\mathfrak{T}_p(M)$ are more effective. Here each module is based on a restricted set of links (and corresponding relations) determined by the index p. (6.7) gives the precise relation between the $\mathfrak{S}$ and $\mathfrak{T}$-sequence, which is not the obvious one. The computations of restricted modules of $S^2 \times S^1$ in section 7 (based on the $\mathfrak{T}$-version) suggest that restricted modules pick up linking numbers whenever these are defined. This is not true for the Hoste-Przytycki-module.

1. Skein modules from local relations

For each oriented 3-manifold M let $\mathfrak{H}(M)$ denote the set of link homotopy classes of unordered oriented smooth links in M. For each r-component link the tuple of homotopy classes of its components is an element of $\pi^{[r]}(M)$, the set of unordered r-tuples of free homotopy classes of loops in M (i.e. for connected M the set of conjugacy classes of elements of the fundamental group). For $\alpha = [\alpha_1, \ldots \alpha_r] \in \pi^{[r]}(M)$ let $|\alpha| = r$ be the *length*. Let $\pi(M) := \bigcup_{r \geq 1} \pi^{[r]}(M)$.

The *wrapping map*:

$$\omega : \mathfrak{H}(M) \longrightarrow \pi(M)$$

collects the homotopy classes of the components of links. Note that ω restricted to knots maps bijectively onto $\pi^{[1]}(M)$.

DEFINITION 1.1. An *unlink section* is a map $\sigma : \pi(M) \to \mathfrak{H}(M)$, such that $\omega \circ \sigma$ is the identity on $\pi(M)$.

Since ω is onto, unlink sections always exist. Throughout we will fix an unlink section σ. Then the image of σ is the set of *unlinks*.

Recall that a crossing (skein triple) in M consists of three links representing elements $K_+, K_-, K_0 \in \mathfrak{H}(M)$, which differ only inside some (oriented) 3-ball, where the picture is the following:

$$K_+ = \quad\quad K_- = \quad\quad K_0 = \quad\quad .$$

In a mixed crossing the two arcs of K_+ are from distinct components.

DEFINITION 1.2. Fix the ring $R = \mathbb{Z}[z]$. Let P be any subset of $\mathfrak{H}(M)$, such that

$$(\mathfrak{C}) \quad\quad K_\pm \in P \implies K_\mp, K_0 \in P \text{ for all mixed crossings}$$

is satisfied. Let $\mathfrak{s}_P$ be the submodule of the free module RP, which is generated by all elements

$$K_+ - K_- - zK_0 \text{ for all } K_+ \in P \text{ and all mixed crossings.}$$

Then

$$\mathfrak{S}_P(M) := RP/\mathfrak{s}_P$$

is the *P-skein module* of M.

EXAMPLES OF P-MODULES.

(a) If $P = \mathfrak{H}(M)$ then $\mathfrak{S}(M) := \mathfrak{S}_P(M)$ is the Hoste-Przytycki skein module.

(b) For $r \geq 1$ let $P = [r]$ be the set of links of $\leq r$ components. Then $(\mathfrak{C})$ holds and $\varinjlim_r \mathfrak{S}_{[r]}(M) = \mathfrak{S}(M)$ (compare section 4).

A set P satisfying $(\mathfrak{C})$ is of the form $\omega^{-1}(p)$ for some set $p \subset \pi(M)$. For each $p \subset \pi(M)$ there is a smallest set $\bar{p} \subset \pi(M)$, the *skein-closure* of p, such that $\bar{p} \supset p$ and $\omega^{-1}(\bar{p})$ satisfies $(\mathfrak{C})$. Obviously $\bar{p} = \bigcup_{\alpha \in p} \bar{\alpha}$.

For $p \subset \pi(M)$ we define skein modules

$$\mathfrak{S}_p(M) := \mathfrak{S}_{\omega^{-1}(\bar{p})}(M).$$

By definition $\mathfrak{S}_p(M) = \mathfrak{S}_{\bar{p}}(M)$. So we have constructed one skein-module for each skein-closed subset of $\pi(M)$.

DEFINITION 1.3. An *elementary module* is a module $\mathfrak{S}_\alpha(M)$ where $\alpha \in \pi(M)$ is any element.

Denote $\mathfrak{H}_p(M) := \omega^{-1}(\bar{p})$ and $\mathfrak{s}_p(M) := \mathfrak{s}_{\omega^{-1}(\bar{p})}$. For $\omega(K) \in \bar{p}$ the image of K in $\mathfrak{S}_p(M)$ is defined and denoted by $[K]_p$. Let $\mathfrak{G}_p(M) \subset \mathfrak{S}_p(M)$ be the set of all $[\sigma(\alpha)]_p$ for $\alpha \in \bar{p}$.

PROPOSITION 1.4. *The module $\mathfrak{S}_p(M)$ is generated $\mathfrak{G}_p(M)$. The inclusion $\sigma(\bar{p}) \subset \mathfrak{H}_p(M)$ induces the isomorphism*

$$\mathfrak{S}_p(M) \cong R\sigma(\bar{p})/(R\sigma(\bar{p}) \cap \mathfrak{s}_p(M)).$$

2. Elementary modules and functorial properties

We describe the set $\bar{\alpha}$ for $\alpha = [\beta, \gamma, \delta, \ldots, \rho] \in \pi^{[r]}(M)$: Define a 1–*descendant* of α to be an element

$$[\beta' \cdot \gamma', \delta, \ldots, \rho] \in \pi^{[r-1]}(M)$$

where β', γ' are lifts of β, γ to $\pi_1(M, *)$ for a basepoint $* \in M$, and $\beta' \cdot \gamma'$ is the image in $\pi^{[1]}(M)$ of the fundamental group product of the two lifts. Note that in each crossing change $K_\pm \to K_\mp$ with $\omega(K) = \alpha$ the wrapping-value of the smoothing is a 1-descendant of α. Finally let a *j–descendant* of α be a 1–descendant of a $(j-1)$–descendant of α. Then for each α of length r

$$\bar{\alpha} = \pi_\alpha^{[1]} \cup \cdots \cup \pi_\alpha^{[r-1]} \cup \{\alpha\} \subset \pi(M),$$

where $\pi_\alpha^{[j]} \subset \pi^{[j]}(M)$ is the set of $(r-j)$-descendants of α.

PROPOSITION 2.1. *Let $\alpha \in \pi(M)$. Then*

$$\mathfrak{S}_\alpha(M) \cong R \oplus \tilde{\mathfrak{S}}_\alpha(M),$$

where R is the submodule which is generated by $\sigma(\alpha)$.

PROOF. Consider the homomorphism induced from $R\mathfrak{H}_\alpha(M) \to R$, which is defined by $K \mapsto 1$ if $\omega(K) = \alpha$, otherwise $K \mapsto 0$. Then $\mathfrak{s}_\alpha(M)$ is contained in the kernel. In fact, in a relation $K_+ - K_- - zK_0$ for $K_+ \in \mathfrak{H}_\alpha(M)$ either: $\omega(K_+) = \alpha$, so also $\omega(K_-) = \alpha$, but $\omega(K_0) \neq \alpha$; or: none of K_+, K_- and K_0 has wrapping-value α. Thus the homomorphism $R \to \mathfrak{S}_\alpha(M)$, which maps 1 to $\sigma(\alpha)$, has a left-inverse. ∎

COROLLARY 2.2. *For each 3-manifold M, and each α of length 1, the module $\mathfrak{S}_\alpha(M)$ is free of rank 1. In particular, $\omega^{-1}(\alpha) \to \mathfrak{S}_\alpha(M)$ is injective.* ∎

The map

$$M \mapsto (\mathfrak{S}_\alpha(M))_{\alpha \in \pi(M)}$$

defines a functor from the category of 3-manifolds and differentiable injections into the category of graded modules. Note that the grading is not absolute, but depends on M. This means more precisely the following: An object in the category of graded modules is a map $A : J(A) \to$ R-modules, $J(A) \ni k \mapsto A_k$ ($J(A)$ is any index set). A morphism from A to B is a map $\iota : J(A) \to J(B)$ together with a sequence of homomorphisms $A_k \to B_{\iota(k)}$.

Note that the index sets $\pi(M)$ are in bijection with $\mathbb{Z}$. A differentiable injection $j : M \hookrightarrow N$ induces maps $j_* : \pi^{[r]}(M) \to \pi^{[r]}(N)$ for $r \geq 1$. This defines homomorphisms

$$\mathfrak{S}_p(M) \to \mathfrak{S}_{j_*(p)}(N)$$

for all $p \subset \pi(M)$ in the obvious way.

Specializing to $p = \{\alpha\}$ we get a morphism between graded modules

$$(\mathfrak{S}_\alpha(M))_{\alpha \in \pi(M)} \to (\mathfrak{S}_\beta(N))_{\beta \in \pi(N)}.$$

Given a link K in M with $\omega(K) = \alpha$, then $[K]_\beta \in \mathfrak{S}_\beta(M)$ is defined for all β with $\alpha \in \bar{\beta}$. The image of a link K in the graded module is put into $\mathfrak{S}_{\omega(K)}(M)$.

3. Algebraic properties of homotopy skein modules

For $p \subset \pi(M)$ let $\mathfrak{A}_p(M)$ be the quotient of the free abelian group on $\mathfrak{H}_p(M)$ by the subgroup $\mathfrak{a}_p$, generated by elements $K_+ - K_-$ for *all* crossings for $K_+ \in \mathfrak{H}_p(M)$.

LEMMA 3.1. *Each unlink section σ induces an isomorphism from the free abelian group on the set $\bar{p}$ onto the group $\mathfrak{A}_p(M)$.*

PROOF. The unlink section

$$\sigma|\bar{p} : \bar{p} \to \mathfrak{H}_p(M)$$

extends to the homomorphism

$$\mathbb{Z}\bar{p} \to \mathbb{Z}\mathfrak{H}_p(M),$$

which induces the epimorphism $\mathbb{Z}\bar{p} \to \mathfrak{A}_p(M)$. Note that each $K \in \mathfrak{H}_p(M)$ is homotopic to $\sigma(\alpha)$ for some $\alpha \in \bar{p}$. Since the corresponding unlink $\sigma(\alpha)$ is determined by K, and the difference is contained in $\mathfrak{a}_p$, there is a well-defined inverse homomorphism $\mathfrak{A}_p(M) \to \mathbb{Z}\bar{p}$. ∎

THEOREM 3.2. *Let $p \subset \pi(M)$. Then the following holds:*

(a) *The set $\mathfrak{G}_p(M)$ is a minimal generating set for $\mathfrak{S}_p(M)$, i.e. no proper subset can generate $\mathfrak{S}_p(M)$.*

(b) *$\mathfrak{S}_p(M)$ is finitely generated if and only if $\bar{p}$ is finite. In particular, $\mathfrak{S}(M)$ is never finitely generated.*

(c) *If $\mathfrak{S}_p(M) \xrightarrow{z\cdot} \mathfrak{S}_p(M)$ is injective (in particular if $\mathfrak{S}_p(M)$ is torsion-free), then $\mathfrak{S}_p(M)$ is free and $\mathfrak{G}_p(M)$ is a basis for each section σ.*

(d) *$Tor_{\mathbb{Z}}\mathfrak{S}_p(M) \subset z\mathfrak{S}_p(M)$ (Here $Tor_{\mathbb{Z}}$ means $\mathbb{Z}$-torsion).*

PROOF. Let $\varepsilon : \mathbb{Z}[z] \to \mathbb{Z}$ be the ring substitution $z \mapsto 0$. It induces the epimorphisms of abelian groups (also denoted ε)

$$R\mathfrak{H}_p(M) \to \mathbb{Z}\mathfrak{H}_p(M) \qquad \text{and} \qquad \mathfrak{S}_p(M) \longrightarrow \mathfrak{A}_p(M).$$

There is the split exact sequence of abelian groups (compare the argument in (c)):

$$0 \to z\mathfrak{S}_p(M) \xrightarrow{\subset} \mathfrak{S}_p(M) \xrightarrow{\varepsilon} \mathfrak{A}_p(M) \to 0.$$

(a) If we could drop a module generator from the set $\mathfrak{B}_p(M) \subset \mathfrak{S}_p(M)$, then the free abelian group $\mathfrak{A}_p(M)$ would be generated (as abelian group) by a proper subset of $\sigma(\bar{p})$. This contradicts (3.1).

(b) The abelian group $\mathfrak{A}_p(M)$ is finitely generated if and only if $\bar{p}$ is finite. But if $\mathfrak{S}_p(M)$ is finitely generated as R-module, then $\mathfrak{A}_p(M)$ is a finitely generated group.

(c) We prove that, for each section σ, the corresponding set $\mathfrak{B}_p(M)$ is linearly independent. Let $\sum \lambda_\beta [\sigma(\beta)]_p$ be a finite vanishing linear combination, $\lambda_\beta \in R$. Then $\sum \varepsilon(\lambda_\beta)\varepsilon([\sigma(\beta)]_p) = 0$. Since $\sigma(\bar{p})$ is in bijection with a basis of $\mathfrak{A}_p(M)$, we get $\varepsilon(\lambda_\beta) = 0$ for all β. Thus z divides λ_β for all β, and we have $z\sum \lambda'_\beta [\sigma(\beta)]_p = 0$ for suitable λ'_β. By assumption $\sum \lambda'_\beta [\sigma(\beta)]_p = 0$. The claim follows now by induction over $\max_\beta deg(\lambda_\beta)$ (Note that only finitely many λ_β are non-zero).

(d) Let $0 \neq w \in Tor_\mathbb{Z}\mathfrak{S}_p(M)$ and $\lambda w = 0$, $\lambda \in \mathbb{Z} \setminus 0$. Then $\varepsilon(\lambda w) = \lambda\varepsilon(w) = 0$, which is only possible if $\varepsilon(w) = 0$. But this is equivalent to $w \in z\mathfrak{S}_p(M)$. ∎

COROLLARY 3.3. *Let $j : M \hookrightarrow N$ be a differentiable injection which induces an injective map $\pi(M) \to \pi(N)$. Then $\mathfrak{S}(N)$ free implies that $\mathfrak{S}(M)$ is free, and $\mathfrak{S}(M) \to \mathfrak{S}(N)$ is injective.*

PROOF. Let $\iota : \pi(N) \to \pi(M)$ be a left-inverse to the injective map $\pi(M) \to \pi(N)$ induced by j. Let $\sigma : \pi(M) \to \mathfrak{H}(M)$ be an unlink section. Then define an unlink section $\tau : \pi(N) \to \mathfrak{H}(N)$ by $\tau(\alpha) := j_*(\sigma\iota(\alpha))$ for each α in the image of j_*, and arbitrarily on the complement. Since $\mathfrak{S}(N)$ is free, $[\ ] \circ \tau : R\pi(N) \to \mathfrak{S}(N)$ is an isomorphism. It follows that the induced homomorphism $j_* : \mathfrak{S}(M) \to \mathfrak{S}(N)$ has left-inverse

$$\mathfrak{S}(N) \xrightarrow{([\ ]\circ\tau)^{-1}\cong} R\pi(N) \xrightarrow{\iota} R\pi(M) \xrightarrow{[\ ]\circ\sigma} \mathfrak{S}(M).$$

Thus $\mathfrak{S}(M)$ is isomorphic to a submodule of the free module $\mathfrak{S}(N)$, so it is torsion-free and finally free by (3.2). ∎

PROOF OF THE THEOREM II. Let $N = M\natural M'$ be a boundary connected sum of two 3-manifolds along disks in their boundaries. The inclusion $M \hookrightarrow M\natural M'$ induces an injective map of the sets of homotopy classes of free loops. In fact $\pi_1(N) = \pi_1(M) * \pi_1(M')$ is a free product, so two elements $x, x' \in \pi_1(M)$ are conjugate in $\pi_1(N)$ only if they are conjugate in $\pi_1(M)$. ∎

REMARK 3.4. Let $j : M \hookrightarrow N$ be a differentiable injection and $\pi_1(M) \to \pi_1(N)$ be injective. This does not imply that $\pi(M) \to \pi(N)$ is injective in general. In fact, let M be a connected oriented 3-manifold, and let $x, y \in im(\pi_1(\partial M) \to \pi_1(M))$ be represented by two smooth simple curves in ∂M. Assume that x, y are not conjugate in $\pi_1(M)$. Let N be the result of glue-ing $S^1 \times I \times I$ to M along $S^1 \times I \times 0$ and $S^1 \times I \times 1$ to neighbourhoods of the two curves (in such a way that N is orientable). Then $\pi_1(N)$ is the HNN-extension $(\pi_1(M), t | txt^{-1} = y)$, such that x and y are conjugate in N, and $\pi_1(M) \to \pi_1(N)$ is known to be injective. (Explicitly let M be the genus 2-handlebody and x, y represent generators of $\pi_1(M)$=free group on x, y. Then N is a genus 2-handlebody again.)

4. Relations between the $\mathfrak{S}$-modules

For $p \subset q \subset \pi(M)$ let

$$\xi_{pq} : \mathfrak{S}_p(M) \to \mathfrak{S}_q(M)$$

be the natural homomorphism. Let

$$\xi_{(p)} : \bigoplus_{\alpha \in p} \mathfrak{S}_\alpha(M) \to \mathfrak{S}_p(M)$$

be defined by $\xi_{(p)}(\sum x_\alpha) = \sum \xi_{\alpha p}(x_\alpha)$.

Next recall the natural functor from graded modules into modules, which maps $(\mathfrak{M}_*)_*$ to $\oplus_* \mathfrak{M}_*$.

PROPOSITION 4.1. Let $\xi_{(\pi(M))} =: \xi_M$. Then $M \to \xi_M$ is a natural transfor-mation $\oplus \mathfrak{S}_* \to \mathfrak{S}$.

PROOF. This is straightforward. ∎

We come back to the general case:

THEOREM 4.2. For $p \subset \pi(M)$ the homomorphism $\xi_{(p)}$ is onto, and the kernel of $\xi_{(p)}$ is generated by all $[K]_\beta - [K]_\gamma$ for $K \in \mathfrak{H}_\beta(M) \cap \mathfrak{H}_\gamma(M)$ and $\beta, \gamma \in p$. Thus, $ker(\xi_{(p)})$ is generated by the set of $[\sigma(\alpha)]_\beta - [\sigma(\alpha)]_\gamma$ for all $\beta, \gamma \in p$ and all $\alpha \in \bar{\beta} \cap \bar{\gamma}$.

PROOF. It is clear that $\xi = \xi_{(p)}$ is onto and $[K]_\beta - [K]_\gamma$ are in the kernel of ξ.

We write K^α to indicate $K \in \mathfrak{H}_\alpha(M) \subset \oplus R\mathfrak{H}_\alpha(M)$. Note that $K^\alpha \neq K^\beta$ in the direct sum if $\alpha \neq \beta$ and $K \in \mathfrak{H}_\alpha(M) \cap \mathfrak{H}_\beta(M)$. Consider the diagram:

$$
\begin{array}{ccc}
\oplus_\alpha R\mathfrak{H}_\alpha(M) & \xrightarrow{\bar{\xi}} & R\mathfrak{H}_p(M) \\
\pi \downarrow & & \pi' \downarrow \\
\oplus_\alpha \mathfrak{S}_\alpha & \xrightarrow{\xi} & \mathfrak{S}_p(M)
\end{array}
$$

where π, π' are the projections, and $\bar{\xi}$ is the obvious lifting of ξ. Note that the kernel of $\bar{\xi}$ is generated by all $L^\beta - L^\gamma$ for $L \in \mathfrak{H}_\beta(M) \cap \mathfrak{H}_\gamma(M)$ and $\beta, \gamma \in p$.

Now, let $w \in \oplus R\mathfrak{H}_\alpha(M)$ such that $\xi \circ \pi(w) = \pi' \circ \bar{\xi}(w) = 0$. Thus

$$\bar{\xi}(w) = \sum \lambda_K (K_+ - K_- - zK_0)$$

for suitable $\lambda_K \in R$ and $K_+ \in \mathfrak{H}_p(M)$. By the note above it follows that

$$w = \sum \lambda_K (K_+^\alpha - K_-^\alpha - zK_0^\alpha) + \sum \mu_L (L^\beta - L^\gamma).$$

The first sum is a linear combination of elements from the sets $\mathfrak{s}_\alpha(M)$ (considered as submodules of $\oplus R\mathfrak{H}_\alpha(M)$) which map trivial into $\oplus \mathfrak{S}_\alpha(M)$. The second sum maps to a linear combination of elements as given in the statement. ∎

COROLLARY 4.3. *Let $j : M \hookrightarrow N$ be a differentiable injection, which induces an isomorphism of graded modules, i.e.*

  (i) $j_* : \pi(M) \to \pi(N)$ *is bijective, and*
  (ii) $j_* : \mathfrak{S}_\alpha(M) \to \mathfrak{S}_{j_*(\alpha)}(N)$ *is an isomorphism for all $\alpha \in \pi(M)$.*
Then $j_* : \mathfrak{S}(M) \to \mathfrak{S}(N)$ is an isomorphism.

PROOF. Because of the isomorphism $j_* : \oplus \mathfrak{S}_\alpha(M) \to \oplus \mathfrak{S}_\beta(N)$, it suffices to observe that j_* maps $\ker \xi_M$ onto $\ker \xi_N$, which follows by (4.2) and (i). ∎

COROLLARY 4.4. *For each 3-manifold M and $p \subset \pi(M)$, $\mathfrak{S}_p(M)$ is free if and only if $\mathfrak{S}_\alpha(M)$ is free for all $\alpha \in p$.*

PROOF. Assume that $\mathfrak{S}_p(M)$ is free. Then it is isomorphic to the free module on $\bar{p}$, the isomorphism being induced by σ (see (3.2)). So the map $\bar{p} \to \mathfrak{S}_\alpha(M)$, which maps $\beta \in \bar{\alpha}$ to $[\sigma(\beta)]_\alpha$, and maps all other $\beta \in \bar{p}$ trivially, induces a left inverse for the canonical homomorphism $\xi_{\alpha p} : \mathfrak{S}_\alpha(M) \to \mathfrak{S}_p(M)$. Then $\mathfrak{S}_\alpha(M)$ is isomorphic to a submodule of $\mathfrak{S}_p(M)$, so it is torsion free, and finally free by (3.2). Conversely, if the modules $\mathfrak{S}_\alpha(M)$ are free for all $\alpha \in p$, then they are isomorphic to the modules $R\bar{\alpha}$ by (3.2) via σ. Now it follows from (4.2) that $\mathfrak{S}_p(M)$ is free. ∎

REMARKS 4.5.

  (a) The family $(\mathfrak{S}_\alpha(M))_{\alpha \in \pi(M)}$ is *not* naturally a directed system of modules. If we define $\beta \prec \alpha$, if $\beta \in \bar{\alpha}$, then the structure homomorphisms $\xi_{\beta\alpha}$ are defined, but $(\pi(M), \prec)$ is not a directed set: For $\beta, \gamma \in \pi(M)$ there is not necessarily an element α with $\beta, \gamma \prec \alpha$. On the other hand, if we define $\beta < \alpha$, if $\beta \in \bar{\alpha}_0$ for any subtuple α_0 of α, then $(\pi(M), <)$ is a directed set. But the *natural* homomorphism $\xi_{\beta\alpha}$ is not always defined. If we complete the sets $\bar{\alpha}$ with respect to sublinks, i.e. consider modules $\mathfrak{S}_{\tilde{\alpha}}(M)$ ($\tilde{\alpha}$ is the set of all subtuples of α), then $\varinjlim_{\tilde{\alpha}} \mathfrak{S}_{\tilde{\alpha}}(M)$ holds.

  (b) The argument in (4.4) proves that if $p \subset q$, and $\mathfrak{S}_q(M)$ is free, then $\mathfrak{S}_p(M)$ is free.

EXAMPLE 4.6.

(a) Decompose $S^2 \times S^1$ into two solid tori glued along its boundaries. The inclusion $j : D^2 \times S^1 \hookrightarrow S^2 \times S^1$ induces a surjective map of link homotopy sets (see [1] or [5]). Thus from the standard unlink section (stack knots, see [3]) σ for $D^2 \times S^1$ we can define the unlink section $\tau := j_* \circ \sigma$ for $S^2 \times S^1$. For each $n \in \mathbb{Z} \cong \pi^{[1]}(D^2 \times S^1)$ let $K(n)$ resp. $L(n)$ be the standard representative links of $\sigma(n-1, 1)$ resp. the result of sliding the $(n-1)$-component of $K(n)$ over an outside disk (handle slide) in $S^2 \times S^1$. So $j(K(n))$ and $j(L(n))$ are isotopic in $S^2 \times S^1$. By a single crossing change in $D^2 \times S^1$ we expand $[L(n)]$ in terms of σ, and get relations for all $n \in \mathbb{Z}$:

$$[\tau(n-1, 1))] = [\tau(n-1, 1))] + z[\tau(n))].$$

So for each knot K in $S^2 \times S^1$ the relation $z[K] = 0$ holds. In particular, for each $n \in \mathbb{Z}$ the kernel of the natural homomorphism

$$R \cong \mathfrak{S}_n(S^2 \times S^1) \to \mathfrak{S}(S^2 \times S^1)$$

is zR (see (2.1)) and (3.2)).

(b) More generally let $\underline{t} := [1, \ldots, 1]$ denote the trivial class of length t in any 3-manifold. If we slide a canonical generator $\sigma(\underline{s})$, $s \leq t$, of $\mathfrak{S}_{\underline{t}}(D^2 \times S^1)$ over the $S^2 \times S^1$-handle, we do not change its link homotopy class in $D^2 \times S^1$. Thus j induces an isomorphism

$$R^t \cong \mathfrak{S}_{\underline{t}}(D^2 \times S^1) \xrightarrow{\cong} \mathfrak{S}_{\underline{t}}(S^2 \times S^1).$$

(The first isomorphism follows from [3] and (4.4).)
Let $K \in \omega^{-1}(\underline{2}) \subset \mathfrak{H}(D^2 \times S^1)$, then

$$[j(K)]_{\underline{2}} = [\tau(\underline{2})]_{\underline{2}} + z \cdot lk(j(K)) \cdot [\tau(\underline{1})]_{\underline{2}} \in \mathfrak{S}_{\underline{2}}(S^2 \times S^1),$$

where lk is the linking number as defined in [10] (compare [5]). But

$$z[\tau(\underline{1})] = 0 \in \mathfrak{S}(S^2 \times S^1).$$

Thus linking numbers are measured in elementary modules, which cannot be measured in the Hoste-Przytycki module.

(c) Let i be the inclusion of a manifold M into the interior connected sum $M \sharp (S^2 \times S^1)$. Since we can replace a 3-manifold N by $N \backslash 3$-ball without changing the homotopy skein module we can consider a boundary connected sum $M' \sharp (S^2 \times S^1 \backslash 3$-ball) instead. It follows from (a) that $i_*[\sigma(\underline{1})]$ is torsion in the homotopy skein module of the connected sum $(i_*[\sigma(\underline{1})]$ is nontrivial and in the kernel of multiplication by z).

5. Obstruction theory from the Conway relation

For $a = (\alpha_1, \ldots, \alpha_r) \in (\pi^{[1]})^r$ let $\mathfrak{D}_a$ be the space of smooth immersions $f = f_1 \amalg \cdots \amalg f_r : S^1 \amalg \cdots \amalg S^1 \to M$ with at most single double-points, such that the homotopy class of f_i is α_i (take the C^∞–topology). Let $\mathfrak{C}_a$ be the subspace of those f with an intersection point of distinct components.

For $\alpha \in \pi^{[r]}$ let $\tilde{\mathfrak{D}}_\alpha := \cup \mathfrak{D}_a$ (where the union is over all possible orderings of α). The symmetric group acts on $\tilde{\mathfrak{D}}_\alpha$ by permuting the S^1 factors in the domain of maps f. This induces a covering $p : \tilde{\mathfrak{D}}_\alpha \to \mathfrak{D}_\alpha$. Let $\mathfrak{C}_\alpha := p(\cup \mathfrak{C}_a)$ be the set of *singular* links in $\mathfrak{D}_\alpha$.

Note that two unordered links with the same wrapping-value α can be joined by a path ρ in $\mathfrak{D}_\alpha$, which is transverse. Here a continuous map $X \to \mathfrak{D}_\alpha$ for a smooth manifold X is called *transverse* if (i) the map lifts to $\tilde{\mathfrak{D}}_\alpha$, and (ii) the resulting trace map $X \times \amalg_r S^1 = \amalg_r (X \times S^1) \to X \times M$ is a self-transverse immersion. We will only consider such maps for $X = I$ or $X = I \times I$. In the first case, transversality implies that for only finitely many $t \in I$, $\rho(t)$ has a mixed crossing (intersection with $\mathfrak{C}_\alpha$).

The natural space of deformation for all links with wrapping-value α (in the spirit of the Conway relation) is

$$\mathfrak{D}(\alpha) := \bigcup_{\beta \in \bar{\alpha}} \mathfrak{D}_\beta.$$

The union of all $\mathfrak{D}_\beta$ with $|\beta| = |\alpha| - j$ is the j-th *layer* of $\mathfrak{D}(\alpha)$; $\mathfrak{D}_\alpha \subset \mathfrak{D}(\alpha)$ is the *top layer*. Note that $\mathfrak{D}(\beta) \subset \mathfrak{D}(\alpha)$ for each $\beta \in \bar{\alpha}$.

DEFINITION 5.1. Let $\alpha \in \pi(M)$. A *scheme* in $\mathfrak{D}(\alpha)$ is a finite family of transverse oriented paths in $\mathfrak{D}(\alpha)$, compatible in the following way:

(i) There is a single path on the top layer. All paths on layers below end in links link homotopic to unlinks.

(ii) For $1 \le j < r$ the starting points of the paths on the j-th layer are the smoothings of the singularity links of paths on the $(j-1)$st layer.

If the top layer path of a scheme begins in K and ends in K' (or links representing K, K'), it is called a (K, K') scheme.

A $(K, \sigma(\alpha))$ scheme is called a K scheme.

A *subscheme* of a (K, K') scheme ρ is an L scheme or (L, K') scheme, where L is a link on a path of ρ (on any layer) as follows: Take the the rest of the path given by ρ from L to the corresponding end-point. Add all paths of ρ emanating from the smoothings and iterated smoothings along this path.

For each (K, K') scheme ρ the (K', K) scheme ρ^{-1} is defined by reversing the direction of the top layer path and keeping the rest.

DEFINITION 5.2. A *homotopy of schemes* in $\mathfrak{D}(\alpha)$ is a finite sequence of elementary moves of schemes $\rho = \rho_0 \to \rho_1$ as follows:

(i) For $t \in [0,1]$ the paths on all layers are deformed by homotopies which keep starting- and end-points away from singular links, and such that at each time $t \in [0,1]$ the union of paths forms a scheme. In particular the set of singular links of the scheme does *not* chang during such a deformation. (ρ_0 and ρ_1 are called *equivalent*.)

(ii) All paths on layers above the j-th layer (for some $j \geq 0$), and all except a single path on the j-th layer, are fixed. A single path on the j-th layer is deformed relative to the endpoints, such that the *deformation* is transverse. Moreover, the paths at different times are transverse, except for a single time parameter t_0. At t_0 a pair of singularity links is created (a) or cancelled (b). (Note that the two smoothed links are link homotopic in M.) We describe the change of the scheme in both cases:

(a) For $t > t_0$ we introduce equivalent schemes for the two smoothings, which we add to the given scheme (equivalence in the sense of (i)).

(b) is the inverse move to (a): This means cancellation of a pair of equivalent schemes emanating from two link homotopic smoothings, which come from cancelling links on the layer above.

A (K, K') scheme ρ and a (K', K'') scheme ρ' can be composed to the (K, K'') scheme (well-defined up to equivalence) $\rho \circ \rho'$ as follows: First replace ρ' by an equivalent scheme, such that the end point of the top-layer path of ρ coincides with the starting point of the top-layer path of ρ'. Then take the usual composition of paths on the top-layer and and collect all paths from the two schemes on layers below.

Scheme homotopy $\simeq$ is an equivalence relation on the set of schemes in $\mathfrak{D}(\alpha)$, which is compatible with composition.

PROPOSITION 5.3. *The set of homotopy classes of $\sigma(\alpha)$ schemes is the group $sc_\alpha(M)$: The neutral element is the constant scheme $*$, which is the constant path in $\sigma(\alpha)$. The homotopy inverse of the scheme ρ is the scheme ρ^{-1}.* ∎

Two K schemes with $\omega(K) = \alpha$ compose to a $\sigma(\alpha)$ scheme (as $\rho_1^{-1} \circ \rho_2$). Conversely, up to homotopy, each two such K schemes differ by a $\sigma(\alpha)$ scheme.

Next we associate to each (K, K') scheme ρ a tree with vertices labelled by signed ($\pm$) link homotopy classes of links as follows:

Consider links on a path of ρ equivalent, if they are joined by a path segment of ρ away from the singularity. Attach a vertex to each equivalence class of links. (Identify vertices with representative links on paths of ρ.) Next for each crossing change and each smoothing on a path of ρ take an edge joining the attached vertices. The direction of edges is induced by the direction of ρ ($K_+ \to K_-, K_0$ or $K_- \to K_+, K_0$). Label the vertex attached to the beginning point of the top layer path by $+K$. For each other vertex v there is a single incoming edge. It joins a vertex v' to v, where v' is the link from which v is formed by either a crossing change or a smoothing. Let $\pm L'$ be the label of v', and L be the link homotopy class of v. Then label v by $\pm L$, if v is either the result of a crossing change of v', or is a smoothing of a crossing, where v' is the K_+. Otherwise label v by $\mp L$. Let the *end-vertices* be the vertices, which are attached to end-points of paths of ρ on all layers except the top layer.

DEFINITION 5.4. Let ρ be a scheme in $\mathfrak{D}(\alpha)$ $(|\alpha| = r)$. From the labels of end-vertices of the associated tree the following formal linear combination is defined:

$$\mathfrak{p}(\rho) = \sum_{j=1}^{r-1} z^{r-j} \sum_{\beta \in \pi_\alpha^{[j]}} \lambda_\beta \beta \in R\bar{\alpha}.$$

The coefficient $\lambda_\beta \in \mathbb{Z}$ is the sum over the signs of the labels of those end-vertices with label $\pm\sigma(\beta)$. If ρ is a (K, K') scheme then $\mathfrak{p}(\rho)$ is the *skein difference* from K to K' along ρ. (This depends on the choice of unlink section σ.)

From the definition of scheme homotopy the following result follows immediately:

PROPOSITION 5.5. *For each scheme ρ in $\mathfrak{D}(\alpha)$, $\mathfrak{p}(\rho) \in R\bar{\alpha}$ only depends on the homotopy class of ρ.* ∎

Write $\mathfrak{p}(K, \rho)$ to indicate that ρ is a K scheme.

The following result shows that the image of $\mathfrak{H}_\alpha(M)$ in $\mathfrak{S}_\alpha(M)$ is "thin".

PROPOSITION 5.6. *For each $K \in \omega^{-1}(\alpha)$ and K scheme ρ the natural epimorphism*

$$R\bar{\alpha} \xrightarrow{\sigma} R\sigma(\bar{\alpha}) \subset R\mathfrak{H}_\alpha(M) \to \mathfrak{S}_\alpha(M)$$

maps

$$\tilde{\mathfrak{p}}(K, \rho) := \alpha + \mathfrak{p}(K, \rho)$$

to the expansion

$$[K]_\alpha = [\sigma(\alpha)]_\alpha + \sum_{j=1}^{r-1} z^{r-j} \sum_{\beta \in \pi_\alpha^{[j]}} \lambda_\beta [\sigma(\beta)]_\alpha \in \mathfrak{S}_\alpha(M)$$

in terms of the generating system $\mathfrak{G}_\alpha(M)$. ∎

PROPOSITION 5.7. *Let σ be an unlink section and $p \subset \pi(M)$. Let $\mathfrak{t}_p(M, \sigma)$ be the submodule of $R\bar{p}$, which is generated by all elements $\mathfrak{p}(\rho)$ for $\rho \in sc_\beta(M)$ and $\beta \in p$ (note that the definition of scheme depends on σ). Let*

$$\mathfrak{T}_p(M, \sigma) := R\bar{p}/\mathfrak{t}_p(M, \sigma).$$

Then for each $K \in \omega^{-1}(p)$ the coset

$$\tilde{\mathfrak{p}}(K, \rho) \mod \mathfrak{t}_p(M, \sigma) \in \mathfrak{T}_p(M, \sigma)$$

is well-defined, i.e. does not depend on the K scheme ρ. ∎

The next result shows that the isomorphism type of the module $\mathfrak{T}_p(M, \sigma)$ does not depend on σ.

PROPOSITION 5.8. *Let σ, τ be unlink sections. Then there exist module isomorphisms*

$$\mathfrak{T}_p(M, \sigma) \to \mathfrak{T}_p(M, \tau)$$

for all $p \subset \pi(M)$.

PROOF. For each $\beta \in \pi(M)$ choose a $(\sigma(\beta), \tau(\beta))$ scheme $\rho(\beta)$. Define the module homomorphism $\mu : R\bar{p} \to R\bar{p}$ by linear extension of the map $\bar{p} \to R\bar{p}$, $\beta \to \beta + \mathfrak{p}(\rho(\beta))$. It is easy to prove that μ is an isomorphism and $\mu(t_p(M, \sigma)) = t_p(M, \tau)$. Thus μ induces an isomorphism between the $\mathfrak{T}_p$-modules. ∎

Because of (5.8) we write $\mathfrak{T}_p(M) := \mathfrak{T}_p(M, \sigma)$, and assume that *all* modules $\mathfrak{T}_p(M, \sigma)$ are defined with respect to the same unlink section.

It is not true that the modules $\mathfrak{T}_p(M)$ and $\mathfrak{T}_{\bar{p}}(M)$ are isomorphic in general (example in (7.3)). The modules $\mathfrak{T}_\alpha(M)$ for $\alpha \in \pi(M)$ are the elementary modules (with respect to the deformation viewpoint).

PROPOSITION 5.9. *For each $p \subset \pi(M)$ there is a natural epimorphism*

$$\eta_p : \mathfrak{T}_p(M) \to \mathfrak{S}_p(M),$$

which maps the image of $K \in \omega^{-1}(p)$ in $\mathfrak{T}_p(M)$ to $[K]_p$.

PROOF. η_p is induced by the epimorphism

$$R\bar{p} \to R\sigma(\bar{p}) \subset R\mathfrak{H}_p(M) \to \mathfrak{S}_p(M),$$

as in (5.6). Since the images of $\mathfrak{p}(\sigma(\beta), \rho)$ $(\beta \in p)$ and $\mathfrak{p}(\sigma(\beta), *) = 0$ in $\mathfrak{S}_p(M)$ coincide, $\sigma(t_p(M)) \subset \mathfrak{s}_p(M)$ holds. ∎

REMARKS 5.10.

(a) The modules $\mathfrak{T}_p(M)$ and $\mathfrak{S}_p(M)$ have a lot of common properties. For example if $p \subset q \subset \pi(M)$ then the natural homomorphism $R\bar{p} \to R\bar{q}$ induces homomorphisms $\mathfrak{T}_p(M) \to \mathfrak{T}_q(M)$.

Also, it is clear from $t_p(M) \subset zR\bar{p}$ that the results of (3.2) also hold for $\mathfrak{T}_p(M)$ (ε induces a group epimorphism $\mathfrak{T}_p(M) \to \mathbb{Z}\bar{p}$). In particular, if $\mathfrak{T}_p(M)$ is free then $\mathfrak{T}_p(M) \cong R\bar{p}$.

Moreover, for $\alpha \in \pi(M)$, $t_\alpha(M) \subset R(\bar{\alpha} \setminus \alpha)$, so

$$\mathfrak{T}_\alpha(M) \cong R \oplus \tilde{\mathfrak{T}}_\alpha(M),$$

where R is the cyclic module generated by α. (The same holds for $\mathfrak{T}_{\bar{\alpha}}(M)$.)

(b) If $t_p(M) \neq 0$ then $\mathfrak{T}_p(M)$ contains torsion. In fact, let $w \neq 1$ be an element of smallest degree in $t_p(M)$, $w = z^k u$ for some $k \geq 1$ and $u \notin t_p(M)$. Then $[w]_p = z^k[u]_p = 0$, but $[u]_p \neq 0$. Note that $t_p(M) \neq 0$ implies that $t_q(M) \neq 0$ for all $q \subset \pi(M)$ with $p \cap q \neq \emptyset$.

(c) If $j : M \hookrightarrow N$, and $\pi(M) \to \pi(N)$ is injective, then sections σ and τ for M and N can be chosen satisfying $\tau \circ j_* = j_* \circ \sigma$. Then j induces an embeddings $\mathfrak{D}(\alpha)(M) \to \mathfrak{D}(j_*(\alpha))(N)$, which maps schemes to schemes. Path independence for N now naturally implies path independence for M. This proves Theorem II for the modules $\mathfrak{T}_{\pi(M)}(M) =: \mathfrak{T}(M)$.

6. Skein linear link homotopy invariants: Comparison of $\mathfrak{S}$ and $\mathfrak{T}$

LEMMA 6.1. *Let $K \in \omega^{-1}(\alpha)$ and $K_+ \in \mathfrak{H}_\alpha(M)$, such that K_+, K_-, K_0 is a mixed skein triple. Then there is a scheme for K, which includes the regular crossing change from $K_\pm$ to $K_\mp$.*

PROOF. This is easy for $K_+ \in \omega^{-1}(\alpha)$: Just take any path on the top layer containing $K_\pm \to K_\mp$. So assume that this is not the case. For each descendant β of α we have to find a scheme for K, which meets the component $\mathfrak{D}_\beta$ of $\mathfrak{D}(\alpha)$. Obviously it is sufficient to be able to descend from a given space $\mathfrak{D}_\gamma$ to each possible 1-descendant of γ. Consider components K_1 and K_2 of a link $K \in \omega^{-1}(\gamma)$, such that in the descendant γ_1 and γ_2 are "fused". Join basepoints on K_1 and K_2 by an arc in M. This defines lifts $\tilde{\gamma}_i$ of γ_i to $\pi_1(M; *)$ for $i = 1, 2$, where $*$ is the midpoint of the path. Assume that the descendant is determined by $\bar{\gamma}_1 \bar{\gamma}_2$; $\tilde{\gamma}_i$ and $\bar{\gamma}_i$ are conjugate in $\pi_1(M, *)$. So we can slide neighbourhoods of the basepoints of K_i along suitable loops, until the result of smoothing along the resulting arc (a regular neighbourhood of the arc is the ball where the crossing change takes place) is a link with wrapping-value equal to the given descendant. ∎

(6.1) immediately generalizes to the following result:

PROPOSITION 6.2. *Let ρ_1, ρ_2 be two L schemes, $L \in \mathfrak{H}_\alpha(M)$, and $K \in \omega^{-1}(\alpha)$. Then there exist K schemes ρ'_1 resp. ρ'_2 containing ρ_1 resp. ρ_2 as subschemes, such that ρ'_2 is formed from ρ'_1 precisely by replacing ρ_1 by ρ_2 and keeping the rest.* ∎

COROLLARY 6.3. *Let ρ_1, ρ_2 be two L schemes, $L \in \mathfrak{H}_\alpha(M)$, and $K \in \omega^{-1}(\alpha)$. Then there exist K schemes ρ'_1, ρ'_2, such that*

$$\mathfrak{p}(\rho'_1) - \mathfrak{p}(\rho'_2) = z^{|\alpha|-j}(\mathfrak{p}(\rho_1) - \mathfrak{p}(\rho_2)),$$

where j is the number of components of L. In particular there is a $\sigma(\alpha)$ scheme ρ, such that

$$\mathfrak{p}(\rho) = z^{|\alpha|-j}(\mathfrak{p}(\rho_1) - \mathfrak{p}(\rho_2)).$$

∎

DEFINITION 6.4. A map $I : \mathfrak{H}_p(M) \to \Lambda$, where Λ is an R–module, is *skein linear* if the natural extension: $R\mathfrak{H}_p(M) \to \Lambda$ factors through $\mathfrak{S}_p(M)$.

Recall that a homomorphism $R\bar{p} \to \Lambda$ is uniquely determined by a map $\bar{p} \to \Lambda$.

THEOREM 6.5. *Let $p \subset \pi(M)$ and let $I : \bar{p} \to \Lambda$ be a map.*
If (P): $I \circ \mathfrak{p}(\rho) = 0$ *for all $\rho \in sc_\alpha(M)$ and $\alpha \in p$, then I induces the map*
$I : \omega^{-1}(p) \to \Lambda$ *by $K \mapsto I \circ \tilde{\mathfrak{p}}(K, \rho)$.*
Furthermore, if either

(a) $I \circ \mathfrak{p}(\rho) = 0$ *for all $\beta \in \bar{p}$ and $\sigma(\beta)$ schemes ρ, or*
(b) Λ *is z-torsion free (i.e. multiplication by z is injective on Λ),*

then I is defined on $\mathfrak{H}_p(M)$, and is skein linear.
(In fact (b) *and* (P) *imply* (a).*)*

PROOF. By (6.3), for two K schemes ρ_1 and ρ_2 with $\omega(K) = \beta \in \bar{\alpha}$ there exists a $\sigma(\beta)$ scheme ρ', such that $\mathfrak{p}(K, \rho_1) - \mathfrak{p}(K, \rho_2) = \mathfrak{p}(\sigma(\beta), \rho')$. It follows from (P), that $I(\tilde{\mathfrak{p}}(K, \rho))$ does not depend on ρ for $K \in \omega^{-1}(p)$. Now we assume that (b) holds and prove that $I \circ \tilde{\mathfrak{p}}$ does not depend on the L scheme ρ' for $L \in \mathfrak{H}_p(M)$ with $\omega(L) \notin p$ and i components. In fact, if this would not be true, then we could find two L schemes ρ'_1 and ρ'_2 with different $I \circ \mathfrak{p}(L, \rho'_j)$ for $j = 1, 2$. By (6.3) there exist K schemes ρ_1 and ρ_2 for $K \in \omega^{-1}(p)$ with r components, such that the difference between $I \circ \mathfrak{p}(K, \rho_j)$ for $j = 1, 2$ is $z^{r-i} I \circ (\mathfrak{p}(L, \rho'_1) - \mathfrak{p}(L, \rho'_2)) \neq 0$.

For skein linearity. it suffices to prove that $I \circ \tilde{\mathfrak{p}}(K_+, \rho_+) = I \circ \tilde{\mathfrak{p}}(K_-, \rho_-) + zI \circ \tilde{\mathfrak{p}}(K_0, \rho_0)$ holds for all mixed skein triples with $K_+ \in \mathfrak{H}_p(M)$ and arbitrary schemes ρ_+, ρ_- and ρ_0. So we can choose for ρ_- and ρ_0 the natural subschemes of ρ_+. Then, actually $\tilde{\mathfrak{p}}(K_+, \rho_+) = \tilde{\mathfrak{p}}(K_-, \rho_-) + z\tilde{\mathfrak{p}}(K_0, \rho_0)$ holds, and the assertion follows by linearity. ∎

THEOREM 6.6. *For $p \subset \pi(M)$ let $I : \bar{p} \to \Lambda = R\bar{p}$ be the inclusion. Then the following are equivalent:*

(i) $\mathfrak{S}_p(M)$ *is free.*
(ii) $I : \mathfrak{H}_p(M) \to \Lambda$ *is defined and skein linear.*
(iii) $\mathfrak{p}(\rho) = 0$ *for all $\rho \in sc_\alpha(M)$ and $\alpha \in p$.*

PROOF.
(iii)$\Longrightarrow$(ii) follows from (6.5) and the fact that Λ is z-torsion free.
(ii) $\Longrightarrow$ (i): Consider the epimorphism $[\ ]_p \circ \sigma : \Lambda \to \mathfrak{S}_p(M)$. The homomorphism $\mathfrak{S}_p(M) \to \Lambda$ induced by I is left-inverse to $[\ ]_p \circ \sigma$, since $I[\sigma(\beta)]_p = \beta$ for all $\beta \in \bar{p}$. Thus $\mathfrak{S}_p(M)$ is free.
(i)$\Longrightarrow$(iii): If $\mathfrak{S}_p(M)$ is free then the coefficients of the expansion K in $\mathfrak{S}_p(M)$ are uniquely determined. Since $\mathfrak{p}(K, \rho)$ maps to this expansion for each ρ, the coefficients in $\dot{\mathfrak{p}}(K, \rho)$ cannot depend on ρ. ∎

THEOREM 6.7. *The epimorphism*

$$\eta_{\bar{p}} : \mathfrak{T}_{\bar{p}}(M) \to \mathfrak{S}_{\bar{p}}(M) = \mathfrak{S}_p(M)$$

is an isomorphism. If one of $\mathfrak{S}_p(M)$ or $\mathfrak{T}_p(M)$ is free then $\mathfrak{T}_{\bar{p}}(M) \cong \mathfrak{T}_p(M) \cong R\bar{p}$, so the epimorphism

$$\eta_p : \mathfrak{T}_p(M) \to \mathfrak{S}_p(M).$$

is an isomorphism.

PROOF. To prove the first result choose $\Lambda := \mathfrak{T}_{\bar{p}}(M)$, $I : \bar{p} \to R\bar{p} \to \mathfrak{T}_{\bar{p}}(M)$, and apply (6.5) based on condition (a).

If $\mathfrak{T}_p(M)$ is free then $\mathfrak{T}_p(M) \cong R\bar{p}$, and $\mathfrak{t}_p(M) = 0$. By (6.6) $\mathfrak{S}_p(M)$ is free and η_p is an isomorphism. Conversely, if $\mathfrak{S}_p(M)$ is free, then $\mathfrak{s}_p(M) \cap R\sigma(\bar{p}) = 0$, thus $\mathfrak{t}_p(M) = 0$ and η_p is an isomorphism.   ∎

EXAMPLE 6.8. If $|\alpha| \leq 2$ then $\mathfrak{S}_\alpha(M) \cong \mathfrak{T}_\alpha(M)$ since $\mathfrak{t}_\alpha(M) = \mathfrak{t}_{\bar{\alpha}}(M)$ ($\mathfrak{t}_\beta(M) = 0$ for $|\beta| = 1$). (7.3) shows that elementary modules are *not* isomorphic in general.

EXAMPLE 6.9.
Let $\Lambda := R$. For α of length r and $1 \leq j < r$ let $I_j : \bar{\alpha} \to R$ be defined by $I(\beta) = 1$ if $|\beta| = r - j$, and $I(\beta) = 0$ otherwise. Then for $K \in \omega^{-1}(\alpha)$ and scheme ρ:

$$I_j \circ \mathfrak{p}(K, \rho) = z^{r-j} lk_j(K, \rho)$$

for integer numbers $lk_j(K, \rho)$. If the coefficient does not depend on the path, then I_j is defined on $\omega^{-1}(\alpha)$, and is called the j–th *total linking number* $lk_j(K)$ of K.

Obviously the jth total linking number $lk_j(K, \rho)$ depends only on that part of the scheme strictly above the j-th layer. For example the *total linking number* $lk(K, \rho) = lk_1(K, \rho)$ only depends on the homotopy class of the path of ρ on the top layer. It is the sum over all signs of crossing changes in the regular homotopy from K to $\sigma(\alpha)$.

Let $\chi : R \to \mathbb{Z}$ be the ring substitution $\chi(\lambda) = \lambda(1)$. For each set X this induces a group homomorphism from the free R-module on a set X to the free abelian group on X by $\sum r_x x \mapsto \sum \chi(r_x)x$. χ induces group homomorphisms from $\mathfrak{S}_p(M)$ resp. $\mathfrak{T}_p(M)$ into abelian groups $\mathfrak{S}_p'(M)$ resp. $\mathfrak{T}_p'(M)$ for all $p \subset \pi(M)$. Here $\mathfrak{S}_p'(M)$ is the quotient of the free abelian group on $\mathfrak{H}_p(M)$ by the subgroup which is generated by all $K_+ - K_- - K_0$ for $\omega(K_+) \in \bar{p}$ (this is just $\chi(\mathfrak{s}_p(M))$). Similarly, $\mathfrak{T}_p'(M)$ is the quotient of $\mathbb{Z}\bar{p}$ by the subgroup $t_p'(M) := \chi(\mathfrak{t}_p(M))$. From (6.3) and (6.7) the following follows immediately:

PROPOSITION 6.10. *For $p \subset \pi(M)$ the groups $\mathfrak{T}_{\bar{p}}'(M)$, $\mathfrak{T}_p'(M)$ and $\mathfrak{S}_p'(M)$ are isomorphic. The natural isomorphism $\mathfrak{T}_p'(M) \to \mathfrak{S}_p'(M)$ maps the image of a link $K \in \omega^{-1}(p)$ in $\mathfrak{T}_p'(M)$ to $[K]_p$.*   ∎

PROPOSITION 6.11. *Let $p \subset \pi(M)$ be bounded (i.e. there exists N, such that $|\alpha| \leq N$ for all $\alpha \in p$). Then there are split group injections $\mathfrak{S}_p'(M) \to \mathfrak{S}_p(M)$ and $\mathfrak{T}_p'(M) \to \mathfrak{T}_p(M)$.*

PROOF. We only prove the first result, the second one is proved similarly. Let N be the upper bound of p. Let $\zeta : \mathbb{Z}\mathfrak{H}_p(M) \to R\mathfrak{H}_p(M)$ be the linear extension of the map $\mathfrak{H}_p(M) \to R\mathfrak{H}_p(M)$, which maps $K \in \mathfrak{H}_p(M)$ to $z^{N-|\omega(K)|}K$. Since a relation $K_+ - K_- - K_0 \in \mathbb{Z}\mathfrak{H}_p(M)$ maps to the relation $z^{(N-|\omega(K_+)|)}(K_+ -$

$K_- - zK_0) \in \mathfrak{s}_p(M)$, ζ induces a group homomorphism $\mathfrak{S}'_p(M) \to \mathfrak{S}_p(M)$ with left inverse induced by χ.   ■

The injections (6.11) depend on N. So the natural image of a link K in the group does not map to its natural image in the module in general. But for $p = \{\alpha\}$ and $N = |\alpha|$ images map to images. So, if $K, K' \in \omega^{-1}(\alpha)$, then the images in $\mathfrak{S}_\alpha(M)$, $\mathfrak{T}_\alpha(M)$ and its group analogs are defined. If the images of K, K' coincide in one of the models, then they coincide in all models. This proves that no new link homotopy invariants can be defined using $\mathfrak{T}$ instead of $\mathfrak{S}$.

7. Computations in $S^2 \times S^1$

The following result is useful for inductive computations.

LEMMA 7.1. *For each $\beta \in \bar\alpha$*

$$\mathfrak{t}_\alpha(M) \supset z^{|\alpha|-|\beta|}\mathfrak{t}_\beta(M).$$

Moreover, if ρ_1, ρ_2 are two $\sigma(\alpha)$ schemes, which agree above the $(|\alpha|-j)$th level (i.e. on the $(|\alpha| - j - 1)$st level and all levels above that, up to the top layer), then $\mathfrak{p}(\rho_1) - \mathfrak{p}(\rho_2)$ is contained in the submodule of $R\bar\alpha$, which is spanned by the set of elements $z^{|\alpha|-j}\mathfrak{p}(\rho)$ for $\rho \in sc_\beta(M)$ and $\beta \in \pi_\alpha^{[j]}$.

PROOF. The first result is a reformulation of (6.3). To prove the second statement compare (5.4).   ■

We consider the special case $S^2 \times S^1$ with $\pi^{[1]}(S^2 \times S^1) \cong \mathbb{Z}$. Without restriction we can replace $S^2 \times S^1$ by $M := S^2 \times S^1 \setminus$ 3-ball, which is diffeomorphic to a 2-handle attached to a solid torus. Let $T \cong D^2 \times S^1$ be the complement of the cocore of the 2-handle. The hight function in the solid torus $I \times (I \times S^1)$ can be used to staple knots (see [3]), and thus define unlink sections for both T and M. Note that the deformation spaces include:

$$\mathfrak{D}(\alpha)(T) \subset \mathfrak{D}(\alpha)(M).$$

Here we identify $\alpha \in \pi(T)$ and the image in $\pi(M)$.

Next consider a $\sigma(\alpha)$ scheme for $\alpha \in \pi(M)$. By a homotopy of schemes we can assume that the path on the top layer traverses only a finite number of handle-slides (transversal crossings with the cocore), and is contained in $\mathfrak{D}_\alpha(T)$ at all other times. Next choose paths in $\mathfrak{D}_\alpha(T)$, which join links between handle-slides to $\sigma(\alpha)$, and choose schemes below. (The existence of these paths is the crucial point. In our case handle sliding does not change the wrapping-value *in T*.)

Using these paths the scheme can be written as a composition of *elementary* schemes, i.e. for which the top layer path traverses only one handle-slide. For those schemes we will furthermore arrange by scheme homotopy that the top layer path first traverses the handle slide, then crossing changes in T.

PROPOSITION 7.2. *Let* $\ell, m \in \mathbb{Z} \cong \pi^{[1]}(S^2 \times S^1)$. *Then*

$$\mathfrak{S}_{[\ell,m]}(S^2 \times S^1) \cong \mathfrak{T}_{[\ell,m]}(S^2 \times S^1) \cong R \oplus (R/tzR),$$

where t *is the greatest common divisor of* ℓ *and* m.

PROOF. By the remark above, to compute $\mathfrak{t}_{[\ell,m]}(M)$, it suffices to consider schemes where the top layer path consists of a single handle slide followed by a regular homotopy in T. Moreover, the scheme below the top layer consists of paths in $\mathfrak{D}_{[\ell+m]}(M)$, which join knots $\subset T$ to $\sigma(\ell + m)$. Without changing the formal expansion (5.4) we can replace these paths by paths of maps in T. Then the scheme is a product of a handle slide followed by a scheme in $\mathfrak{D}_{[\ell,m]}(T)$. By [3] the homotopy scheme modules of T are free (thus all $\mathfrak{t}_p(T) = 0$). So we can compute $\mathfrak{t}_{[\ell,m]}(M)$ from the expansions in T of the results of single handle-slides of generators. By definition the result of a single handle-slide of a link K in T is an oriented band connected sum of K with a meridian (positively or negatively oriented) of T. There are many different bands, but an easy calculation shows that the image of a resulting link in a homotopy skein module of T does not depend on the band, but only upon the component to which it is attached (compare [5]). Also it is sufficient to restrict to positive handle slides (consider the inverse of an elementary scheme). Then $\mathfrak{t}_{[\ell,m]}(M)$ is generated by $zl[l + m]$ and $zm[l + m]$, and the result follows (apply (6.8)). ∎

PROPOSITION 7.3. *Let* $\ell, m, n \in \mathbb{Z} \cong \pi^{[1]}(S^2 \times S^1)$. *Then*

$$\mathfrak{S}_{[\ell,m,n]}(S^2 \times S^1) \cong R \oplus A[\ell, m, n] \oplus (R/tzR)$$
$$\mathfrak{T}_{[\ell,m,n]}(S^2 \times S^1) \cong R \oplus A[\ell, m, n] \oplus (R/tz^2R),$$

where t *is the greatest common divisor of* ℓ, m, n.

For ℓ, m, n *pairwise distinct*, $A[\ell, m, n]$ *is the quotient of* $R^3 \cong R[\ell + m, n] \oplus R[\ell + n, m] \oplus R[m + n, \ell]$ *by the submodule which is generated by the rows of the matrix*

$$z \begin{pmatrix} \ell & 0 & n \\ m & n & 0 \\ 0 & \ell & m \end{pmatrix}.$$

$A[\ell, \ell, n]$ *is the quotient of* $R^2 \cong R[2\ell, n] \oplus R[\ell + n, \ell]$ *by the submodule generated by the rows of*

$$z \begin{pmatrix} \ell & n \\ 0 & 2\ell \end{pmatrix}.$$

Finally $A[\ell, \ell, \ell] \cong R/2\ell zR$.

PROOF. This is proved similarly to (7.2). To compute $\mathfrak{t}_{[\ell,m,n]} = \mathfrak{t}_{[\ell,m,n]}(M)$ we use (7.1),i.e. $\mathfrak{t}_{[\ell,m,n]} \supset z(\mathfrak{t}_{[\ell+m,n]} \cup \mathfrak{t}_{[l+n,m]} \cup \mathfrak{t}_{[m+n,\ell]})$, and the difference between contributions from schemes, which agree on the top layer, is contained in the submodule generated by the right hand side. Thus we only have to take care of arbitrary top layer paths, and can choose schemes below freely. In this way we have reduced the problem to the computation of the effect of single handle-slides of $[\ell, m, n]$ in the homotopy skein module of the torus. This is straightforward. ∎

EXAMPLE 7.4. In the notation of (7.3), let $\Lambda := \mathbb{Z}_{2t}[z]$ (considered as R-module), and $I : \overline{[\ell, m, n]} \to \Lambda$ be defined by $I(\beta) = 1$ if and only if $|\beta| = 2$. Then I induces a skein linear link homotopy invariant (compare section 6) $\mathfrak{H}_{[\ell, m, n]}(S^2 \times S^1) \to \Lambda$. This maps each link K with wrapping-value $[\ell, m, n]$ to $lk_{(2t)}(K)z$, where $lk_{(2t)} \in \mathbb{Z}_{2t}$ is the total linking number, which is defined mod $2t$. For further results about the role of linking numbers in the study of homotopy skein modules see [5] and [8].

REFERENCES

1. N. Habegger, *Link homotopy in simply connected 3-manifolds*, Preprint 1992.
2. N. Habegger and X.S. Lin, *The classification of links up to homotopy*, Journal of the American Mathematical Society **3** (1990), 389–419.
3. J. Hoste and J. Przytycki, *Homotopy skein modules of orientable 3-manifolds*, Math. Proc. Cambridge Phil. Society **108** (1990), 475–488.
4. J. Hoste and J. Przytycki, *The $(2, \infty)$-skein module of Lens spaces: a generalization of the Jones polynomial*, will appear in Journal of Knot Theory and its Ramifications.
5. U. Kaiser, *Linking numbers and homotopy skein modules*, in preperation 1993.
6. L. Kauffman, *State models for link polynomials*, L'Enseignment Math. **t.36** (1990), 1–37.
7. J. Milnor, *Link groups*, Annals of Mathematics **59** (1954), 177–195.
8. J. Przytycki, *q-analog of the homotopy skein module*, Preprint 1992.
9. J. Przytycki, *Skein modules of 3-manifolds*, Bulletin of the Polish Academy of Sciences Mathematics **39**, 91–100.
10. H. Seifert and W. Threlfall, *Lehrbuch der Topologie* (1947), Chelsea, New York.
11. V. A. Vassiliev, *Cohomology of knot spaces*, Theory of singularities and its Applications (V. I. Arnold, ed.), Amer. Math. Soc., Providence (1990).

UNIVERSITÄT-GH SIEGEN, MATHEMATIK V, HÖLDERLINSTR. 3, 57068 SIEGEN, GERMANY

E-mail address: kaiser@ hrz.uni-siegen.d400.de

Contemporary Mathematics
Volume **164**, 1994

BORDISM OF LINK MAPS AND SELFINTERSECTIONS

Ulrich Koschorke

Universität-GH Siegen

ABSTRACT. Given a framed link map $f : N'^p \amalg N''^q \longrightarrow \mathbb{R}^m$, we discuss three equivalent geometric descriptions of the (total) γ–invariant defined by the author in 1987. Via canonical (homotopy theoretical) isomorphisms it turns out to correspond to a suspension homomorphism. Hence in the "stable range"

$$2\,(p + q + 2) \;\leq\; 3m$$

γ induces an isomorphism in framed link bordism theory. As a consequence, in this range spherical link maps are completely determined (up to framed link bordisms) by their α–invariant; we construct examples to show that the dimension requirement is sharp.

INTRODUCTION

In this paper we study *framed link maps*

$$(I.1) \qquad f \;=\; f' \amalg f'' \;:\; N' \amalg N'' \;\longrightarrow\; \mathbb{R}^m$$

i.e. N' and N'' are closed smooth *framed* (=stably parallelized) manifolds and the continuous maps f' and f'' are only required to have disjoint images, $f'(N') \cap f''(N'') = \emptyset$; we will assume throughout this paper that $p = \dim N'$ and $q = \dim N''$ do not exceed $m - 2$.

A *framed link bordism* between two such link maps f_+ and f_- is defined in the obvious way as a map

$$F \;:\; B' \amalg B'' \;\longrightarrow\; \mathbb{R}^m \times [-1, +1]$$

such that $F(B') \cap F(B'') = \emptyset$, $F|F^{-1}(\mathbb{R}^m \times \{ {}_{(-)}^{+} 1 \}) = f_{{}_{(-)}^{+}} \times \{ {}_{(-)}^{+} 1 \}$, and the framed manifolds B' and B'' induce the original framings on their boundaries $N'_+ \amalg (-N'_-)$ and $N''_+ \amalg (-N''_-)$. A very special case is a *link homotopy* when $B' = N' \times [-1, +1]$, $B'' = N'' \times [-1, +1]$, and F preserves the levels in the interval $[-1, +1]$.

1991 *Mathematics Subject Classification:* primary: 57 Q 45, 57 R 42; secondary 55 P 35
This paper is in final form.

In the mid-eighties higher dimensional link homotopy classification questions attracted enormous attention, and the resulting big progress in a large "triple point free" dimension range was set off largely by the construction of appropriate double point invariants (see e.g. [FR], [Ki 1], [Ki 2], [Ko 1], [Ko 2]). So it became very natural to search also for invariants which involve higher order selfintersections. In [Ko 3] a whole family of link bordism invariants

$$(I.2) \qquad \gamma_{r,s}(f) \quad \in \quad \Omega_{u_{r,s}}(B\Sigma_r \times B\Sigma_s; \Phi_{r,s}) \qquad (r,s \geq 1)$$

was introduced which can be interpreted as follows. Using Hirsch theory [H], we may assume after an approximation that the maps $f' = (\underline{f'}, \overline{f}')$ and $f'' = (\underline{f''}, \overline{f}'')$ into $\mathbb{R}^m = \mathbb{R}^{m-1} \times \mathbb{R}$ project to framed (self–)transverse immersions $\underline{f}'$ and $\underline{f}''$ into $\mathbb{R}^{m-1}$. Then $\gamma_{r,s}(f)$ is represented by the overcrossing locus where "r branches of f' are stacked on top of s branches of f'' ". More precisely, let the framed submanifold $\widetilde{K}_{r,s}$ of $(N')^r \times (N'')^s$ consist of all $(r+s)$–tuples $(y_1, \ldots, y_r; z_1, \ldots, z_s)$ with pairwise distinct components such that

$$\underline{f}'(y_1) = \cdots = \underline{f}'(y_r) = \underline{f}''(z_1) = \cdots = \underline{f}''(z_s)$$

and

$$\overline{f}'(y_i) > \overline{f}''(z_j) \quad \text{for all} \quad 1 \leq i \leq r, \ 1 \leq j \leq s.$$

It comes with an obvious free action of the product $\Sigma_r \times \Sigma_s$ of permutation groups. The resulting quotient manifold

$$(I.3) \qquad\qquad K_{r,s} \quad := \quad \widetilde{K}_{r,s} \ / \ \Sigma_r \times \Sigma_s$$

has dimension

$$(I.4) \qquad u_{r,s} \quad := \quad m - 1 \ - r(m-p-1) \ - s(m-q-1)$$

and is naturally endowed with a map

$$g \ : \quad K_{r,s} \quad \longrightarrow \quad B\Sigma_r \times B\Sigma_s$$

and with a description of the (stable) normal bundle of $K_{r,s}$ as the pullback $g^*(\Phi_{r,s})$. (Here

$$(I.5) \qquad\qquad \Phi_{r,s} \quad := \quad (m-p-1)\lambda_r \times (m-q-1)\lambda_s \ ,$$

where λ_t denotes the canonical t–plane bundle over the classifying space $B\Sigma_t$). We define

$$(I.6) \qquad\qquad \gamma_{r,s}(f) \quad := \quad [K_{r,s}]$$

to be the class of the resulting "$\Phi_{r,s}$–manifold" in the "normal bordism group" of all such manifolds. (Other geometric descriptions will be discussed in section 1).

In this paper we compare these geometric γ–invariants with homotopy theoretically defined γ–invariants which were inspired by an argument of B. Sanderson ([Sa], §6). They turn out to correspond just to a suspension homomorphism (via canonical Ganea and James-Hopf isomorphisms). In section 3 we will prove

Theorem 1. *The geometric and homotopy theoretical γ–invariants coincide. Thus in the "stable" ("Freudenthal") dimension range*

$$(F) \qquad 2(p+q+2) \quad \leq \quad 3m$$

the total γ–invariant

$$\gamma \quad := \quad \bigoplus_{r,s \geq 1} \gamma_{r,s} \quad : \quad \widetilde{LB}_{p,q}^{m} \longrightarrow \bigoplus_{r,s \geq 1} \Omega_{u_{r,s}}(B\Sigma_r \times B\Sigma_s; \Phi_{r,s})$$

is an isomorphism.

Here $\widetilde{LB}_{p,q}^{m}$ denotes the ("reduced") group of all framed link bordism classes of link maps as in (I.1) such that N' and N'' are *framed nulbordant* manifolds.

Of course, it is of particular interest to describe the subgroup of $\widetilde{LB}_{p,q}^{m}$ formed by *spherical link maps* (i.e. where N' and N'' are spheres). We attack this question in section 2 and obtain

Theorem 2. *In the Freudenthal range*

$$(F) \qquad 2(p+q+2) \quad \leq \quad 3m$$

the wellknown generalized linking number $\alpha = \gamma_{1,1}$ establishes an isomorphism from framed link bordism classes of spherical link maps onto the framed bordism group $\pi_{p+q+1-m}^{S}$.

Our dimension requirement is sharp in the sense that the injectivity claim in theorem 2 can already fail to hold when $2(p+q+2) = 3m+1$. Indeed, we have

Theorem 3. *There exists a link map*

$$f \quad : \quad S^7 \quad \mathrm{II} \quad S^5 \quad \longrightarrow \quad \mathbb{R}^9$$

such that $\alpha(f) = 0$, but $\gamma_{2,2}(f) \neq 0$.

The surjectivity part of theorem 2 is due to N. Habegger and U. Kaiser (cf. [HK]). Recently these authors have also proved that in the range $2(p+q+2) < 3m$ the α–invariant establishes a bijection from the set of link *homotopy* classes of spherical link maps onto $\pi_{p+q+1-m}^{S}$; hence here link homotopy and framed link bordism are the same. For this claim the dimension requirement is also sharp.

Example: $p = q = 2, m = 4$. This is the only nontrivial case where both $p = q = m - 2$ and the Freudenthal condition (F) are satisfied. The isomorphism

$$\gamma = \alpha \oplus \gamma_{1,2} \oplus \gamma_{2,1} \quad : \quad \widetilde{LB}_{2,2}^{4} \quad \xrightarrow{\cong} \quad \Omega_1^{fr} \oplus (\Omega_0(P^\infty; \lambda))^2 = \mathbb{Z}_2 \oplus \mathbb{Z}_2 \oplus \mathbb{Z}_2$$

of theorem 1 was already established by B. Sanderson (see [Sa], 6.1). The subgroup of framed link *bordism* classes of link maps

$$f \quad : \quad S^2 \quad \mathrm{II} \quad S^2 \quad \longrightarrow \quad \mathbb{R}^4$$

is generated by the Fenn-Rolfsen link map (cf. [FR]) and corresponds to the diagonal $\mathbb{Z}_2(1,1,1)$. However, P. Kirk has shown that his refined double point invariant σ maps spherical link *homotopy* classes onto an infinitely generated free abelian group.

Notations and conventions. An immersion (always assumed to be smooth) is called *framed* if its normal bundle is trivialized; it is called *(self–) transverse* if all iterated (self–)intersections in sight are transverse. Σ_j denotes the permutation group in j elements. Base points of topological spaces are usually denoted by $*$; in the case of a sphere $S^i = \mathbb{R}^i \cup \{\infty\}$ we pick $* = \infty$.

§ 1. The geometric γ–invariants

In this section we discuss three equivalent definitions 1.4, 1.7 and I.6 of the geometric γ–invariants. Much of this material is already inherent in [Ko 3].

First, given any framed link map

$$f \; = \; f' \amalg f'' \quad : \quad N' \; \amalg \; N'' \; \longrightarrow \; \mathbb{R}^m$$

we associate to it a whole double sequence of "generalized link maps"

$$(1.1) \qquad f_{r,s} = f'_r \amalg f''_s \quad : \quad N'_r \; \amalg \; N''_s \; \longrightarrow \; \mathbb{R}^m \qquad (r, s \geq 1)$$

(starting with $f_{1,1} = f$) as follows.

Pick any relative bordism

$$\widetilde{f'} \quad : \quad B' \; \longrightarrow \; \mathbb{R}^m$$

of f', i.e. the boundary of the framed bordism B' consists of N' and of other components which $\widetilde{f'}$ maps close to a faraway point $c \in \mathbb{R}^m$. Since $p = \dim N'$ and $q = \dim N''$ are strictly smaller than $m - 1$, we may assume (after an approximation) that $\widetilde{f'}$ is a selftransverse framed immersion. Consider the boundary of the r–tuple point manifold of $\widetilde{f'}$: it consists of a "faraway part" (which gets mapped close to c and which we neglect) and of the "nearby part" N'_r which gets mapped into a neighborhood of $f'(N')$. After smoothing corners N'_r is what we call an $(m - p - 1)\lambda_r$–manifold of dimension

$$\dim N'_r \; = \; m - 1 - r(m - p - 1) \; ,$$

where λ_r denotes the canonical r–plane bundle over the classifying space $B\Sigma_r$. In other words, N'_r is equipped with a map

$$g \quad : \quad N'_r \; \longrightarrow \; B\Sigma_r$$

and with a (stable) trivialization

$$(1.2) \qquad \overline{g} \; : \; TN'_r \oplus g^*((m - p - 1)\lambda_r) \; \overset{\cong}{\longrightarrow} \; N'_r \times \mathbb{R}^{m-1} \; ;$$

as usual, this "normal structure" is derived from the action of the symmetric group Σ_r permuting r branches of a selftransverse immersion.

Similarly we define the $(m - q - 1)\lambda_s$–manifold N''_s of dimension

$$\dim N''_s \; = \; m - 1 - s(m - q - 1)$$

which gets mapped into a neighborhood of $f''(N'')$. Thus f restricts to a "generalized link map" $f_{r,s}$ as in 1.1, where N'_r and N''_s are equipped with the normal structures described above (see [Ko 1], 1.9 for the corresponding obvious notion of (normal) link bordisms).

Proposition 1.3. *For every pair* $r, s \geq 1$ *of integers the (normal) link bordism class of* $f_{r,s}$ *depends only on the framed link bordism class of* f.

This result follows from standard techniques. It leads to the welldefined invariant

$$(1.4) \qquad \gamma_{r,s}(f) \; := \; \alpha(f_r' \amalg f_s'') \; \in \; \Omega_{u_{r,s}}(B\Sigma_r \times B\Sigma_s; \; \Phi_{r,s})$$

which also depends only on the framed link bordism class of f (compare I.4 and I.5). This generalized α–invariant is represented by the $(m-p-1)\lambda_r \times (m-q-1)\lambda_s$–manifold

$$(1.5) \qquad\qquad\qquad \widetilde{f}_r' \quad \pitchfork \quad f_s''$$

obtained by intersecting f_s'' with some relative bordism $\widetilde{f}_r'$ of f_r' (see [Ko 1], section 1). In particular, the r–tuple point manifold of $\widetilde{f}'$ is such a relative bordism, and in this case $\widetilde{f}_r' \pitchfork f_s$ is easily seen to be the r–tuple point manifold of the immersion

$$(1.6) \qquad\qquad i_s \quad : \quad \widetilde{f}' \pitchfork f_s'' \quad \looparrowright \quad N_s''.$$

Thus in the terminology of [Ko 3], we have

$$(1.7) \qquad\qquad \gamma_{r,s}(f) \quad = \quad h_r(\gamma_s(f))$$

where $\gamma_s(f) = [i_s]$ is the "joint bordism class" of this immersion and h_r denotes the *r–tuple point Hopf invariant homomorphism*.

Finally, by Hirsch theory we may assume from the very beginning that

$$f' = (\underline{f}', \overline{f}') \quad : \quad N' \quad \longrightarrow \quad \mathbb{R}^m = \mathbb{R}^{m-1} \times \mathbb{R}$$

and similarly $f'' = (\underline{f}'', \overline{f}')$ project to smooth (self–)transverse immersions $\underline{f}'$ and $\underline{f}''$ into $\mathbb{R}^{m-1}$. Moreover, we may pick $\widetilde{f}'$ to be the homotopy which moves f' downwards in the negative x_m–direction. Then f_r' is given by the immersed r–tuple selfintersection manifold of $\underline{f}'$ in $\mathbb{R}^{m-1}$, lifted in $\mathbb{R}^{m-1} \times \mathbb{R}$ to the minimal $\overline{f}'$– level of its r branches. Similarly, f_s'' can be given by the s–fold selfintersection manifold of $\underline{f}''$, lifted to the maximal $\overline{f}''$–level. We deduce the description of $\gamma_{r,s}(f)$ given in the introduction (see I.6).

§ 2. Spherical link maps

In this section we prove results concerning the existence or nonexistence of relations which our γ–invariants must satisfy on spherical link maps. This will allow us to deduce theorems 2 and 3 of the introduction.

If in the previous discussion $N'' = N_1''$ is a sphere, we have the very unusual situation that the joint bordism class of the immersion

$$i_1 \quad : \quad \widetilde{f}' \pitchfork f'' \quad \looparrowright \quad S^q$$

(see 1.6) depends only on the framed bordism class of its domain, i.e. on $\alpha(f) = [\widetilde{f}' \pitchfork f''] \in \Omega^{fr}_{p+q+1-m}$; from 1.7 we get the sphericity relation

$$(2.1) \qquad\qquad \gamma_{r,1}(f) \;=\; h_r^{(q)}(\alpha(f)) \qquad \text{for all } r \geq 1 \;,$$

where $h_r^{(q)}$ denotes the r–tuple point Hopf invariant for framed immersions into $\mathbb{R}^q$.

Similarly, if $N' = S^p$ then

$$(2.2) \qquad\qquad \gamma_{1,s}(f) \;=\; \pm h_s^{(p)}(\alpha(f)) \qquad \text{for all } s \geq 1$$

since $\alpha(f'' \amalg f') = \pm\alpha(f' \amalg f'')$ (see e.g. [Ko 1], 1.3).

In particular, if both N' and N'' are spheres and if the Freudenthal condition (F) holds, then $\gamma_{r,1}(f)$ and $\gamma_{1,s}(f)$ depend only on $\alpha(f)$ for $r,s \geq 1$, and the remaining γ–invariants lie in bordism groups of strictly negative dimensions. Thus the injectivity claim in theorem 2 follows from theorem 1. ∎

In contrast, we will show next that $\gamma_{2,2}$ is *not* always a homomorphic image of α. We start from an immersed link map

$$f \;=\; f' \amalg f'' \;:\; S^p \amalg S^q \;\looparrowright\; \mathbb{R}^m$$

and a (base point preserving) map $g : S^{q+\ell} \longrightarrow S^q$. A standard k–fold suspension procedure for $k \geq 1$ (see e.g. [Ko 1], 2.1) and precomposition then yield the new link map

$$(2.3) \qquad \widehat{f} \;=\; S^k f' \amalg f'' \circ g \;:\; S^{p+k} \amalg S^{q+\ell} \;\longrightarrow\; \mathbb{R}^{m+k}.$$

We will now calculate $\gamma_{2,2}(\widehat{f})$.

Let $\widetilde{f}'_2$ be defined by the double points of a relative bordism of f' in $\mathbb{R}^m$ as in 1.5f. The corresponding map $S^k \widetilde{f}'_2$ for $S^k f'$ into $\mathbb{R}^{m+k}$ has product form $\widetilde{f}'_2 \times id$ near $\mathbb{R}^m \times \{0\}$. If

$$(2.4) \qquad\qquad 2p + 2q + 3 \;\leq\; 3m \;,$$

then generically both maps intersect the immersion f'' only in its "embedded part" $E(f'') \subset \mathbb{R}^m$. The resulting intersections

$$\widetilde{f}'_2 \pitchfork f'' \;=\; S^k \widetilde{f}'_2 \pitchfork f''$$

(in $\mathbb{R}^m$ and $\mathbb{R}^{m+k}$, resp.) represent

$$\gamma_{2,1}\,(f) \;=\; \gamma_{2,1}(S^k f' \amalg f'').$$

Furthermore, after a suitable homotopy g restricts to a smooth product fibration over $S^q - \{*\}$. If also

$$(2.5) \qquad\qquad 0 \leq \ell \;<\; m - q - 1 + k \;,$$

we may pick a framed selftransverse immersion i of the resulting fiber M^ℓ into $\mathbb{R}^{m-q-1+k}$; denote its double point manifold by D. In a small tubular neighborhood $E(f'') \times \mathbb{R}^{m-q-1+k}$ of $E(f'')$ in $\mathbb{R}^{m+k-1}$ we can then approximate $f'' \circ g$ by the immersion $f'' \times i$ (at least away from $* \in S^q$), and its double point manifold $E(f'') \times D$ turns out to be the relevant part of the manifold N_2'' (when constructed for $f'' \circ g$ and for a relative homotopy into the positive x_{m+k-} direction, see section 1). Its intersection in $\mathbb{R}^{m+k}$ with $S^k \widetilde{f_2'}$ is the product of D with the intersection $\widetilde{f_2'} \pitchfork f''$ in $\mathbb{R}^m$. Finally observe that the framed manifold M^ℓ represents the stable suspension $E^\infty[g] \in \pi_\ell^S \cong \Omega_\ell^{fr}$ and hence D represents its value $h_2^{(m-q-1+k)}(E^\infty[g])$ under the indicated double point Hopf invariant (compare 2.1). We conclude

Proposition 2.6. *If 2.4 and 2.5 hold, then*

$$\gamma_{2,2}(\widehat{f}) = \gamma_{2,1}(f) \cdot h_2^{m-q-1+k}(E^\infty[g])$$
$$= h_2^{(q)}(\alpha(f)) \cdot h_2^{m-q-1+k}(E^\infty[g])$$

and

$$\alpha(\widehat{f}) = \alpha(f) \cdot E^\infty[g].$$

∎

Now we know from the table 3.7 in [Ko 1], and from the surjectivity of the Hopf invariant homomorphisms $h_2^{(6)}$ and $h_2^{(2)}$ from π_3^S and π_1^S, resp., into $\Omega_0(P^\infty; 3\lambda) = \Omega_0(P^\infty; \lambda) = \mathbb{Z}_2$, that there exist a link map

$$f : S^4 \amalg S^6 \longrightarrow \mathbb{R}^8$$

and a map $g : S^7 \longrightarrow S^6$ such that both $h_2^{(6)}(\alpha(f))$ and $h_2^{(2)}(E^\infty[g])$ are nontrivial. Therefore, the $\gamma_{2,2}-$ invariant of

$$\widehat{f} = Sf' \amalg f'' \circ g : S^5 \amalg S^7 \longrightarrow \mathbb{R}^9$$

is also nontrivial whereas $\alpha(\widehat{f})$ lies in the trivial group π_4^S . Theorem 3 of the introduction follows from the symmetry theorem in [Ko 3].

§ 3. The homotopy theoretical γ–invariant

In this section we outline the proof of theorem 1. Along the way we obtain a very simple standard form for all framed link maps (up to bordism; see theorem 3.9).

The discussion will center around the following diagram (which will turn out to commute).

(3.1)

$$
\widetilde{LB}_{p,q}^m \xrightarrow{\ \gamma=\oplus\gamma_{r,s}\ } \bigoplus_{r,s\geq 1}\Omega_{u_{r,s}}(B\Sigma_r\times B\Sigma_s;\phi_{r,s})
$$

$$
\text{S}\| \uparrow PT \qquad\qquad\qquad\qquad \text{S}\|\ \uparrow PT
$$

$$
\widetilde{\pi}_m(QS^{m-p}\vee QS^{m-q})
$$

$$
\text{S}\| \uparrow G \qquad\qquad \bigoplus_{r,s\geq 1}\pi_{m-1+t}(S^v D_r S^{m-p-1}\wedge S^w D_s S^{m-q-1})
$$

$$
\pi_m(SQS^{m-p-1}\wedge QS^{m-q-1})
$$

$$
\text{S}\| \uparrow \qquad\qquad\qquad\qquad \text{S}\| \Big\downarrow{\scriptstyle \oplus_{r,s}(h_r\wedge h_s)_*}
$$

$$
\pi_m(SF_kC_\ell S^{m-p-1}\wedge F_kC_\ell S^{m-q-1}) \xrightarrow{\ E^\infty\ } \pi_{m-1+t}(S^t F_kC_\ell S^{m-p-1}\wedge F_kC_\ell S^{m-q-1}).
$$

Here $k,\ell,v,w>>0$, $t=v+w$, and $QX:=\Omega^\ell S^\ell X$ denotes the ℓ–fold loop space of the ℓ–th reduced suspension of any space X with base point $*$. Often X will be a sphere $S^j=\mathbb{R}^j\cup\{\infty\}$ with base point $*=\infty$.

Given a framed n–manifold M^n, recall that a map $g:M\to QS^j$ yields (after a suitable approximation of its adjoint $G:M\times\mathbb{R}^\ell\to S^{\ell+j}$) a map

(3.2) $$N^{n-j}\ =\ G^{-1}\{0\}\ \longrightarrow\ M^n$$

of framed manifolds. In particular, if we restrict a map $g:\mathbb{R}^m\cup\{\infty\}\to QS^i\vee QS^j$ to the disjoint open subsets $g^{-1}(QS^i-\{*\})$ and $g^{-1}(QS^j-\{*\})$ of $\mathbb{R}^m$, we obtain a framed link map into $\mathbb{R}^m$. This well-known procedure defines the first (Pontrjagin-Thom) isomorphism PT on the left hand side in diagram 3.1.

Here we denote the kernel of the canonical homomorphism

$$\pi_m(QS^{m-p}\vee QS^{m-q})\ \longrightarrow\ \pi_m(QS^{m-p}\times QS^{m-q})$$

by $\widetilde{\pi}_m(QS^{m-p}\vee QS^{m-q})$. A result of Ganea (see [B], 3. 1. 20) gives the indicated second vertical isomorphism between this reduced homotopy group and

$$\pi_m(S\Omega QS^{m-p}\wedge\Omega QS^{m-q})\ =\ \pi_m(SQS^{m-p-1}\wedge QS^{m-q-1}).$$

In order to understand its geometric meaning, let us consider any base point preserving map

$$h:\mathbb{R}^m\cup\{\infty\}\ \longrightarrow\ S^1\wedge QS^{m-p-1}\wedge QS^{m-q-1}$$

and its restriction

$$h|=(h_0,h',h''):\mathbb{R}^m-h^{-1}\{*\}\ \longrightarrow\ \mathbb{R}\times(QS^{m-p-1}-\{*\})\times(QS^{m-q-1}-\{*\}).$$

After suitable approximations and homotopies h takes the following *standard form*

(3.3) a) there is a compact smooth framed $(m-1)$–dimensional submanifold $M\subset$
$\mathbb{R}^m$ (essentially $M=h_0^{-1}\{0\}$);

b) the boundary ∂M consists of two parts,

$$\partial M \;\;=\;\; \partial' M \,\cup\, \partial'' M \; ,$$

and has corners along the 1–codimensional intersection $\partial' M \cap \partial'' M$;

c) $h' \,:\, M \,\to\, QS^{m-p-1}$ (and $h'' \,:\, M \,\to\, QS^{m-q-1}$) map a whole collar neighborhood of $\partial' M$ (and of $\partial'' M$, resp.) into the basepoint; and

d) on a "collar neighborhood" $M \times \mathbb{R} \;\subset\; \mathbb{R}^m$ of M the map h is defined by

$$h(x,t) = [t, h'(x), h''(x)] \in S^1 \wedge QS^{m-p-1} \wedge QS^{m-q-1}$$

while the complement $\mathbb{R}^m - M \times \mathbb{R}$ is mapped to the base point.

Similarly, homotopies of h can be made into "standard form bordisms" of these data (in $\mathbb{R}^m \times I$).

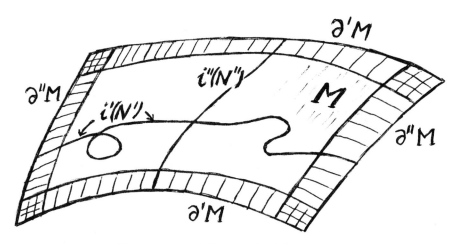

FIGURE 3.4. The standard form of h.

Note that h' and h'' correspond to framed maps

$$(3.5) \qquad \begin{aligned} i' \;&:\; (N'^p, \partial N'^p) \longrightarrow (M, \partial'' M) \qquad \text{and} \\ i'' \;&:\; (N''^q, \partial N''^q) \longrightarrow (M, \partial' M). \end{aligned}$$

It follows from Smale-Hirsch theory [H] that these maps are homotopic to framed immersions. In other words, in the discussion above we can replace the iterated loop spaces QS^j by the "Thom spaces for immersions"

$$(3.6) \qquad F_k C_\ell S^j \;\;=\;\; (\coprod_{r=1}^{k} (\widetilde{C}_{\ell,r} \times (S^j)^r)/\Sigma_r)/\sim$$

(see e. g. [M], [Se] and compare with [KS]). Here the permutation group Σ_r acts in the obvious way on the ordered configuration space $\widetilde{C}_{\ell,r}$ of r distinct points in $\mathbb{R}^\ell$; i.e. we are dealing with unordered configurations of r points in $\mathbb{R}^\ell$, each labelled by an element of S^j , and the equivalence relation $\sim$ drops all occurences of the basepoint $\infty \in S^j$. Thus the third vertical isomorphism to the left in diagram 3.1 can be obtained from Hirsch theory.

By definition the Ganea isomorphism G is induced by a Whitehead product. We will now describe it geometrically. Given M and i', i'' as in 3.3a, 3.3b and 3.5, consider the framed maps

$$(3.7) \qquad \begin{aligned} I' &= i' \times \text{incl} \; : N' \times [0,2] \; \longrightarrow \; M \times \mathbb{R} \subset \mathbb{R}^m, \quad \text{and} \\ I'' &= i'' \times \text{incl} \; : N'' \times [1,3] \; \longrightarrow \; M \times \mathbb{R} \subset \mathbb{R}^m. \end{aligned}$$

Restrict these maps to the boundaries of their domains and smooth the corners. The resulting framed link map $\partial I' \amalg \partial I''$ into $\mathbb{R}^m$ represents $PT \circ G([h])$. The following consequence seems to be worth noting.

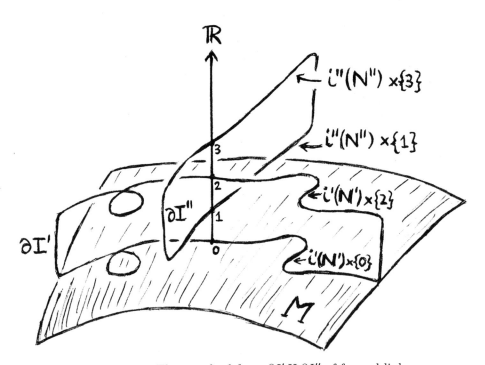

FIGURE 3.8. The standard form $\partial I' \amalg \partial I''$ of framed links.

Theorem 3.9. *Every element of $\widetilde{LB}_{p,q}^m$ can be represented in this standard form (for some M, i', i'' as in 3.3a, 3.3b and 3.5).*

Next we turn to the ("stable") right hand side of diagram 3.1.

Given a sphere $S^j = \mathbb{R}^j \cup \{\infty\}$, note that $F_r C_\ell S^j$ (see 3.6) is the disjoint union of $F_{r-1} C_\ell S^j$ and of the total space $(\widetilde{C}_{\ell,r} \times (\mathbb{R}^j)^r)/\Sigma_r$ of the direct sum of j copies of the canonical r–plane bundle λ_r over the configuration space $C_{\ell,r}$. Hence

$$(3.10) \qquad D_r S^j \quad := \quad F_r C_\ell S^j / F_{r-1} C_\ell S^j$$

is the Thom space of $j\lambda_r$. Similarly $D_r S^{m-p-1} \wedge D_s S^{m-q-1}$ is the Thom space of the vector bundle $(m-p-1)\lambda_r \times (m-q-1)\lambda_s$ over $C_{\ell,r} \times C_{\ell,s}$ (which is basically our coefficient bundle $\Phi_{r,s}$), and we obtain the upper right-hand Pontrjagin-Thom isomorphism in diagram 3.1.

The lower right-hand isomorphism is induced by the smash product of two homotopy equivalences of the form

$$(3.11) \qquad \oplus h_r \quad : \quad S^t(F_k C_\ell S^j) \longrightarrow \bigvee_{r=1}^{k} S^t(D_r S^j) \quad ,$$

where the summands are (restricted) James-Hopf maps (see [CMT], definition 5.5 and theorem 7.1).

Now we define the *homotopy theoretical γ–invariant* to be the suspension homomorphism E^∞, composed with all the vertical isomorphisms in diagram 3.1.

Theorem 3.12. *The homotopy theoretical and geometric γ–invariants coincide. In particular, γ is an isomorphism in the stable range $p + q + 2 \leq \frac{3}{2}m$.*

Proof. Given $r, s \geq 1$, let an element

$$[h] \in \pi_m(SF_k C_\ell S^{m-p-1} \wedge F_k C_\ell S^{m-q-1})$$

be represented in standard form as in 3.3.

We may assume that the maps i', i'' in 3.5 are (self–)transverse immersions into M which are decompressed by embeddings in $M \times \mathbb{R}^v$ and $M \times \mathbb{R}^w$, resp.. The effect e.g. of the James-Hopf map h_r is to replace i' by a similar decompressed version of the immersion i'_r of the r–tuple point manifold of i' into M. Thus if we apply $h_r \wedge h_s$ and the Pontrjagin-Thom procedure, we obtain a submanifold of $M \times \mathbb{R}^v \times \mathbb{R}^w$ which projects to the immersed intersection $i'_r \pitchfork i''_s$ in M .

On the other hand, $PT \circ G([h])$ is represented by the link map $\partial I' \amalg \partial I''$ in standard form (see 3.7). We can obtain its $\gamma_{r,s}$–value by intersecting $i'_r \times \{2\}$ with the s–tuple points of I'' in $M \times \mathbb{R} \subset \mathbb{R}^m$ or, equivalently, by intersecting i'_r with i''_s in M. We conclude that our diagram 3.1 commutes.

The remaining claim follows from the Freudenthal suspension theorem (see e.g. [W], p. 369). Indeed, the space $SF_k C_\ell S^{m-p-1} \wedge F_k C_\ell S^{m-q-1}$ is $(2m-p-q-2)$–connected since its i–th homotopy group is isomorphic to $\widetilde{LB}^i_{i-m+p, i-m+q}$ and hence vanishes by a simple transversality argument when

$$(i - m + p) + (i - m + q) + 1 \quad < \quad i.$$

■

REFERENCES

[B]   H. Baues, *Obstruction theory*, Lect. Notes in Math. **628** (1977), Springer.

[CMT]   F. R. Cohen, J. P. May and L. R. Taylor, *Splitting of certain spaces CX*, Math. Proc.
         Camb. Phil. Soc. **84** (198), 465–496.

[FR]   R. Fenn, D. Rolfsen, *Spheres may link homotopically in 4-space*, J. Lond. Math. Soc. **34**
         (1986), 177–184.

[H]   M. Hirsch, *Immersions of manifolds*, Trans. Am. Math. Soc. **93** (1959), 242–276.

[HK]   N. Habegger, U. Kaiser, *Link homotopy in the 2-metastable range*, preprint (1992), Siegen
         University.

[Ki 1]   P. Kirk, *Link maps in the 4-sphere*, Proc. Siegen Topology Symp., Lect. Notes in Math.
         1350 (1988), Springer, 31–43.

[Ki 2]   _____, *Link homotopy with one codimension two component*, Trans AMS **319, No. 2**
         (1990), 663–688.

[Ko 1]   U. Koschorke, *Link maps and the geometry of their invariants*, Manuscr. Math. **61** (1988),
         383–415.

[Ko 2]   _____, *On link maps and their homotopy classification*, Math. Ann. **286** (1990), 753–
         782.

[Ko 3]   _____, *Multiple point invariants*, Proc. Siegen Topology Symp., Lect. Notes in Math.
         1350 (1988), Springer, 44–86.

[KS]   U. Koschorke, B. Sanderson, *Selfintersections and higher Hopf invariants*, Topology **17**
         (1978), 283–290.

[M]   J. P. May, *The geometry of iterated loop spaces*, Lect. Notes in Math **271** (1972), Springer.

[Sa]   B. Sanderson, *Bordism of links in codimension 2*, J. London Math. Soc. **35 (2)** (1987),
         367–376.

[Se]   G. Segal, *Configuration spaces and iterated loop spaces*, Inv. math. **21** (1973), 213–221.

[W]   G. W. Whitehead, *Elements of homotopy theory*, Grad. Texts Math. (1978), Springer.

UNIVERSITÄT-GH SIEGEN, MATHEMATIK V, HÖLDERLINSTR. 3, SIEGEN, GERMANY
E-mail address: koschorke@ hrz.uni-siegen.d400.de

Contemporary Mathematics
Volume **164**, 1994

The monodromy of the Brieskorn bundle

A. LEIBMAN *   D. MARKUSHEVICH [†]

Abstract

The lower central series (l.c.s.) of the fundamental group of the complement of the reflecting hyperplanes of the D_n Coxeter group (that is, the generalized pure braid group PD_n) is investigated in terms of the Brieskorn bundle. The monodromy of the Brieskorn bundle is calculated and applied to prove the triviality of the nilpotent residue of PD_n. Another group-theoretic tool for the investigation of the l.c.s. of the fundamental group of the total space of a fiber type bundle is given, generalizing a Falk - Randell result for the case when the monodromy acts trivially on the one-dimensional homology.

0. Introduction

There are a number of papers devoted to the investigation of the lower central series (l.c.s.) of the fundamental groups of complements of arrangements of hyperplanes in $\mathbf{C}^n$, see, e.g. [FR-1], [FR-2], [FR-3], [Ko-1]. The paper [H] shows the importance of the property of the nilpotent approximability of fundamental groups, which is our main concern in the present paper. The nilpotent residue of the fundamental group is an obstruction to the existence of a formal analytic solution to the generalized Riemann-Hilbert problem on recovery of a Pfaff system of PDE with regular singularities from its monodromy representation. Of special interest are the fundamental groups of complements of classical arrangements of hyperplanes which are complexified reflecting hyperplanes of the Coxeter groups A_n, B_n, D_n (the C_n arrangement coincides with the B_n one). The fundamental group of the complement of the A_n arrangement is the classical pure braid group P_{n+1}. It is a well-studied object. Many group-theoretic facts about P_{n+1} are known, e.g. a presentation in terms of generators and relations

---

1991 mathematical subject classification. Primary 20F36, 32C40

*Supported by BTS.

[†]Research at MSRI supported in part by NSF grant # DMS 8505550.

This paper is in final form and no version of it will be submitted for publication elsewhere.

(that of Burau), normal form of elements, the structure of the lower central series (see [MKS], [M], [Bi], [Lin]). Certain matrix representations of P_{n+1} by monodromy of Knizhnik-Zamolodchikov equations are investigated ([Ko-2], [TK], [L]). Much less has been done for the other arrangements. The forthcoming paper [Lei] provides presentations in terms of generators and relations of PB_n, PD_n, where we denote by PX_n the generalized pure braid group of the Coxeter group X_n, defined as the fundamental group of the complement of the complexified reflecting hyperplanes of X_n (so that $P_{n+1} = PA_n$ in this notation); [Mar] gives a different presentation for PD_n. The study of the l. c. s. of PA_n, PB_n is covered by Falk-Randell, as A_n and B_n are fiber type.

An arrangement of hyperplanes $\bigcup_{1 \le j \le N} L_j \subset \mathbf{C}^n$ passing through the origin is said to be fiber type if its complement M admits a linear projection onto the complement of an arrangement of hyperplanes N in $\mathbf{C}^{n-1}$ which introduces a structure of a fiber bundle on M with punctured complex line $\mathbf{C} \setminus \{a_1, \ldots, a_r\}$ as a fiber, and if this construction can be iterated up to the one-dimensional base $M_1 = \mathbf{C}^* := \mathbf{C} \setminus \{0\}$. So, the complement of a fiber type arrangement $M = M_n \subset \mathbf{C}^n$ admits a tower of fiber bundles

$$M_n \xrightarrow{p_{n-1}} M_{n-1} \xrightarrow{p_{n-2}} \ldots \xrightarrow{p_1} M_1 = \mathbf{C}^*$$

where each M_k is the complement of an arrangement of hyperplanes passing through the origin in $\mathbf{C}^k$, and the fiber F_k of each bundle map p_k is $\mathbf{C} \setminus \{a_1, \ldots, a_{r_k}\}$. On each stage of the tower we have the split exact sequence:

$$1 \longrightarrow \pi_1(F_k) \xrightarrow{\alpha_k} \pi_1(M_{k+1}) \underset{\sigma_k}{\overset{\beta_k}{\rightleftarrows}} \pi_1(M_k) \longrightarrow 1$$

where $\pi_1(F_k)$ is a free group with r_k generators. It is easily seen that the maps p_k admit a cross-section (that is why the exact sequence is split), and that the monodromy action of $\pi_1(M_k)$ on the homology $H_1(F_k) = \pi_1(F_k)/[\pi_1(F_k), \pi_1(F_k)]$ is trivial. Let us denote by $\gamma_n(G)$ the n-th term of the l.c.s. of a group G, defined by $\gamma_1(G) = G$ and $\gamma_q(G) = [\gamma_{q-1}(G), G]$. Thus, the following Falk-Randell result applies [FR-1]:

Falk–Randell Lemma. *Let*

$$1 \longrightarrow A \xrightarrow{\alpha} B \underset{\sigma}{\overset{\beta}{\rightleftarrows}} C \longrightarrow 1$$

be a split exact sequence of groups. Suppose that the induced action of C on $A/\gamma_2 A$ is trivial, or equivalently, that C acts on A by conjugations in A. Then the maps α, β, σ restricted to the q-th terms of the l.c.s. fit into the split exact sequence

$$1 \longrightarrow \gamma_q(A) \xrightarrow{\alpha} \gamma_q(B) \underset{\sigma}{\overset{\beta}{\rightleftarrows}} \gamma_q(C) \longrightarrow 1$$

for every $q \ge 1$.

This lemma easily implies the triviality of the nilpotent residue γ_∞ of $\pi_1(M)$, defined as the intersection of the terms of the l.c.s.:

$$\gamma_\infty(G) := \bigcap_{q=1}^{\infty} \gamma_q(G)$$

for every group G.

Thus, for the A_n and B_n arrangements the triviality of γ_∞ follows from the Falk-Randell Lemma. We give two other approaches to the problem of computing γ_∞. The first one applies specifically to the D_n arrangement, using the structure of a fiber bundle on its complement induced not by a linear projection, but by a quadratic map (see Sect. 1.1) discovered by Brieskorn [Br]. The base of the Brieskorn bundle is the complement of a fiber type arrangement, so that we have again a tower of fiber bundles, and the Falk- Randell Lemma applies for every stage of the tower except for the last one, which is the Brieskorn bundle.

Our first main result is the explicit computation of the monodromy action of the fundamental group of the base N on that of the fiber F of the Brieskorn bundle; see formulas (9). This includes in particular a certain nice presentation for $\pi_1(F)$. To give an idea what this nice presentation is, we should say that $\pi_1(F)$ is identified with the commutator subgroup of the free product of n copies of the cyclic group of order 2:

$$\pi_1(F) = [E, E], \ E := \mathbf{Z}_2 * \ldots * \mathbf{Z}_2.$$

In the case $n = 3$ it is a free group of rank 5, what can be seen also from geometric considerations, as F is an elliptic curve with four points deleted. It is quite common to use a standard basis of the fundamental group of a Riemann surface of genus g with l punctures in the form $\{a_1, b_1, \ldots, a_g, b_g, c_1, \ldots, c_l\}$ subject to one relation $[a_1, b_1] \ldots [a_g, b_g] c_1 \ldots c_l = 1$. This idea is misleading here, as the connection between the natural basis of $[E, E]$ and the standard geometric basis seems to be incomprehensible.

In fact, the cumbersome formulas (9) are obtained by restricting an action of $\pi_1(N)$ on E to that on $[E, E]$. The action on E is pretty simple, given by formulas (8), and it is an action by conjugations in E. This enables us to give a cheap proof of the triviality of the nilpotent residue of PD_n in applying the Falk-Randell Lemma not to PD_n itself, which is an extension of $[E, E]$ by $\pi_1(N)$, but to a group $\widetilde{PD_n}$ without any topological interpretation and having PD_n as its subgroup, defined as an extension of E by $\pi_1(N)$. This is the second main result (Theorem 1.8).

The second part of the article gives another approach to the proof of $\gamma_\infty = 1$ in a more general setting. The following result is proved:

Generalized Falk–Randell Lemma 1. *Let*

$$1 \longrightarrow A \xrightarrow{a} B \underset{s}{\overset{b}{\underset{\longleftarrow}{\longrightarrow}}} C \longrightarrow 1$$

be a split exact sequence of groups, $\alpha \geq 2$, $\beta \geq 1$ fixed integers, and assume that *either* (i) $\gamma_n(A)/\gamma_{n+1}(A)$ *is free abelian for every* n,

or (ii) $\gamma_n(A)/\gamma_{n+1}(A)$ *is finitely generated and has no torsions of order prime to* α.

Suppose that the induced action of C *on* $\gamma_1(A)/\gamma_2(A) \otimes_{\mathbf{Z}} (\mathbf{Z}/l\mathbf{Z})$ *is nilpotent of index* $\leq \beta$, *that is*

$$\forall x \in A, c_1, \ldots, c_\beta \in C, \ \exists y \in A : [\ldots [x, c_1], \ldots, c_\beta] \equiv y^\alpha \bmod \gamma_2(A).$$

Then there exists an integer valued function $k = k(m, \beta)$, $\lim_{m \to \infty} k(m, \beta) = \infty$, *such that for all* $m \geq 1$

$$\gamma_m(B) \bigcap A \subset \bigcup_{q=1}^{k} \{g^{\alpha^{m-q}} h \mid g \in \gamma_q(A), \ h \in \gamma_{q+1}(A)\}.$$

This condition implies the triviality of $\gamma_\infty(B)$, whenever $\gamma_\infty(A)$ and $\gamma_\infty(C)$ are trivial (Theorem 2.2). We also provide a separate formulation of Generalized Falk Randell Lemma for the case when the action of C on $\gamma_1(A)/\gamma_2(A)$ is nilpotent over $\mathbf{Z}$ (that is, the case $\alpha = 0$):

Generalized Falk–Randell Lemma 2. *Let*

$$1 \longrightarrow A \xrightarrow{a} B \underset{s}{\overset{b}{\underrightarrow{\longrightarrow}}} C \longrightarrow 1$$

be a split exact sequence of groups. Assume that the Lie action of C *on the abelian group* $\gamma_1(A)/\gamma_2(A)$ *is nilpotent of order* $\beta \geq 1$, *that is*

$$\forall x \in A, \ c_1, \ldots, c_\beta \in C \ \ [\ldots [x, c_1], \ldots, c_\beta] \in \gamma_2(A).$$

Then there exists an integer valued function $k = k(m, \beta)$, $\lim_{m \to \infty} k(m, \beta) = \infty$, *such that for all* $m \geq 1$

$$B_m \bigcap A \subset A_k.$$

This condition also implies the triviality of $\gamma_\infty(B)$ as soon as $\gamma_\infty(A) = \gamma_\infty(C) = 1$ (Theorem 2.3).

Unfortunately, we could not verify the assumptions of the above lemmas in the D_n case in dimensions greater than 3. For $n = 3$, Lemma 1 applies with $a = b = 2$, thus giving an alternative proof of $\gamma_\infty(PD_3) = 1$.

1. Brieskorn bundle

1.1. Denote the complement of the complexification of the reflecting hyper-planes of the Coxeter group D_n in $\mathbf{C}^n$ by M:

$$M = \mathbf{C}^n \setminus \bigcup_{1 \leq i < j \leq n} \{(x_1, \ldots, x_n)|x_i = \pm x_j\}$$

Consider the mapping $p' : \mathbf{C}^n \to \mathbf{C}^{n-1}$ defined by

$$(1) \qquad y_i = p'_i(x_1, \ldots, x_n) = x_i^2 - x_n^2 \; , i = 1, \ldots, n-1$$

This mapping sends our hyperplanes onto the arrangement of hyperplanes

$$L = (\bigcup_{1 \leq i < j \leq n-1} U_{i,j}) \bigcup (\bigcup_{1 \leq i \leq n-1} U_{i,n})$$

where $U_{i,j} = \{y \in \mathbf{C}^{n-1}|y_i = y_j\}$, $U_{i,n} = \{y \in \mathbf{C}^{n-1}|y_i = 0\}$.

On their complement M the mapping $p = p'|_M$ is a locally trivial bundle: $p : M \to N$, $N = \mathbf{C}^{n-1} \setminus L$.

Moreover, p has a continuous section:

$$(2) \qquad \eta(y_1, \ldots, y_{n-1}) = (\sqrt{u + y_1}, \ldots, \sqrt{u + y_{n-1}}, \sqrt{u}),$$

where $u = \mathrm{i}(\max_{1 \leq k \leq n-1} |y_k| + 1)$, and the values of all the roots are chosen in the upper half-plane.

All this enables us to write the split exact sequence of fundamental groups

$$(3) \qquad 1 \longrightarrow \pi_1(F_y) \longrightarrow \pi_1(M) \underset{\eta_*}{\overset{p_*}{\rightleftarrows}} \pi_1(N) \longrightarrow 1$$

where $F_y = p^{-1}(y_1, \ldots, y_{n-1})$ is the fiber over any point $y \in N$.

1.2. The arrangement of hyperplanes L in $\mathbf{C}^{n-1}$ is fiber-type. The fiber-type structure is given by the projection along the direction $y_1 = y_2 = \ldots = y_{n-1}$, restricted to N. This projection is locally trivial and has the complement of the arrangement of hyperplanes of the form $\{y_i = y_j\}$ as its base. The latter is the complexification of the reflecting hyperplanes of the Coxeter group A_{n-1} in $\mathbf{C}^{n-1}$. It is well-known that this arrangement is fiber-type.

This proves that the nilpotent residue of $\pi_1(N)$ is trivial (see [FR-1]).

1.3. Now we will investigate the fundamental group of the fiber $H = \pi_1(F_y)$, $y \in N$. Let $\psi_y : F_y \to \mathbf{C}^1$ be defined by the formula

$$\psi_y(x_1, \ldots, x_n) = -x_n^2$$

It is a 2^n-sheeted ramified covering of $\mathbf{C}^1$: $x_k = \pm\sqrt{y_k - s}$, $x_n = \pm\sqrt{-s}$ for $s \in \mathbf{C}^1$. The ramification locus consists of the n points

$$a_1 = y_1, \ldots, a_{n-1} = y_{n-1}, \quad a_n = 0.$$

Denote $B_y = \mathbf{C}^1 \setminus \{a_1, \ldots, a_n\}$.

The sheets stick together pairwise over each a_k, forming 2^{n-1} pairs. If we delete from F_y the ramification points, $\tilde{F}_y = F_y \setminus \psi_y^{-1}(\{a_1, \ldots, a_n\})$, we will obtain an unramified covering $\tilde{\psi}_y : \tilde{F}_y \to B_y$ over the plane with n punctures B_y. The fundamental group $\tilde{H} = \pi_1(\tilde{F}_y)$ of the covering space $\tilde{F}_y$ can be embedded into the free group $\mathcal{F}_n$ which is the fundamental group of B_y and is freely generated by n simple loops around the punctures $a_1, \ldots, a_n$. We will denote these loops and their classes in $\pi_1(B_y)$ by the same symbols $a_1, \ldots, a_n$. The image of $\tilde{H}$ consist of the classes of the loops in B_y which can be lifted to closed loops in $\tilde{F}_y$.

Passing along the simple loop a_k around the puncture a_k leads to the change of the sign of the corresponding coordinate x_k of the reference point of the fiber. Consequently, we will find ourselves after passing along the loop $\alpha = \prod_{k=1}^r a_{i(k)}^{h(k)}$ at the same point of the fiber if and only if $\sum_{k:i(k)=j} h(k)$ is even for every $j = 1, \ldots, n$.

Such elements α form the group $\tilde{H}$. As concerns H, it is obtained from $\tilde{H}$ by gluing up the holes in the covering. Gluing up the holes over the point a_k produces the relations in H of the form $Aa_k^2A^{-1} = 1$, and holes in different sheets correspond to different parities of occurencies of the letters a_i, $i \neq k$, in the word A. So, we have

$$H = \tilde{H}/(\{Aa_k^2A_{-1}, \ k = 1, \ldots, n \ , \ A \in \mathcal{F}_n\}).$$

The group $\tilde{H}$ is free as a subgroup of a free group, and its generators can be easily written out with the help of the standard procedure (see, for example, [MKS]). These are the elements of one of the following 3 forms:

(4)
$$a_{i_1} \ldots \widehat{a_{i_p}} \ldots a_{i_k} a_{i_p} a_{i_k}^{-1} \ldots a_{i_p}^{-1} \ldots a_{i_1}^{-1} ,$$
$$a_{i_1} \ldots a_{i_p} \ldots a_{i_k} a_{i_p} a_{i_k}^{-1} \ldots a_{i_p}^{-1} \ldots a_{i_1}^{-1} ,$$
$$a_{i_1} \ldots a_{i_k}^2 a_{i_{k-1}}^{-1} \ldots a_{i_1}^{-1} ,$$
$$\text{for } i_1 < \ldots < i_p < \ldots < i_k, \ 2 \le k \le n, \ 1 \le p < k.$$

When we pass to H, the first two groups of the generators are identified, the 3rd one vanishes, and H stays free with the $2^{n-1}(n-2) + 1$ generators of the form

(5)
$$a_{i_1} \ldots \widehat{a_{i_p}} \ldots a_{i_k} a_{i_p} a_{i_k} \ldots a_{i_p} \ldots a_{i_1} ,$$
$$\text{for } i_1 < \ldots < i_p < \ldots < i_k, \ 2 \le k \le n, \ 1 \le p < k.$$

(note, that a_i^{-1} can be replaced in the words presenting the elements of H by a_i).

The way the elements of H are expressed in terms of these generators is clear: the form of the generators gives us a possibility to rearrange the symbols a_i forming a word in H to obtain the natural ordering, as in the following example:

$$
\begin{aligned}
a_2a_1a_3a_1a_2a_3 &= (a_2a_1a_2a_1)a_1a_2a_3a_1a_2a_3 \\
&= (a_2a_1a_2a_1)(a_2a_3a_1a_3a_2a_1)^{-1}a_2a_3a_2a_3 \\
&= (a_2a_1a_2a_1)(a_2a_3a_1a_3a_2a_1)^{-1}(a_3a_2a_3a_2)^{-1}
\end{aligned}
$$

We will denote the generator of the form (5) by $(i_1 \ldots i_k i_p)$; in this notation the previous example acquires the form

$$213123 = (21)(231)^{-1}(32)^{-1}$$

1.4. The conclusion that H is free of rank $2^{n-1}(n-2)+1$ agrees to the algebraic-geometric one: the generic fiber F_y is an intersection of $n-1$ quadrics in $\mathbf{P}^n$ in the general position with infinite points deleted; so, F_y is an algebraic curve of genus

$$g = \frac{\deg \omega_F}{2} + 1 = \frac{2^{n-1}(n-3)}{2} + 1 = 2^{n-2}(n-3) + 1$$

with 2^{n-1} punctures.
Therefore, $H = \pi_1(F_y)$ is free of rank

$$2g + 2^{n-1} - 1 = 2^{n-1}(n-2) + 1.$$

1.5. Our next goal is description of the action on H of the monodromies of the elements of $\pi_1(N)$. The construction used to obtain the presentation of $\pi_1(F_y)$ depends continuously on the choice of the point $y \in \pi_1(N)$, that is it can be extended to bundles. We are going to describe a construction of such an extension.

Denote by $\tilde{M}$ the manifold obtained from M by deleting the hyperplanes given by the equation $x_k = 0$, $k = 1, \ldots, n$. (Note, that $\tilde{M}$ is the complement of the complexification of the reflecting hyperplanes of the Coxeter group B_n in $\mathbf{C}^n$:

$$\tilde{M} = \mathbf{C}^n \setminus (\bigcup_{1 \le i < j \le n} \{x_i = \pm x_j\} \bigcup_{1 \le i \le n} \{x_i = 0\})).$$

Let $\tilde{p} = p|_{\tilde{M}}$. Then $\tilde{p}$ is a locally trivial bundle as well; its fiber $\tilde{F}_y$ is obtained from the fiber F_y of p by deleting $n2^{n-1}$ points, in which F_y meets the deleted hyperplanes.

Denote by E the complement in $\mathbf{C}^n$ of the arrangement of hyperplanes

$$(\bigcup_{1\leq i<j\leq n} \{y \in \mathbf{C}^n | y_i = y_j\}) \bigcup (\bigcup_{1\leq i\leq n} \{y \in \mathbf{C}^n | y_i = 0\})$$

and consider the projection $q : E \to N$ defined by

$$q(y_1,\ldots,y_n) = (y_1,\ldots,y_{n-1}) \in \mathbf{C}^{n-1}.$$

It is a locally trivial bundle whose fiber B_y over a point $y = (y_1,\ldots,y_{n-1}) \in N$ is the complex line with n punctures $y_1,\ldots,y_{n-1}, 0$.

Define the mapping $\tilde{\psi} : \tilde{M} \to E$ by the formula

$$\tilde{\psi}(x_1,\ldots,x_n) = (x_1^2 - x_n^2,\ldots,x_{n-1}^2 - x_n^2,\ -x_n^2).$$

It is easy to check that $\tilde{\psi}$ really maps $\tilde{M}$ onto E and is a 2^n-sheeted unramified covering.

We have the following two morphisms of the bundles, that is the commutative diagram of continuous mappings:

(6)
$$\begin{array}{ccc} M \xhookleftarrow{i} \tilde{M} \xrightarrow{\tilde{\psi}} E \\ {}_{p}\searrow \quad \downarrow_{\tilde{p}} \quad \swarrow_{q} \\ N \end{array}$$

and over every $y \in N$ the fibers F_y, $\tilde{F}_y$ and B_y together with the restricted mappings $i_y = i|_{\tilde{F}_y}$, $\tilde{\psi}_y = \tilde{\psi}|_{\tilde{F}_y}$, realize the construction described in 1.3.

Furthermore, the formula (2) defines a section $\eta : N \longrightarrow \tilde{M}$ of $\tilde{p}$, and therefore, we have the consistent sections $i \circ \eta$ and $\tilde{\psi} \circ \eta$ of p and q respectively.

This shows, that the actions of the elements of $\pi_1(N)$ by monodromies on the fundamental groups of the fibers $\pi_1(F_y)$, $\pi_1(\tilde{F}_y)$ and $\pi_1(B_y)$ are consistent, that is for every $z \in \pi_1(N)$ the diagram of homomorphisms

(7)
$$\begin{array}{ccc} \pi_1(F_y) & \xleftarrow{i_{y*}} & \pi_1(\tilde{F}_y) & \xrightarrow{\tilde{\psi}_{y*}} & \pi_1(B_y) \\ \downarrow_{z*} & & \downarrow_{z*} & & \downarrow_{z*} \\ \pi_1(F_y) & \xleftarrow{i_{y*}} & \pi_1(\tilde{F}_y) & \xrightarrow{\tilde{\psi}_{y*}} & \pi_1(B_y) \end{array}$$

is commutative.

This enables us to limit ourselves by investigation of the action of $\pi_1(N)$ on the fundamental group $\pi_1(B_y) \simeq \mathcal{F}_n$ of the fiber B_y of the bundle $q : E \longrightarrow N$.

1.6. Choose the reference point s in B_y at $i(+\infty)$, the loops a_k passing round the corresponding punctures counterclockwise (see Fig 1). Remember, that over the point $y = (y_1,\ldots,y_{n-1}) \in N$ we have $a_k = y_k$, $a_n = 0$.

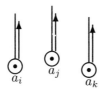

Figure 1:

Choose the reference point of $\pi_1(N)$ at $b = (n-1, \ldots, 1) \in N$ and consider the ray of a real line through it:

$$R : y(t) = b - vt,$$
$$t \in \mathbf{R}_+, \ v = (v_1, \ldots, v_{n-1}) \in \mathbf{R}^{n-1}, \ 0 \ll v_{n-1} \ll \ldots \ll v_1.$$

If we increase t from $t = 0$, the punctures a_k of $B_{y(t)}$ will move along the real line of the fiber as follows: over $t = 0$ we have $a_1 > a_2 > \ldots > a_n = 0$, then a_1 passes through $a_2, \ldots, a_n$ into the negative part of the line; after that a_2 passes through $a_3, \ldots, a_n$, and so on, until the punctures will rearranged in the order $a_1 < \ldots < a_{n-1} < a_n = 0$.

At the moment when the ray R meets the hyperplane $u_{k,l} = \{y_k = y_l\}$, a_k coincides with a_l; at the moment when the ray meets the hyperplane $u_{k,n} = \{y_k = 0\}$, a_k coincides with a_n. Every point of intersection $c_{k,l} = R \bigcap u_{k,l}$ ($1 \le k < l \le n$) provides a generator of $\pi_1(N)$, defined as the product $\alpha_{k,l}\rho_{k,l}\alpha_{k,l}^{-1}$, where $\alpha_{k,l}$ is the path going along R from b to a point $c'_{k,l}$ in a small neighborhood of $c_{k,l}$ and bypassing all other $c_{i,j}$ it meets before $c_{k,l}$ along small arcs in the counterclockwise direction in the complex line $\mathbf{C} \cdot R$ (these are the points $c_{i,j}$ for $i < k$ and $c_{k,j}$ for $k < j < l$), and $\rho_{k,l}$ is the small loop in $\mathbf{C} \cdot R$ starting at $c'_{k,l}$ and passing round $c_{k,l}$ in the clockwise direction (see Fig.2). We will

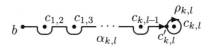

Figure 2: $u_{k,l}$

denote these generators by the same symbols $u_{k,l}$.

The transportation of the loops a_i surrounding the punctures a_i along the loops $u_{k,l}$ will entangle them, and the resulting loops will be the images of a_i under the action of the monodromy of $u_{k,l}$. Instead of this, we will transport the loops a_i along $u_{k,l}^{-1}$ and express the initial loops in terms of their transports.

It is easy to see that the action of $u_{k,l}$ on a_i for $i < k$ or $i > l$ is trivial, as the corresponding loops are not involved into the process of entangling, or if they are, they cancel out. That is, when moving along $\alpha_{k,l}$, all the loops

a_j, $k \le j \le l$, acquire conjugations by a_i for $i < k$, but on the way back they receive conjugations by a_i^{-1}, so that all the conjugations kill each other.

Therefore, we get non-trivial conjugations only in the following way:

1. a_k acts by conjugation on all the a_j, $k < j < l$:
   $a_j \mapsto a_k^{-1} a_j a_k$ (see Fig.3);

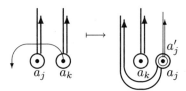

Figure 3: $a_j' = a_k^{-1} a_j a_k$

2. a_k is acted on by the conjugation with respect to a_l:
   $a_k \mapsto a_k^{-1} a_l^{-1} a_k a_l a_k$, $a_l \mapsto a_k^{-1} a_l a_k$ (see Fig.4);

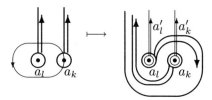

Figure 4: $a_l' = a_k^{-1} a_l a_k$, $a_k' = a_k^{-1} a_l^{-1} a_k a_l a_k$

3. a_k comes back, all the $a_k^{-1} a_j a_k$ receiving the action of $a_k^{-1} a_l^{-1} a_k^{-1} a_l a_k$:
   $a_k^{-1} a_j a_k \mapsto a_k^{-1} a_l^{-1} a_k a_l a_k a_k^{-1} a_j a_k a_k^{-1} a_l^{-1} a_k^{-1} a_l a_k$.

Thus, the action of $u_{k,l}$ is given by the following formulas:

$$(8) \qquad u_{k,l} : a_i \mapsto \begin{cases} a_i \, , \ i < k \\ a_k^{-1} a_l^{-1} a_k a_l a_k \, , \ i = k \\ a_k^{-1} a_l^{-1} a_k a_l a_i a_l^{-1} a_k^{-1} a_l a_k \, , \ k < i < l \\ a_k^{-1} a_l a_k \, , \ i = l \\ a_i \, , \ i > l \end{cases}$$

1.7. Now we can write out the action of $\pi_1(N)$ on H in terms of generators. The fundamental group $\pi_1(N)$ is generated by $u_{k,l}$. The group H is generated by the $2^{n-1}(n-2) + 1$ generators. Each of them we denote by an increasing sequence of integers between 1 and n and one additional integer which is not a member of this sequence (see 1.3).

Denote by capitals increasing sequences of integers, possibly empty. Use small letters for single integers. The following table describes all the possible arrangements of the two integers k, l which are the subscripts of $u_{k,l}$ together with the sequence of integers representing the generators of H, and the results of the action of $u_{k,l}$ and $u_{k,l}^{-1}$ on the corresponding generators.

(9)   • $\underline{\text{Action of } u_{k,l},\ 1 \le k < l \le n}$

1) $A_1 < r < A_2 < k < B < l < C$

$$(A_1A_2BCr) \;\mapsto\; (A_1A_2lk)^{-1}(A_1A_2Blk)(A_1A_2Bk)^{-1}(A_1A_2BCr)$$
$$(A_1rA_2Bk)(A_1rA_2Blk)^{-1}(A_1rA_2lk)$$
$$(A_1A_2kBCr) \;\mapsto\; (A_1A_2Blk)^{-1}(A_1A_2Bk)(A_1A_2kBCr)(A_1rA_2Bk)^{-1}$$
$$(A_1rA_2Blk)$$
$$(A_1A_2BlCr) \;\mapsto\; (A_1A_2lk)^{-1}(A_1A_2BlCr)(A_1rA_2lk)$$
$$(A_1A_2kBlCr) \;\mapsto\; (A_1A_2kBlCr)$$

2) $A < r = k < B < l < C$

$$(ABCk) \;\mapsto\; (Alk)^{-1}(ABlk)(ABk)^{-1}(ABCk)(AkBCl)(ABlCk)^{-1}$$
$$(ABCl)^{-1}(ABCk)(ABk)^{-1}(ABlk)$$
$$(ABlCk) \;\mapsto\; (Alk)^{-1}(ABlCk)(AkBCl)^{-1}(ABCk)^{-1}(ABCl)(ABlCk)$$

3) $A < k < B_1 < r < B_2 < l < C$

$$(AB_1B_2Cr) \;\mapsto\; (Alk)^{-1}(AB_1B_2lk)(AB_1B_2k)^{-1}(AB_1B_2Ck)(AkB_1B_2Cl)$$
$$(AB_1B_2lCk)^{-1}(AB_1B_2Cl)^{-1}(AB_1B_2Cr)(AB_1rB_2Cl)$$
$$(AB_1rB_2lCk)(AkB_1rB_2Cl)^{-1}(AB_1rB_2Ck)^{-1}(AB_1rB_2k)$$
$$(AB_1rB_2lk)^{-1}(Alk)$$
$$(AkB_1B_2Cr) \;\mapsto\; (AB_1B_2lk)^{-1}(AB_1B_2k)(AB_1B_2Ck)^{-1}(AB_1B_2Cl)$$
$$(AB_1B_2lCk)(AkB_1B_2Cl)^{-1}(AkB_1B_2Cr)(AkB_1rB_2Cl)$$
$$(AB_1rB_2lCk)^{-1}(AB_1rB_2Cl)^{-1}(AB_1rB_2Ck)(AB_1rB_2k)^{-1}$$
$$(AB_1rB_2lk)$$
$$(AB_1B_2lCr) \;\mapsto\; (Alk)^{-1}(AB_1B_2lCk)(AkB_1B_2Cl)^{-1}(AB_1B_2Ck)^{-1}$$
$$(AB_1B_2Cl)(AB_1B_2lCr)(AB_1rB_2Cl)^{-1}(AB_1rB_2Ck)$$
$$(AkB_1rB_2Cl)(AB_1rB_2lCk)^{-1}(Alk)$$
$$(AkB_1B_2lCr) \;\mapsto\; (AB_1B_2lCk)^{-1}(AB_1B_2Cl)^{-1}(AB_1B_2Ck)(AkB_1B_2Cl)$$
$$(AkB_1B_2lCr)(AkB_1rB_2Cl)^{-1}(AB_1rB_2Ck)^{-1}(AB_1rB_2Cl)$$
$$(AB_1rB_2lCk)$$

4) $A < k < B < r = l < C$

$$(ABCl) \;\mapsto\; (Alk)^{-1}(ABlk)(ABk)^{-1}(ABCk)(AkBCl)(ABlCk)^{-1}(Alk)$$
$$(AkBCl) \;\mapsto\; (ABlk)^{-1}(ABk)(ABCk)^{-1}(ABCl)(ABlCk)$$

5) $A < k < B < l < C_1 < r < C_2$

$$(ABC_1C_2r) \mapsto (Alk)^{-1}(ABlk)(ABk)^{-1}(ABC_1C_2r)(ABk)(ABlk)^{-1}(Alk)$$
$$(AkBC_1C_2r) \mapsto (ABlk)^{-1}(ABk)(AkBC_1C_2r)(ABk)^{-1}(ABlk)$$
$$(ABlC_1C_2r) \mapsto (Alk)^{-1}(ABlC_1C_2r)(Alk)$$
$$(AkBlC_1C_2r) \mapsto (AkBlC_1C_2r)$$

•Action of $u_{k,l}^{-1}$, $1 \le k < l \le n$

1) $A_1 < r < A_2 < k < B < l < C$

$$(A_1A_2BCr) \mapsto (A_1A_2lk)(A_1A_2Bk)(A_1A_2Blk)^{-1}(A_1A_2BCr)(A_1rA_2Blk)$$
$$(A_1rA_2Bk)^{-1}(A_1rA_2lk)^{-1}$$
$$(A_1A_2kBCr) \mapsto (A_1A_2Bk)^{-1}(A_1A_2Blk)(A_1A_2kBCr)(A_1rA_2Blk)^{-1}$$
$$(A_1rA_2Bk)$$
$$(A_1A_2BlCr) \mapsto (A_1A_2lk)(A_1A_2BlCr)(A_1rA_2lk)^{-1}$$
$$(A_1A_2kBlCr) \mapsto (A_1A_2kBlCr)$$

2) $A < r = k < B < l < C$

$$(ABCk) \mapsto (Alk)(ABk)(ABlk)^{-1}(ABCl)(ABlCk)(AkBCl)^{-1}(ABlk)^{-1}$$
$$(ABk)$$
$$(ABlCk) \mapsto (Alk)(ABCl)^{-1}(ABCk)(AkBCl)$$

3) $A < k < B_1 < r < B_2 < l < C$

$$(AB_1B_2Cr) \mapsto (Alk)(AB_1B_2k)(AB_1B_2lk)^{-1}(AB_1B_2Cl)(AB_1B_2lCk)$$
$$(AkB_1B_2Cl)^{-1}(AB_1B_2Ck)^{-1}(AB_1B_2Cr)(AB_1rB_2Ck)$$
$$(AkB_1rB_2Cl)(AB_1rB_2lCk)^{-1}(AB_1rB_2Cl)^{-1}(AB_1rB_2lk)$$
$$(AB_1rB_2k)^{-1}(Alk)^{-1}$$
$$(AkB_1B_2Cr) \mapsto (AB_1B_2k)^{-1}(AB_1B_2lk)(AkB_1B_2Cl)(AB_1B_2lCk)^{-1}$$
$$(AB_1B_2Cl)^{-1}(AB_1B_2Ck)(AkB_1B_2Cr)(AB_1rB_2Ck)^{-1}$$
$$(AB_1rB_2Cl)(AB_1rB_2lCk)(AkB_1rB_2Cl)^{-1}(AB_1rB_2lk)^{-1}$$
$$(AB_1rB_2k)$$
$$(AB_1B_2lCr) \mapsto (Alk)(AB_1B_2Cl)^{-1}(AB_1B_2Ck)(AkB_1B_2Cl)$$
$$(AB_1B_2lCk)^{-1}(AB_1B_2lCr)(AB_1rB_2lCk)(AkB_1rB_2Cl)^{-1}$$
$$(AB_1rB_2Ck)^{-1}(AB_1rB_2Cl)(Alk)^{-1}$$
$$(AkB_1B_2lCr) \mapsto (AkB_1B_2Cl)^{-1}(AB_1B_2Ck)^{-1}(AB_1B_2Cl)(AB_1B_2lCk)$$
$$(AkB_1B_2lCr)(AB_1rB_2lCk)^{-1}(AB_1rB_2Cl)^{-1}(AB_1rB_2Ck)$$
$$(AkB_1rB_2Cl)$$

4) $A < k < B < r = l < C$

$$(ABCl) \mapsto (Alk)(ABk)(ABlk)^{-1}(ABCl)(ABlCk)(AkBCl)^{-1}(ABCk)^{-1}$$
$$(ABCl)(Alk)^{-1}$$
$$(AkBCl) \mapsto (ABk)^{-1}(ABlk)(AkBCl)(ABlCk)^{-1}(ABCl)^{-1}(ABCk)$$
$$(AkBCl)$$

$$5)\ A < k < B < l < C_1 < r < C_2$$

$$(ABC_1C_2r) \mapsto (Alk)(ABk)(ABlk)^{-1}(ABC_1C_2r)(ABlk)(ABk)^{-1}(Alk)^{-1}$$
$$(AkBC_1C_2r) \mapsto (ABk)^{-1}(ABlk)(AkBC_1C_2r)(ABlk)^{-1}(ABk)$$
$$(ABlC_1C_2r) \mapsto (Alk)(ABlC_1C_2r)(Alk)^{-1}$$
$$(AkBlC_1C_2r) \mapsto (AkBlC_1C_2r)$$

1.8. Our next goal is to prove the triviality of the nilpotent residue of $\pi_1(M) = PD_n$.

Theorem. $\gamma_\infty(PD_n) = 1$.

Proof.
If the action of $\pi_1(N)$ on $H = \pi_1(F_y)$ were reduced to conjugations (that is if it were *trivial* modulo $[H, H]$) then the Falk-Randell lemma applied to the exact sequence (3) would have solved the problem (see [FR-1]).

This is not the case, but the extension of this action to the group generated by a_k, $k = 1, \ldots, n$ is given by conjugations. This is clear a priori, without precise computations, as all the punctures a_k return upon acting by elements of $\pi_1(N)$ to their own places.

We can describe H in another way: H is the subgroup of the group $\mathcal{G}_n = \mathcal{F}_n/(a_k^2, \ k = 1, \ldots, n)$ consisting of all the words with even numbers of oc-curencies of each generator a_k. In fact, H coincides with the commutator group $[\mathcal{G}_n, \mathcal{G}_n]$.

Therefore, we can embed the exact sequence (3) into the commutative diagram

$$(10) \qquad \begin{array}{ccccccc} 1 & \longrightarrow & H & \longrightarrow & \pi_1(M) & \rightleftarrows & \pi_1(N) & \longrightarrow & 1 \\ & & \big\uparrow & & \big\uparrow & & \big\Vert \\ 1 & \longrightarrow & \mathcal{G}_n & \longrightarrow & \overline{M} & \rightleftarrows & \pi_1(N) & \longrightarrow & 1 \end{array}$$

where the second line is exact as well, and $\overline{M} = \mathcal{G}_n \ltimes \pi_1(N)$ is defined by the action of $\pi_1(N)$ on the generators a_k of $\mathcal{G}_n$ described by (8); this action is well-defined on $\mathcal{G}_n$ and is *trivial* modulo $[\mathcal{G}_n, \mathcal{G}_n]$.

The mentioned lemma of Falk and Randell ensures the triviality of nilpotent residue of $\overline{M}$, and consequently, that of $\pi_1(M)$, if that of $\mathcal{G}_n$ is trivial.

1.9. It only remains to prove, that $\gamma_\infty(\mathcal{G}_n) \overset{def}{=} \bigcap_{k=1}^\infty \gamma_k(\mathcal{G}_n) = \{1\}$.

Proposition. Let $\mathcal{G}_n = \{a_1, \ldots, a_n | a_1^2, \ldots, a_n^2\}$. Then $\gamma_\infty(\mathcal{G}_n) = \{1\}$.

Proof.
We follow the proof of the similar fact for free groups given in [LS].

Let A be the ring of the polynomials with coefficients in $\mathbf{Z}_2$ in the non-commutative variables $a_1, \ldots, a_n$ with monomials free from squares of a_k for all k, i.e.

$$A = \mathbf{Z}_2 \mathcal{F}_n / (\{a_k^2, \ k = 1, \ldots, n\}).$$

Define the homomorphism $\varphi : \mathcal{G}_n \longrightarrow A$ on the generators by the formulas $\varphi(a_k) = 1 + a_k$. The only relations in $\mathcal{G}_n$, which are of the form $a_k^2 = 1$, are also satisfied by their images:

$$(1 + a_k)^2 = 1 + 2a_k + a_k^2 = 1.$$

Hence, the homomorphism φ is well defined. It is injective, as for every reduced word $x \in \mathcal{G}_n$ the image $\varphi(x)$ contains the only one monomial identical to x; it is the monomial of the greatest degree.

Denote by $\mathrm{mdeg}(p)$ for $p \in A$ the minimal degree of the monomials of p. Let $A_r \subset A$ be the set of polynomials of the form $1 + p_r$, where $\mathrm{mdeg}(p_r) \geq r$.

Lemma. $\varphi(\gamma_r(\mathcal{G}_n)) \subseteq A_r$.

Proof. First, if $\varphi(x) = 1 + p_r \in A_r$, then $\varphi(x^{-1}) = 1 + p_r + q_{2r}$, where $\mathrm{mdeg}(q_{2r}) \geq 2r$. It follows from the equality $(1 + p_r)(1 + p_r + q_{2r}) = 1$, which gives $q_{2r} = -p_r^2 - p_r q_{2r}$. It is evident that $\varphi(\gamma_1(\mathcal{G}_n) = \mathcal{G}_n) \in A_1$. If we already have $\varphi(x_r) = 1 + p_r \in A_r$ for $x_r \in \gamma_r(\mathcal{G}_n)$ by induction, then

$$\varphi([p_r, a_k]) = (1 + p_r + q_{2r})(1 + a_k)(1 + p_r)(1 + a_k) =$$
$$= 1 + 2p_r + 2a_k + (\text{terms of degree} \geq r + 1) \in A_{r+1}.$$

This proves the lemma.

The fact that $\bigcap_{r=1}^{\infty} A_r = \{1\}$ ends the proof of the proposition.

1.10. Remark. The analogy with $\mathcal{F}_n$ ends with this proposition. Magnus (see, e.g., [MKS]) has proved that in the $\mathcal{F}_n$ case the minimal degree of monomials mdeg of the images of the elements of $\gamma_k(\mathcal{F}_n) \setminus \gamma_{k+1}(\mathcal{F}_n)$ is equal to k. This statement is wrong for $\mathcal{G}_n$; a counterexample is given by the element $x = [[a_1, a_2], a_3]$ of $\gamma_3(\mathcal{G}_2)$: its image in A belongs to A_4, but $x = (a_1 a_2)^4$, while $\gamma_4(\mathcal{G}_2)/\gamma_5(\mathcal{G}_5)$ is generated by $(a_1 a_2)^8$, and hence $x \notin \gamma_4(\mathcal{G}_2)$.

2. A generalization of the Falk-Randell lemma

2.1. We have proved the triviality of the nilpotent residue of the group $\pi_1(M)$ by embedding it into another group. Can this result be obtained directly, like it is done in [FR-1] for the arrangements of fiber type? The group $\pi_1(M)$ is a semidirect product $H \rtimes \pi_1(N)$ and the action of $\pi_1(N)$ on H is known and described by formulas (9).

Consider the simplest nontrivial case n=3. The action of $\pi_1(N)$ on H is described by the formulas

$$
\begin{aligned}
u_{1,2}: \quad (21) &\mapsto (21)\\
(31) &\mapsto (31)(132)(231)^{-1}(32)^{-1}(31)(21)\\
(32) &\mapsto (31)(132)(231)^{-1}(21)\\
(132) &\mapsto (21)^{-1}(31)^{-1}(32)(231)\\
(231) &\mapsto (21)^{-1}(231)(132)^{-1}(31)^{-1}(32)(231)
\end{aligned}
$$

$$
\begin{aligned}
u_{1,3}: \quad (21) &\mapsto (31)^{-1}(231)\\
(31) &\mapsto (31)\\
(32) &\mapsto (32)\\
(132) &\mapsto (31)^{-1}(132)(21)^{-1}(231)\\
(231) &\mapsto (31)^{-1}(231)(21)^{-1}
\end{aligned}
$$

(11)

$$
\begin{aligned}
u_{2,3}: \quad (21) &\mapsto (32)^{-1}(21)(132)\\
(31) &\mapsto (32)^{-1}(31)(132)\\
(32) &\mapsto (32)\\
(132) &\mapsto (132)\\
(231) &\mapsto (231)
\end{aligned}
$$

where $u_{1,2}, u_{1,3}, u_{2,3}$ are the generators of $\pi_1(N)$ and $(21),(31),(32),(132),(231)$ are the generators of H introduced in 1.3.

For the application of the Falk-Randell lemma, it is necessary that the action of $\pi_1(N)$ on the abelian group $\hat{H} = H/[H,H]$ be trivial or equivalently, that the action of $\pi_1(N)$ by Lie brackets be identically zero. It is not the case; the action of $\pi_1(N)$ is given in the basis formed by the chosen generators of H by the following matrices:

(12)

$$
v_{1,2} = \begin{pmatrix} 0 & 1 & 1 & -1 & -1 \\ 0 & 1 & 1 & -1 & -1 \\ 0 & -1 & -1 & 1 & 1 \\ 0 & 1 & 1 & -1 & -1 \\ 0 & -1 & -1 & 1 & 1 \end{pmatrix} \quad
v_{1,3} = \begin{pmatrix} -1 & 0 & 0 & -1 & -1 \\ -1 & 0 & 0 & -1 & -1 \\ 0 & 0 & 0 & 0 & 0 \\ 0 & 0 & 0 & 0 & 0 \\ 1 & 0 & 0 & 1 & 1 \end{pmatrix}
$$

$$
v_{2,3} = \begin{pmatrix} 0 & 0 & 0 & 0 & 0 \\ 0 & 0 & 0 & 0 & 0 \\ -1 & -1 & 0 & 0 & 0 \\ 1 & 1 & 0 & 0 & 0 \\ 0 & 0 & 0 & 0 & 0 \end{pmatrix}, \quad
\begin{aligned}
&\text{where } v_{1,2} = u_{1,2} - \mathbf{1}\,,\\
&v_{1,3} = u_{1,3} - \mathbf{1}\,,\ v_{2,3} = u_{2,3} - \mathbf{1}
\end{aligned}
$$

The images of the corresponding operators lie in the 2-dimensional space spanned by $x = (132) - (32)$, $y = (231) - (31) - (21)$, and

$$
\begin{array}{lll}
v_{1,2}: x \mapsto 2y - 2x & v_{1,3}: x \mapsto y & v_{2,3}: x \mapsto 0\\
\phantom{v_{1,2}:} y \mapsto 2y - 2x & \phantom{v_{1,3}:} y \mapsto 0 & \phantom{v_{2,3}:} y \mapsto -2x
\end{array}
$$

It is easily seen, that after we apply any 3 such operators successively, $\hat{H}$ is taken into $2\hat{H}$.

We are going to prove in this section the generalization of the Falk-Randell lemma which can be used in this situation.

2.2. Theorem. *Let* $1 \longrightarrow A \longrightarrow B \overset{\varphi}{\underset{\longleftarrow}{\longrightarrow}} C \longrightarrow 1$ *be a split exact sequence of group homomorphisms, and let the nilpotent residues of A and C be trivial. Assume furthermore, that the Lie action of the group C on the commutative group $A/[A,A]$ is nilpotent of order β modulo α, where $\alpha, \beta \in \mathbf{N}$, $\alpha \geq 2$, that is*

$$\forall x \in A, \ c_1, \ldots, c_\beta \in C \ \exists y \in A : [\ldots [x, c_1], \ldots, c_\beta] \equiv y^\alpha \mathrm{mod}[A, A].$$

and one of the following two properties holds:
A) $\gamma_n(A)/\gamma_{n+1}(A)$ *is a free abelian group for every n,*
        or
B) $\gamma_n(A)/\gamma_{n+1}(A)$ *is finitely generated for every n and does not have for every n any torsions which are mutually prime to α.*
Then the nilpotent residue of B is trivial.

2.3. The case $\alpha = 0$ is a separate one.
Theorem. *Let* $1 \longrightarrow A \longrightarrow B \overset{\varphi}{\underset{\longleftarrow}{\longrightarrow}} C \longrightarrow 1$ *be a split exact sequence of group homomorphisms, and let the nilpotent residues of A and C be trivial. Assume furthermore, that the Lie action of the group C on the group $A/[A,A]$ is nilpotent of order $\beta \in \mathbf{N}$, that is*

$$\forall x \in A, \ c_1, \ldots, c_\beta \in C \ [\ldots [x, c_1], \ldots, c_\beta] \in [A, A].$$

Then the nilpotent residue of B is trivial.

2.4. We need a great number of technical definitions. Let B be a group, A its subgroup.

The *commutator over B* is a formal expression defined inductively. If $b \in B$, then the word "b" is a commutator over B; its *weight* is $w(b) = 1$, its *value* is $v(b) = b \in B$, its *relative weight over A* is $w_A(b) = \begin{vmatrix} 1, & b \in A \\ 0, & b \notin A \end{vmatrix}$. The element b is a *member* of the commutator b, its *depth* is $h_b(b) = 0$. The commutator b *takes part* in the commutator b. If P, Q are commutators over B, then $[P, Q]$ is a commutator over B, and P and Q *belong to the same bracket*. Furthermore, $w([P, Q]) = w(P) + w(Q)$, $w_A([P, Q]) = w_A(P) + w_A(Q)$, $v([P, Q]) = v(P)^{-1}v(Q)^{-1}v(P)v(Q) \in B$. If b is a member of P or Q of depth h, then b is a member of $[P, Q]$ of depth $h_{[P,Q]}(b) = h + 1$. All the commutators taking part in P or Q and $[P,Q]$ itself take place in $[P,Q]$. Note that if A is normal in B and $w_A(P) \geq 1$, so $v(P) \in A$.

The element $b \in B$ is a *linear commutator over* B; its *center* $c(b) = b$. If P is a linear commutator and $b \in B$, then $[P,b]$ and $[b,P]$ are also linear commutators; their centers $c([P,b]) = c([b,P]) = c(P)$. Note, that the center of a linear commutator is not uniquely defined. We will say that the center of a linear commutator P lies in A and write $c(P) \in A$ if one of its centers is contained in A.

The linear commutator P with center in A is an A-*oriented commutator* of A-*weight* $Aw(P) = 1$. If P and Q are A-oriented, then $[P,Q]$ is A-oriented too, and $Aw([P,Q]) = Aw(P) + Aw(Q)$. The A-weight of a commutator is not uniquely defined and depends on the way it was built.

Every commutator P is a *linear commutator from* P; its *weight over* P is $w^P(P) = 0$, its *relative weight over* P is $w_A^P = 0$. If P is a linear commutator from Q and $b \in B$, so $[P,b]$ and $[b,P]$ are linear commutators from Q too, and $w^Q([P,b]) = w^Q([b,P]) = w^Q(P) + 1$, $w_A^Q([P,b]) = w_A^Q([b,P]) = $

$$\left| \begin{array}{l} w_A^Q(P) + 1, \ b \in A \\ w_A^Q(P), \ b \notin A \end{array} \right. .$$

Define the action of *conjugation* on the commutators: $b^a = a^{-1}ba \in B$ for $a, b \in B$; $[P,Q]^a = [P^a, Q^a]$ for commutators P, Q. It is clear, that the weight, the property to be linear and all the other properties of a commutator P defined above which are independent of A, are invariant under conjugations; if A is a normal subgroup, then all the properties which depend on A are invariant as well. In particular, $v(P^c) = (v(P))^c$.

Define the operation of *change* of a commutator Q_1, taking part in a commutator P_1, by a commutator Q_2. If $P_1 = Q_1$, the result of this change is $P_2 = Q_2$. If $P_1 = [R_1, S]$ and Q_1 takes part in R_1, so $P_2 = [R_2, S]$, where R_2 is the result of the change of Q_1 by Q_2 in R_1. It is clear, that if Q_1 and Q_2 have the same properties, the properties of P_1 and P_2 coincide.

Every commutator is a *commutator expression*; if U and W are commutator expressions, then UW, U^W, U^n, $n \in \mathbf{Z}$ and $[U,W]$ are commutator expressions. Their values are equal, respectively, to $v(U)v(W)$, $v(U)^{v(W)}$, $v(U)^n$ and $v(U)^{-1}v(W)^{-1}v(U)v(W)$. We will often omit the symbol of value "v" in equations including commutator expressions.

We will say, that the commutator expression U *is represented in the form of* the commutator expression W, or U *is equal to* W, and write $U = W$ if $v(U) = v(W)$; we will sometimes denote equal commutator expressions in the same way, if it does not lead to a confusion.

Call formal finite product of commutators $U = P_1^{n_1} \ldots P_m^{n_m}$, $n_i \in \mathbf{N}$, a *commutator product* and define its *weight* $w(U) = \min_i w(P_i)$ and its *relative weight* $w_A(U) = \min_i w_A(P_i)$.

Let $\alpha \in \mathbf{N}$. Call commutator product U of the form $U = P_1^{\alpha^{n_1}} P_2^{\alpha^{n_2}} \ldots P_m^{\alpha^{n_m}}$, $n_1, \ldots, n_m \in \mathbf{Z}_+$, an α-*product* ; its α-*rank* is $r^\alpha(U) = \min_i(n_i + w(P_i))$ and its *relative* α-*rank* is $r_A^\alpha(U) = \min_i(n_i + w_A(P_i))$.

We will also use the following known commutator identities:

(13) $$[PQ, R] = [P, R]^Q[Q, R] \quad [P, QR] = [P, R][P, Q]^R$$

(14) $$[PQ, R] = [P, R][[P, R], Q][Q, R] \quad [P, QR] = [P, R][R, [Q, P]][P, Q]$$

(15) $[[P, Q], R] = [\tilde{P}, [Q, \tilde{R}]][\tilde{Q}, [\tilde{R}, P]]$, where $\tilde{R} = R^{P^{-1}}, \tilde{P} = P^Q, \tilde{Q} = Q^{\tilde{R}}$

$[P, [Q, R]] = [[\tilde{P}, Q], \tilde{R}][[R, \tilde{P}], \tilde{Q}]$, where $\tilde{P} = P^{Q^{-1}}, \tilde{R} = R^{\tilde{P}}, \tilde{Q} = Q^R$

(16) $$PQ = QP[P, Q]$$

(17) $$[P_1, P_2][Q_1, Q_2] = [[P_2, P_1], [Q_2, Q_1]][Q_1, Q_2][P_1, P_2]$$

We will often consider only one of the possible imbeddings of commutator without notice, as the equations (13)-(17) are symmetric with respect to occurencies of their members. For example, we will often say that a commutator P *has the form* $[R, Q]$ if $P = [\tilde{R}, \tilde{Q}]$ or $P = [\tilde{Q}, \tilde{R}]$, where R and Q have the forms $\tilde{R}$ and $\tilde{Q}$ respectively.

2.5. Lemma. *Let a commutator Q taking part in a commutator P be represented in the form of a commutator product $Q = Q_1 Q_2 \ldots Q_n$. Then P is represented in the form of the commutator product $P = P_1 P_2 \ldots P_n$, where every P_i is obtained from P by the change of Q by $Q_{\sigma(i)}$, where σ is some permutation, and, furthermore, by some conjugations of its members.*

<u>Proof.</u> We have to prove, that there exist the $P_i, i = 1, \ldots, n$, for which $v(P) = v(P_1) \ldots v(P_n)$. If we change the commutator Q in P by its value $v(Q)$, find the required P_i obtained from P by the change of $v(Q)$ by $v(Q_i)$ and some conjugations, and apply the change of $v(Q_i)$ by Q_i, we will be done. Therefore, we can assume $Q = b$ and $Q_i = b_i$ are some elements of B and, moreover, $b = b_1 b_2$ by induction. In this case the statement is proved by induction on the depth $h_P(b)$. The base of the induction is trivial, the step is given by the formulas (13):

$$[b_1 b_2, R] = [b_1, R]^{b_2}[b_2, R], \quad [R, b_1 b_2] = [R, b_2][R, b_1]^{b_2}$$

2.6. Lemma. *Let A be a normal subgroup of B, P a commutator over B, $w(P) = w$, $w_A(P) \geq 1$. Then P is represented in the form of the commutator product $Q_1 Q_2 \ldots Q_m$, where the commutators Q_i are linear and $c(Q_i) \in A$, $w(Q_i) = w$ for every i.*

<u>Proof.</u> We will argue by induction on the weight of the linear commutator L with the center $c(L) \in A$ taking part in P. Such a commutator of weight 1 exists by the assumption; if $w(L) = w$ then $P = L$ and the lemma is proved. The second induction is done on the weight of the commutator R which belongs to the same bracket as L; if $w(R)=1$, R is a single element $b \in B$, and we can add it to L to increase the weight of the latter: $L' = [L, b]$ (or $[b, L]$), $w(L') = w(L) + 1$ and

this gives us the step of the first induction. Let $R = [R_1, R_2]$. Then $[L, R] = [[\tilde{L}, R_1], \tilde{R}_2][[R_2, \tilde{L}], \tilde{R}_1]$ by the formula (15); the tilde denotes a conjugation. We have $w([[\tilde{L}, R_1], \tilde{R}_2]) = w([[R_2, \tilde{L}], \tilde{R}_1) = w(L) + w(R_1) + w(R_2) = w([L, R])$; $\tilde{L}$ is linear and $c(\tilde{L}) \in A$, as A is normal. At the same time, the weights of the commutators which belong to the same brackets as $\tilde{L}$ decrease: $w(R_1) < w(R)$, $w(R_2) < w(R)$. Using Lemma 2.5, represent P in the form of the product $P = P_1 P_2$ of the commutators P_1 and P_2 obtained by the change of $[L, R]$ in P by $[[\tilde{L}, R_1], \tilde{R}_2]$ and $[[R_2, \tilde{L}], \tilde{R}_1]$ respectively and, in addition, by conjugations of its members; they contain, consequently, the linear commutators $\tilde{L}_1$ and $\tilde{L}_2$, for which $c(\tilde{L}_1) \in A$, $c(\tilde{L}_2) \in A$, $w(\tilde{L}_1) = w(\tilde{L}_2) = w(\tilde{L})$, and the commutators which belong to the same brackets have smaller weights than R. By the inductive assumption, P_1 and P_2 can be represented in the required form, hence P does.

2.7. Lemma. *Let $B = A \ltimes C$, $U = P_1 \ldots P_n$ be a commutator product over B, $w(U) \geq w$. Then U can be represented in the form of the commutator product $U = R_1 \ldots R_l S_1 \ldots S_m$, where R_i are commutators over B, $w(R_i) \geq w$, $w_A(R_i) \geq 1$; S_i are commutators over C, $w(S_i) \geq w$.*

<u>Proof.</u> Prove the lemma for the case when $U = P$ is a single commutator by induction on $w(P)$. If $w(P) = 1$, that is $P = b \in B$, then $b = ac$, $a \in A$, $c \in C$.

Let $P = [Q, T]$; then, by the inductive assumption, $Q = R_1' \ldots R_{l'}' S_1' \ldots S_{m'}'$, $T = R_1'' \ldots R_{l''}'' S_1'' \ldots S_{m''}''$, where S_i', S_i'' are commutators over C of weight $\geq w(P)$ and $\geq w(T)$ respectively, R_i', R_j'' are commutators over B of weight $\geq w(Q)$ and $\geq w(T)$ respectively, containing a member from A. Multiple application of the identity (14) allows us to represent P in the form of commutator product composed from commutators over C of the form $[S_i', S_j'']$ of weight $\geq w(Q) + w(T) \geq w$ and from commutators containing a member from A, and of weight $\geq w$.

Let $S = [S_1, S_2]$ be a commutator of the first type, $R = [R_1, R_2]$ of the second. Then the identity (17), applied to SR, gives $SR = [[S_2, S_1], [R_2, R_1]]RS$. The first commutator has weight $\geq w$ and contains a member from A, that is, it belongs to the second type. Multiple application of this identity to the pairs of the form SR occurring in the expression of P enables us to transport all the commutators of the first type to the right; and this proves the statement.

If now $U = P_1 \ldots P_n$ is a commutator product, decompose every P_i and apply the same algorithm to gather all the commutators over C in this decomposition of U on the right; all the commutators remaining on the left will contain a member from A, that, is their relative weight $w_A \geq 1$.

2.8. Lemma. *Let A be a normal subgroup of B, $\alpha \in \mathbf{N}$, $a \in A$, $b \in B$, U be an α-product of linear commutators with centers in A, $w(U) \geq w$, $r^\alpha(U) \geq r$, $w_A(U) \geq w_A$, $r_A^\alpha(U) \geq r_A$. Then the commutator expressions $[U, b]$, $[U, a]$ and U^α can be represented in the form of α-products U_b, U_a and U_α respectively consisting of linear commutators with their centers in A, such that*

a) $w(U_b) \geq w + 1$, $r^\alpha(U_b) \geq r + 1$, $w_A(U_b) \geq w_A$, $r_A^\alpha(U_b) \geq r_A$

b) $w(U_a) \geq w + 1$, $r^\alpha(U_a) \geq r + 1$, $w_A(U_a) \geq w_A + 1$, $r_A^\alpha(U_a) \geq r_A + 1$

c) $w(U_\alpha) \geq w$, $r^\alpha(U_\alpha) \geq r + 1$, $w_A(U_\alpha) \geq w_A$, $r_A^\alpha(U_\alpha) \geq r_A + 1$

Proof. Prove b) and c) simultaneously using the double induction on increasing of r and decreasing of w_A.

If $w_A = r$, decompose all the powers in the expression of U, that is represent U in the form

$$U = P_1 P_2 \ldots P_k, \ w(P_i) \geq w_A(P_i) \geq r \geq r_A \ \text{for every } i.$$

Applying Lemma 2.5, we obtain b). In order to obtain c), permute the commutators P_i in the commutator product $U^\alpha = P_1 \ldots P_k P_1 \ldots P_k \ldots P_1 \ldots P_k$ so that for every i all the commutators P_i get together; a possibility of such permutation is given by a modification of the formula (16): $Q P_i = P_i Q [v(Q), P_i]$. The new commutator which arises in the result of this transposition is linear and, if $v(Q) \in A$, so $w_A(Q') = w_A(P_i) + 1$, $v(Q') \in A$.

After this transposition of the commutators of U^α, we will obtain: $U^\alpha = U_\alpha = P_1^\alpha \ldots P_k^\alpha Q_1 \ldots Q_m$, where all Q_j have the form $Q_j = [P_{i(j)}, a_j]$, $a_j \in A$. But $r^\alpha(Q_j) = w(Q_j) \geq w + 1 = r + 1$ and $r_A^\alpha(Q_j) \geq r_A + 1$ for every j; $r^\alpha(P_i^\alpha) = w(P_i) + 1 \geq r + 1$ and $r_A^\alpha(P_i^\alpha) \geq r_A + 1$ for every i, hence $w(U_\alpha) \geq w$, $r^\alpha(U_\alpha) \geq r + 1$, $w_A(U_\alpha) \geq w_A$, $r_A^\alpha(U_\alpha) \geq r_A + 1$.

Let now $U = P_1^{\alpha^{n_1}} \ldots P_k^{\alpha^{n_k}}$, all P_i be linear with their centers in A and $w(A_i) \geq w$, $r^\alpha(A_i) \geq r$, $w_A(A_i) \geq w_A$, $r_A^\alpha(A_i) \geq r_A$ for every i. Applying Lemma 2.5, represent $[U, a]$ in the form $[U, a] = [\tilde{P}_1^{\alpha^{n_1}}, a_1] \ldots [\tilde{P}_k^{\alpha^{n_k}}, a_k]$. Consequently, to prove b) it suffices to consider the case $k = 1$, $U = P^{\alpha^n}$.

If $n = 0$ the statement is trivial; otherwise, using formula (14), decompose

$$[P^{\alpha^n}, a] = [P^{\alpha^{n-1}}, a][[P^{\alpha^{n-1}}, a], v(P^{\alpha^{n-1}(\alpha-1)})][P^{\alpha^{n-1}(\alpha-1)}, a],$$

then decompose the last bracket on the right hand side, and continue this process. As the result we will obtain

$$[P^{\alpha^n}, a] = [P^{\alpha^{n-1}}, a][[P^{\alpha^{n-1}}, a], a_1][P^{\alpha^{n-1}}, a] \ldots [P^{\alpha^{n-1}}, a].$$

Permute the commutators on the right hand side as it was done in the first part of the proof; we will have $U_a = [P^{\alpha^{n-1}}, a]^\alpha Q_1 \ldots Q_m$, where all Q_i have the form $Q_i = [[P^{\alpha^{n-1}}, a], a_i]$ for some $a_i \in A$.

The expression $P' = P^{\alpha^{n-1}}$ is an α-product of linear commutators with centers in A, and $w(P') \geq w$, $r^\alpha(P') \geq r - 1$, $w_A(P') \geq w_A$, $r_A^\alpha(P') \geq r_A - 1$. Because of the second inequality, the part b) of the inductive assumption permits to represent $P'' = [P', a]$ in the form of a similar product with the parameters $w(P'') \geq w + 1$, $r^\alpha(P'') \geq r$, $w_A(P'') \geq w_A + 1$, $r_A^\alpha(P'') \geq r_A$. Because of the third inequality, we can apply the part b) once more and obtain the demanded presentation for $Q_i = [[P', a], a_i]$ with the parameters $w(Q_i) \geq w + 2$, $r^\alpha(Q_i) \geq r + 1$, $w_A(Q_i) \geq w_A + 2$, $r_A^\alpha(Q_i) \geq r_A + 1$.

In just the same way, the parts b) and c) of the inductive assumption, applied consequently to the commutator expression $[P^{\alpha^{n-1}}, a]^\alpha$, permit to factorize it into the demanded product with the parameters $w+1$, $r+1$, w_A+1, r_A+1. The step of the induction in the proof of b) has been performed.

To prove c), rearrange $P_i^{\alpha^{n_i}}$ in the product $U_\alpha = P_1^{\alpha^{n_1}} \dots P_k^{\alpha^{n_k}} P_1^{\alpha^{n_1}} \dots P_k^{\alpha^{n_k}}$ $\dots P_1^{\alpha^{n_1}} \dots P_k^{\alpha^{n_k}}$ so that they get together for every i. We have: $U^\alpha = P_1^{\alpha^{n_1+1}}$ $\dots P_k^{\alpha^{n_k+1}} Q_1 \dots Q_m$, where every Q_j has the form $Q_j = [P_i^{\alpha^{n_i}}, a_j]$, $a_j \in A$. This bracket can be represented in the form of an α-product of linear commutators with centers in A with the parameters $w+1$, $r+1$, w_A+1, r_A+1 by part b) of the lemma, which has been already proved; hence, the part c) has been proved too.

We will prove the part a) of the lemma with by induction on r. By virtue of Lemma 2.5, we can assume $U = P^{\alpha^n}$, where P is a linear commutator with center in A. In this case the base of the induction $r = 1$ and the case $n = 0$ are trivial.

Factorize $[P^{\alpha^n}, b] = [P^{\alpha^{n-1}}, b]^\alpha Q_1 \dots Q_m$ as it was done above, where all Q_i have the form $Q_i = [[P^{\alpha^{n-1}}, b], a_i]$, $a_i \in A$. Apply to the bracket $[P^{\alpha^{n-1}}, b]^\alpha$ the inductive assumption a) and the part c) of the lemma, to the bracket Q_i the inductive assumption a) and the part b) of the lemma; then we will obtain the wanted presentation with the parameters $w(U_b) \geq w+1$, $r^\alpha(U_b) \geq r+1$, $w_A(U_b) \geq w_A$, $r_A^\alpha(U_b) \geq r_A$.

2.9. Lemma. *Let A be a normal subgroup of B, $U = P_1^{\alpha^{n_1}} \dots P_k^{\alpha^{n_k}}$ be an α-product of linear commutators with centers in A . Then*
a) U^{-1} can be represented in the form of a similar product with the same parameters;
b) if σ is an arbitrary permutation of the elements of the set $\{1, \dots, k\}$, U can be represented in the form $U = P_{\sigma(1)}^{\alpha^{n_{\sigma(1)}}} \dots P_{\sigma(k)}^{\alpha^{n_{\sigma(k)}}} U'$, where U' is an α-product of the same kind and $w(U') \geq w(U)+1$, $r^\alpha(U') \geq r^\alpha(U)+1$, $w_A(U') \geq w_A(U)+1$, $r_A^\alpha(U') \geq r_A^\alpha(U)+1$.

Proof.
a) $U^{-1} = (P_k^{-1})^{\alpha^{n_k}} \dots (P_1^{-1})^{\alpha^{n_1}}$. If $w(P_i) = 1$, that is $P_i = a_i \in A$, then $P_i^{-1} = a_i^{-1}$. If $w(P_i) > 1$, that is $P_i = [Q_i, R_i]$ for some Q_i, R_i, then $P_i^{-1} = [R_i, Q_i]$, and we have a representation of U^{-1} in the form of α-product of linear commutators with centers in A and the same parameters.
b) We will transport the commutators $P_i^{\alpha^{n_i}}$ to the left hand side of U to obtain the needed order; that is, at first we transport $P_{\sigma(1)}^{\alpha^{n_{\sigma(1)}}}$, then $P_{\sigma(2)}^{\alpha^{n_{\sigma(2)}}}$ so that it find itself on the right of $P_{\sigma(1)}^{\alpha^{n_{\sigma(1)}}}$, and so on. The transpositions are described by the variant of the formula (16): $QP_i^{\alpha^{n_i}} = P_i^{\alpha^{n_i}} Q[v(Q), P_i^{\alpha^{n_i}}]$.

Thus, we obtain the wanted order, and the new factors which arose in U have the form $[a_i, P_i^{\alpha^{n_i}}]$, $a_i \in A$, hence they can be factorized into an α-product of the desired form by Lemma 2.8 b).

2.10. Lemma. *Let A be a normal subgroup of B, P be a linear commutator from an A-oriented commutator Q. Then P can be represented in the form of a product of A-oriented commutators of the same weight as P and of A-weight $\geq w_A^Q(P) + Aw(Q)$.*

Proof. The proof is produced by induction on the weight $w^Q(P)$ of P over Q. Let $P = [\dots [Q, b_1], \dots b_k], b]$. By the inductive assumption $[\dots [Q, b_1] \dots b_k] = Q_1 \dots Q_n$, where all Q_i are A-oriented, and $Aw(Q_i) \geq Aw(Q) + w_A^Q(P) - w_A(b)$, $w(Q_i) = w(P) - 1$ for every i.

Using Lemma 2.5, factorize $P = [Q_1 \dots Q_n, b] = [\tilde{Q}_1, c_1] \dots [\tilde{Q}_n, c_n]$, where $\tilde{Q}_i$ is conjugate to some Q_j, all c_i are conjugate to b. All $\tilde{Q}_i$ are A-oriented commutators of the same weight and A-weight as Q_i, all c_i lie or do not lie in A simultaneously. It is sufficient, consequently, to consider the case $P = [Q, b]$. If $b \in A$, P is A-oriented and $Aw(P) = Aw(Q) + 1$. If $b \notin A$, then $w_A(b) = w_A^Q(P) = 0$. Use the induction on $Aw(Q)$. If $Aw(Q) = 1$, that is $Q = [\dots [a, b_1], \dots, b_k]$, $a \in A$, then $P = [Q, b] = [[\dots [a, b_1], \dots, bk], b]$ is A-oriented and $Aw(P) = Aw(Q) = 1$. Else, $Q = [Q_1, Q_2]$, Q_1 and Q_2 are A-oriented and $[Q, b] = [\tilde{Q}_1, [Q_2, \tilde{b}]][\tilde{Q}_2, [\tilde{b}, Q_1]]$ by the formula (15). As $Aw(Q_1) < Aw(Q)$ and $Aw(Q_2) < Aw(Q)$, the commutators $[Q_2, \tilde{b}]$ and $[\tilde{b}, Q_1]$ can be factorized into products of A-oriented commutators of the same weights, and, applying Lemma 2.5, we obtain the statement.

2.11. Proposition. *Let $B = A \ltimes C$ and the Lie action of C on $A/[A, A]$ be nilpotent of order β modulo α, where $\alpha, \beta \in \mathbf{N}$. Let $k \in \mathbf{N}$, $m = \frac{k(k-1)}{2}(\beta - 1) + k$, $x \in \gamma_m(B) \bigcap A$. Then x can be represented in the form of an α-product of linear commutators over A of α-rank $r^\alpha = r_A^\alpha \geq k$.*

Proof. By Lemma 2.7, x can be represented in the form of the commutator product $x = R_1 \dots R_l S$, where $w(R_i) \geq m$, $w_A(R_i) \in 1$ for every i, S is a commutator product over C. As $v(R_i) \in A$ and $x \in A$, so $v(S) \in A$ and, consequently, $v(S) = 1$.

We can assume, therefore, that x is a commutator, $w_A(x) \geq 1$. Applying Lemma 2.6, we can even assume that x is a linear commutator and $c(x) \in A$. Thus, let $x = [\dots [a, b_{m-1}], \dots, b_1]$.

We will argue by induction on k; the statement is trivial for $k = 1$. Factorize the inner bracket $x' = [\dots [a, b_{m-1}], \dots], b_{(k-1)(\beta-1)+2}]$ into an α-product of linear commutators over A using the inductive assumption: we have $r^\alpha(x') = r_A^\alpha(x') \geq k - 1$, as $m - (k-1)(\beta - 1) - 1 = \frac{(k-1)(k-2)}{2}(\beta - 1) + k - 1$.

Applying Lemma 2.8 a) a few times , represent the commutator expression $x = [\dots [x', b_{(k-1)(\beta-1)+1}], \dots, b_1]$ in the form of an α-product of linear commutators with centers in A : $x = P_1^{\alpha^{n_1}} \dots P_l^{\alpha^{n_l}}$, $r^\alpha(x) \geq (k-1)(\beta - 1) + 1 + k - 1$, $r_A^\alpha(x) \geq k - 1$. We can assume now that $x = P^{\alpha^n}$, P is linear, $c(P) \in A$, $w(P) + n \geq (k-1)\beta + 1$, $w_A(P) + n \geq k - 1$.

As P is linear over its center, which belongs to A, we can apply Lemma 2.10 and decompose P into a product of A-oriented commutators $P = Q_1 \dots Q_l$,

$w(Q_i) = w(P) \geq (k-1)\beta - n + 1$, $Aw(Q_i) \geq w_A(P) \geq k - n - 1$ for every i. Consider one of Q_i, say Q.

Q is A-oriented; if $Aw(Q) \geq k - n$, replace all the linear commutators with the centers in A by their values; we will obtain a commutator Q'' over A of the weight $w(Q'') = Aw(Q) \geq k - n$. If $Aw(Q) = k - n - 1$, then $w(Q)/Aw(Q) \geq \frac{(k-1)\beta - n + 1}{(k-1)-n} > \beta$, and, consequently, one of the linear commutators taking part in Q has weight $\geq \beta + 1$. Replace the other $k - n - 2$ commutators taking part in Q by their values, lying in A; the remaining commutator has the form $S = [\ldots [a, d_1], \ldots, d_\beta]$. Decompose every $d_i = a_i c_i$, $a_i \in A$, $c_i \in C$. Then $[a, d_1] = [a, c_1][a, a_1]^{c_1} \equiv [a, c_1] \bmod \gamma_2(A)$. Continuing this process, we obtain, that $v(S) \equiv [\ldots [a, c_1], \ldots, c_\beta] \bmod \gamma_2(A)$; by the assumption, there exists $y \in A$ for which $v(S) \equiv y^\alpha \bmod \gamma_2(A)$. It shows, that $Q \equiv (Q')^\alpha \bmod \gamma_{k-n}(A)$, that is $Q = (Q')^\alpha Q''$, where Q', Q'' are commutator products over A, $w(Q') = k-n-1$, $w(Q'') \geq k - n$.

By Lemma 2.6, we can assume, that Q' and Q'' are products of some linear commutators over A. By Lemma 2.8 b), $r^\alpha(Q) = r_A^\alpha(Q) \geq k - n$.

After we transform all Q_i in the expression of P in the same way and apply n time Lemma 2.8 c) to $x = P^{\alpha^n}$, we will obtain $r^\alpha(x) = r_A^\alpha(x) \geq k$.

2.12. From this proposition we immediately obtain
Proof of Generalized Falk-Randell Lemma 1.

Denote $k(m, \beta) = [\frac{\beta + 1 + \sqrt{(\beta+1)^2 + 8(\beta-1)m}}{2(\beta-1)}]$. Then, by Proposition 2.11, we can represent every $x \in \gamma_m(B) \cap A$ in the form of an α-product of linear commutators over A of α-rank $\geq k$: $x = P_1^{\alpha^{n_1}} \ldots P_l^{\alpha^{n_l}}$. Let $q = w(x)$; then $n_i \geq k - q$ for every i.

Define $g = P_1^{\alpha^{n_1 - (k-q)}} \ldots P_l^{\alpha^{n_l - (k-q)}}$; then $g \in \gamma_q(A)$ as $w(P_i) \geq q$ for every i, and $x \equiv g^{\alpha^{k-q}} \bmod \gamma_{q+1}(A)$.

2.13. Proof of Theorem 2.2.

Let $x \in \bigcap_1^\infty \gamma_m(B)$. Then $\varphi(x) \in \bigcap_1^\infty \gamma_m(C) = \{1\}$, and $x \in A$. According to Proposition 2.11, x can be represented in the form of an α-product of linear commutators over A of an arbitrarily large α-rank . We will assume later all commutators to be defined over A.

Part A.

Let $x \in \gamma_N(A)$ and x be represented in the form of an α-product of linear commutators of α-rank $r \geq N$. We want to show by induction on r, that x can be represented in the form of an α-product of linear commutators of α-rank $\geq r$ and of weight $\geq N$; if $r = N$ the statement is trivial. When r is fixed, we use induction on decreasing of the weight of the α-product $x = P_1^{\alpha^{n_1}} \ldots P_l^{\alpha^{n_l}}$ from N to 1.

Let $w = w(x) < N$. Transfer all the factors $P_i^{\alpha^{n_i}}$ of the weight w to the left side of the expression using Lemma 2.9 b). Then x acquires the form $x = x'x''$, where x' and x'' are α-product of linear commutators , $r^\alpha(x') \geq r$, $r^\alpha(x'') \geq r$,

$w(x'') \geq w + 1$. Let $x' = P_1^{\alpha^{n_1}} \dots P_k^{\alpha^{n_k}}$. As $w(P_i) = w < N$ for every $i \leq k$, all $n_i \geq 1$. Define the α-product $U = P_1^{\alpha^{n_1-1}} \dots P_k^{\alpha^{n_k-1}}$. Then, by Lemma 2.9 b), $U^\alpha = (P_1^{\alpha^{n_1-1}} \dots P_k^{\alpha^{n_k-1}}) \dots (P_1^{\alpha^{n_1-1}} \dots P_k^{\alpha^{n_k-1}}) = x'U'$, where U' is an α-product of linear commutators , $w(U') \geq w + 1$, $r^\alpha(U') \geq r(U) + 1 \geq r$.

By Lemma 2.9 a), $x' = U^\alpha U''$, where U'' is an α-product of linear commutators with the same parameters as U' and $x = U^\alpha U'' x''$. The weights of U'' and x'' are $\geq w + 1$ and their α-ranks $\geq r$; the only problem is U^α. But $v(U) \in \gamma_N(A) \subseteq \gamma_{w+1}(A)$, $v(U''x'') \in \gamma_{w+1}(A)$. Therefore, $v(U^\alpha) = v(U)^\alpha \in \gamma_{w+1}(A)$. We have $v(U) \in \gamma_w(A)$, $\gamma_w(A)/\gamma_{w+1}(A)$ is free abelian, so $v(U) \in \gamma_{w+1}(A)$. Therefore, as $r^\alpha(U) \geq r-1$, the assumption of the induction on r can be applied to U. Consequently, U can be represented in the form of an α-product of linear commutators of weight $\geq w + 1$ and of α-rank $\geq r - 1$. Lemma 2.8 c) permits to represent U^α in the form of an α-product of linear commutators of weight $\geq w + 1$ and of α-rank $\geq r$, and the inductive step is accomplished.

Let $x \in \gamma_N(A)$. Then x can be represented in the form of an α-product of linear commutators of weight N and of an arbitrarily large α-rank , that is

$$x \bmod \gamma_{N+1}(A) \in \bigcap_{k=1}^{\infty} \alpha^k(\gamma_N(A)/\gamma_{N+1}(A)) = \{0 \bmod \gamma_{N+1}(A)\},$$

as $\gamma_N(A)/\gamma_{N+1}(A)$ is free. Therefore, $x \in \gamma_{N+1}(A)$ too. So, we have $x \in \bigcap_{N=1}^{\infty} \gamma_N(A) = \{1\}$.

Part B.

Define e_k, $k \in \mathbf{N}$, as the maximal integer such that there is a torsion of order α^{e_k} in $\gamma_k(A)/\gamma_{k+1}(A)$, that is

$$e_k = \max\{n | \exists y \in \gamma_k(A) : y^{\alpha^n} \in \gamma_{k+1}(A), y^{\alpha^{n-1}} \notin \gamma_{k+1}(A)\}.$$

All the e_k are finite, as $\gamma_k(A)/\gamma_{k+1}(A)$ are finitely generated abelian groups.

Let $x \in \gamma_N(A)$ and x be represented in the form of an α-product of weight $N - 1$ and of α-rank $r > e_{N-1}$. We are going to prove by induction on r that x can be represented in the form of an α-product of weight N and of α-rank $r - e_{N-1}$. The statement is trivial for $r = N + e_{N-1}$. Let $x = P_1^{\alpha^{n_1}} \dots P_l^{\alpha^{n_l}}$; using Lemma 2.9, reorder the factors of x so that all the factors of weight $N - 1$ will find themselves in the left, that is represent x in the form $x = x'x''$, $r^\alpha(x') \geq r$, $r^\alpha(x'') \geq r$, $w(x'') \geq N$ and, hence, $v(x') \in \gamma_n(A)$. Let $x' = P_1^{\alpha^{n_1}} \dots P_k^{\alpha^{n_k}}$; then for every $i \leq k$ we have $n_i \geq r - w(P_i) > N + e_{N-1} - (N - 1) = e_{N-1} + 1$.

Define $W = P_1^{\alpha^{n_1-e_{N-1}-1}} \dots P_k^{\alpha^{n_k-e_{N-1}-1}}$. Then $v(W)^{\alpha^{e_{N-1}+1}} \equiv v(X') \equiv 0 \bmod \gamma_N(A)$, and, by the definition of e_k, $v(W)^{\alpha^{e_{N-1}}} \in \gamma_N(A)$. Let $U = P_1^{\alpha^{n_1-1}} \dots P_k^{\alpha^{n_k-1}}$. Then $r^\alpha(U) \geq r - 1$, $v(U) \equiv v(W)^{\alpha^{e_{N-1}}} \bmod \gamma_N(A)$, so $v(U) \in \gamma_{N+1}(A)$, therefore, we can apply the inductive assumption and represent U in the form of an α-product of linear commutators of weight $w(U) \geq N$ and of α-rank $r^\alpha(U) \geq r - 1 - e_{N-1}$. But, as it was shown in the proof of part A of the theorem, $x = U^\alpha U'' x''$, where $w(U'') \geq N$, $r^\alpha(U'') \geq r$, $w(x'') \geq N$,

$r^\alpha(x'') \geq r$, and the application of Lemma 2.8 c) to U^α gives us the step of the induction.

Let $x \in \gamma_N(A)$. Then, as x can be represented in the form of an α-product of linear commutators of α-rank $r > \sum_{k=1}^{N-1} e_k$, it can be represented also in the form of an α-product of linear commutators of weight N and of α-rank $r - \sum_{k=1}^{N-1} e_k$, that is, similar to the above,

$$x \bmod \gamma_{N+1}(A) \in \bigcap_{k=1}^{\infty} \alpha^k(\gamma_N(A)/\gamma_{N+1}(A)).$$

But, by the assumption of the theorem, $\gamma_N(A)/\gamma_{N+1}(A)$ does not have any torsions which are mutually prime to α, therefore, this intersection is $0 \bmod \gamma_{N+1}(A)$, that is $x \in \gamma_{N+1}(A)$. Hence, $x \in \bigcap_{N=1}^{\infty} \gamma_N(A) = \{1\}$.

Proof of Theorem 2.3 and Generalized Falk-Randell Lemma 2.

The statement of Theorem 2.3 corresponds to the case $\alpha = 0$; in this case the α-rank and the relative α-rank coincide with the weight and the relative weight of commutator product respectively. The proof of Proposition 2.11 is translated for this case almost literally, and we obtain the statement of the Lemma 2 for $k(m, \beta) = \left[\frac{\beta+1+\sqrt{(\beta+1)^2+8(\beta-1)m}}{2(\beta-1)}\right]$. Hence, if $x \in (\bigcap_1^\infty \gamma_n(B)) \bigcap A$, then x can be represented in the form of a product of commutators over A of arbitrarily large weight. It shows, that $\bigcap_1^\infty \gamma_n(B) = (\bigcap_1^\infty \gamma_n(A))(\bigcap_1^\infty \gamma_n(C)) = 1$.

2.14. The following lemma is useful for applications:

Lemma. *Let C be a subgroup of group B, A be a normal subgroup, M be a subgroup of $\hat{A} = A/[A, A]$. Let C° be a set of generators of C as a semigroup and the action of this generators on $\hat{A}$ be nilpotent of order β modulo M, that is*

$$\forall a \in \hat{A}, \ c_1, c_2, \ldots, c_\beta \in C^\circ \ \ [\ldots [a, c_1], c_2], \ldots, c_\beta] \bmod [A, A] \in M.$$

Then the action of C on $\hat{A}$ is also nilpotent of order β modulo M.

Proof. Consider the commutator $P = [\ldots [Q, b_1] \ldots, b_k]$, $b_i \in C$, where $Q = [\ldots [a, c_1] \ldots, c_l]$, $c_j \in C^\circ$, $k + l \geq \beta$. Prove by induction on k that $v(P) \in M$. It is true for $k = 0$, by the assumptions. If $k \neq 0$, factorize b_1 into the product of generators from C°: $b_1 = e_1 e_2 \ldots e_n$, $e_k \in C^\circ$. Then, by the formula (14), $[Q, b_1] = [Q, e_1 \ldots e_n] = [Q, e_2 \ldots e_n][e_2 \ldots e_n, [Q, e_1]][Q, e_1]$. Using the induction on the number of the generators forming b_1, we obtain $[Q, b_1] = Q_1 \ldots Q_r$, where every Q_i has the form $Q_i = [\ldots [Q, e_{i_1}] \ldots, e_{i_s}]$, $s \geq 1$. Applying the formula (14) to $[Q_1 \ldots Q_r, b_2]$, we will have
$[[Q, b_1], b_2] = [Q_1, b_2][[Q_1, b_2], Q_2 \ldots Q_r][Q_2 \ldots Q_r, b_2]$.
The middle bracket is equal to 0 in $A/[A, A]$; so, continuing this process, we will obtain $[[Q, b_1], b_2] \equiv [Q_1, b_2] \ldots [Q_r, b_2]$ in $\hat{A}$, and
$P \equiv [\ldots [Q_1, b_2], \ldots, b_k][\ldots [Q_2, b_2], \ldots, b_k] \ldots [\ldots [Q_r, b_2], \ldots, b_k]$.
If we apply now the inductive assumption to each factor, we will obtain $v(P) \in M$.

2.15. As it was shown, the action of the generators of $\pi_1(N)$ on H is nilpotent of order 3 modulo 2 in the case $n = 3$ (i.e. for D_3), and it can be easily checked, that it is also true for their inverses. All of the groups $\gamma_k(H)/\gamma_{k+1}(H)$ are free abelian, as H is free, so, Lemma 2.14 and Theorem 2.2 show that the nilpotent residue of the group $\pi_1(M) = H \ltimes \pi_1(N)$ is trivial in this case.

We could not establish a similar fact for $n > 3$; though it is seen from the formulas (9), that $(u_{k,l} - 1)^2$ are equal to zero modulo 2 for all k, l.

References

[Bi] Birman, J. 'Braids, links, and mapping class groups', *Ann. Math. Stud.* **82** (1975).

[Br] Brieskorn, E. 'Die Fundamentalgruppen des Raumes der regulären Orbits einer endlichen komplexen Spiegelungsgruppe', *Invent. Math.* **12**(1971), 57-61.

[FR-1] Falk, M. and Randell, R. 'The lower central series of a fiber type arrangement', *Invent. Math.* **82**(1985), 77-88.

[FR-2] Falk, M. and Randell, R. 'The lower central series of generalized pure braid groups', in: *Geometry and Topology*(Athens,Ga., 1985), *Lecture Notes in Pure and Appl. Math.* **105**, Dekker, New York, 1987, p. 102-108.

[FR-3] Falk, M. and Randell, R. 'Pure braid groups and products of free groups', *Contemp. Math.* **78**(1988), 217-228.

[H] Hain, R. 'On a generalization of Hilbert 21st problem', *Ann. ENS* **49** (1986), 609-627.

[Ko-1] Kohno, T. 'Série de Poincaré - Koszul associée aux groupes de tresses pures', *Invent. Math.* **82**(1985), 57-75.

[Ko-2] Kohno, T. 'Monodromy representations of braid groups and Yang-Baxter equations', *Ann. Inst. Fourier (Grenoble)* **37**, 4(1987), 139-160.

[L] Lawrence, R.J. 'Braid group representations associated with $\mathcal{SL}_m$', Harvard preprint, 1991.

[Lei] Leibman, A. 'Fiber bundles with degenerations and their applications to computing fundamental groups', to appear.

[Lin] Lin, V.Ya. 'Artin braids, and groups and spaces connected with them', Itogi nauki i tehniki VINITI AN SSSR. Ser. *Algebra, Topologiya, Geometriya* **17**, VINITI, Moscow, 1979, p. 159-228 (in Russian).

[LS] Lyndon, R., Schupp, P. *Combinatorial Group Theory*, Springer-Verlag, Berlin, Heidelberg, New York, 1977.

[M] Markov, A.A. 'Fundamentals of the Algebraic Theory of Braids', *Trudy MIAN SSSR* **16**(1945), 1-53 (in Russian).

[Mar] Markushevich, D.G. 'The D_n generalized pure braid group', *Geometriae Dedicata*, **40**(1991), 73-96.

[MKS] Magnus, W., Karrass, A. and Solitar, D. *Combinatorial Group Theory: Presentation of Groups in Terms of Generators and Relations*, Wiley, New York, London, Sydney, 1966.

[TK] Tsuchiya, A. and Kanie., Y. 'Vertex operators in conformal field theory on $\mathbf{P}^1$ and monodromy representations of braid groups', *Adv. Studies in Pure Math.* **16** (1988), 297-372. Erratum ibid **19**(1990), 675-682.

A.Leibman. Department of Mathematics, Israel Institute of Technology, Technion, 32000 Haifa, Israel.
E-mail address: mat2136@technion.technion.ac.il

D.Markushevich. Department of Mathematics, U.R.A. 746, bat.101, Universite Claude Bernard (Lyon 1) 43 bd du 11 novembre 1918, 69622 Villeurbanne, France.
E-mail address: markushe@geometrie.univ-lyon1.fr

Contemporary Mathematics
Volume **164**, 1994

Verlinde Formulae for Surfaces with Spin Structure

by G. Masbaum and P. Vogel

January 26, 1993

This is an introduction to the construction of Topological Quantum Field Theories (in the sense of Atiyah [A1]) from knot polynomials, in the particular case of the Kauffman bracket.[1]

We begin by describing a series of $3-$manifold invariants $< >_p$, p a positive integer. We then show how to use them to construct a module $V_p(\Sigma)$ for every closed surface Σ. These modules turn out to be finite-dimensional, and we explain in what sense they satisfy a TQFT. We also explain briefly why the rank of these modules is given by Verlinde's formula. We then describe in more detail an action of a Heisenberg type group associated to a surface Σ on the $V-$modules of Σ. Using this action, one obtains, for example, a natural decomposition of the modules $V_{8k}(\Sigma)$ as a direct sum of submodules $V_{8k}(\Sigma, q)$ canonically associated to spin structures q on Σ. The ranks of these submodules can be computed, thereby giving a refinement of Verlinde's formula.

---

[1]This paper is a slightly expanded version of talks given by the authors, and essentially expository in nature. More results, details, and complete proofs can be found in [BHMV3]. All results presented here are joint work with C. Blanchet and N. Habegger. They were obtained in 1991 when all four of us were at the University of Nantes, France.

Acknowledgements: The first author wishes to acknowledge the hospitality of MSRI at Berkeley, the Sonderforschungsbereich "Geometrie und Analysis" at Göttingen, and the Isaac Newton Institute at Cambridge.

This paper is in final form and no version of it will be submitted for publication elsewhere.

1. The 3−manifold invariant

We begin by describing a series of 3−manifold invariants $< >_p$, $p \geq 1$.

A *banded link* K in a closed 3-manifold M is an oriented submanifold homeomorphic to a finite disjoint union of annuli $S^1 \times \mathbf{I}$ in M. [2] A p_1−*structure* is a weak form of framing, discussed below.

By a 3−*manifold with structure*, we mean a (compact oriented) 3−manifold together with a banded link and a p_1−structure.

1.1. Theorem. *[BHMV3] There is a series of invariants $< >_p$, p a positive integer, on the set of closed oriented 3−manifolds with structure, and taking values in a commutative ring k_p with involution, satisfying the following five properties:*

(i) (Multiplicativity)

$$< M_1 \amalg M_2 >_p = < M_1 >_p < M_2 >_p \quad and \quad < \emptyset >_p = 1$$

(ii) (Orientation reversal)

$$< -M >_p = \overline{< M >_p}$$

(iii) (Kauffman bracket relations) [Ka] There is an invertible element $A \in k_p$ such that the following skein relations hold:

$$\diagdown\!\!\!\!\!\diagup = A \;\;)(+ A^{-1} \asymp$$

$$L \cup \bigcirc = \delta \, L$$

(Here, $\delta = -A^2 - A^{-2}$. By convention, a line always represents a band parallel to the plane, with orientation induced from the plane. The first equation means the following: Assume that we have three banded links K, K_0, and K_∞, in a closed 3−manifold M, which are locally (in a ball) given by the three diagrams appearing in the

---

[2]Only the bands, and not their cores (which form an ordinary link in M), are assumed to be oriented.

first equation. Then the three corresponding $< >_p$ −invariants are related as indicated. The second equation means that a banded link consisting of an unknotted component, lying in a ball disjoint from the rest of the link, may be removed at the cost of the factor $-A^2 - A^{-2}$.)

(iv) (Index One Surgery.) There is an invertible element $\eta \in k_p$ such that the following holds. If M' is obtained from M by index one surgery, then

$$< M' >_p = \eta^{-1} < M >_p$$

(v) (Index Two Surgery.) There exist finitely many banded links L_i in the solid torus $-(S^1 \times D^2)$ and elements $\lambda_i \in k_p$ such that the following holds. Let $\phi : -(S^1 \times D^2) \to M$ be an embedding corresponding to a framed knot $K \subset M$ (disjoint from the given banded link in M). Let M' be the result of index two surgery along K (equipped with the same banded link as M). Then

$$< M' >_p = \sum_i \lambda_i < M_i >_p$$

where M_i is the manifold M with $\phi(L_i)$ adjoined to the banded link in M.

1.2. Remarks.

1. The invariant $< >_p$ is constructed from (and essentially equivalent to) the invariant θ_p (on 3−manifolds without p_1−structures) of [BHMV1]. In the case where $p = 2r$ is even, the latter invariant is related to the invariant $\tau_r(M)$ constructed by Reshetikhin and Turaev [RT] and Kirby and Melvin [KM] from the representation theory of the quantum group $U_q SU(2)$ at $q = e^{2\pi i/r}$, and by Lickorish [Li1,2] using the Temperley-Lieb algebra. If p is odd, the invariant θ_p is related to the refined invariant $\tau_r'(M)$ of [KM]. (See [BHMV2] for details.)

2. As is well known, the Kauffman bracket is essentially equivalent to Jones' original V−polynomial [J2]. But for our purposes, the Kauffman bracket is more convenient.

3. The coefficient ring k_p is defined as

$$k_p = \mathbf{Z}[\frac{1}{d}, A, \kappa]/(\varphi_{2p}(A), \kappa^6 - u)$$

where $\varphi_{2p}(A)$ is the $2p$-th cyclotomic polynomial in the indeterminate A, and $d = p$ and $u = A^{-6-\frac{p(p+1)}{2}}$ except for a few low values of p (see [BHMV3] for the exceptional cases.) The element η is given for $p \geq 3$ by

$$\eta = \kappa^3 A^{3+\frac{p(p-1)}{2}}(A^2 - A^{-2})\frac{g(p,1)}{p}$$

where $g(p,1)$ is the Gauss sum $\sum_{m=1}^{p}(-1)^m A^{m^2}$. The involution is defined by sending A to A^{-1} and κ to κ^{-1}.

Thus, the invariant is determined by the choice of p, A, and κ, where p is a positive integer, A is a primitive $2p-$th root of unity, and κ is a sixth root of a certain power of A.

4. One has

$$< S^3 >_p = \eta$$

$$< S^2 \times S^1 >_p = 1$$

(where S^3 and $S^2 \times S^1$ are equipped with the standard [3] p_1-structure.)

5. There is also a *uniqueness theorem* [BHMV3]: If an invariant of closed 3−manifolds with structure [4] satisfies properties (i) - (v), then it is one of the $< >_p$, up to change of coefficients.

6. The index two surgery formula (v) means that surgery along a framed knot K may be replaced, for the purpose of computing the invariant, by cabling K by a certain linear combination $\omega = \sum_i \lambda_i L_i$ of banded links in the solid torus.

7. The construction of the invariants $< >_p$ goes roughly as follows: One observes that an invariant satisfying properties (i) - (v) is essentially determined by $\omega = \sum_i \lambda_i L_i$, and that the possible ω's can be viewed as solutions of a certain algebraic equation coming from the Kirby calculus. This algebraic equation was completely solved in [BHMV1] in the case of invariants satisfying the Kauffman bracket relations. (See [BHMV3] for details.)

1.3. The framing anomaly and p_1-structures.

From section 2 onwards, we shall only need an invariant satisfying properties (i) - (v) of theorem 1.1, and the fact that mani-

---

[3]i.e. a p_1-structure with $\sigma-$invariant zero, see 1.3.

[4]and taking values in an integral domain

folds with structure can be cut and pasted just as ordinary manifolds. Note that properties (i) - (v) do not explicitly mention the
p_1−structure. (Hence the reader may well skip the following discussion of p_1−structures at first reading.)

For us, p_1−structures are just a convenient way to resolve the
so-called *framing anomaly*. By the latter, we mean the following: If
the knot K in the index two surgery formula (v) is the unknot with
writhe $+1$, contained in a ball disjoint from the link in M, then the
underlying manifolds M and M' are diffeomorphic, but

$$< M' >_p = \kappa^3 < M >_p$$

Thus the invariant *does* depend on the p_1−structure (which is a weak
form of framing.)

Formally, p_1−structures are defined as follows: Let X be the
homotopy fiber of the map $p_1 : BO \rightarrow K(\mathbf{Z}, 4)$ corresponding to the
first Pontryagin class of the universal stable bundle γ over BO. Let
γ_X be the pull-back of γ over X. A p_1-*structure* on a manifold M is
a fiber map from the stable tangent bundle of M, τ_M, to γ_X.

In fact, the invariants $< >_p$ depend only on the homotopy class
of the p_1−structure. Moreover, there is a $\mathbf{Z}$−valued homotopy invariant, $\sigma(\alpha)$, of p_1−structures α on closed 3−manifolds, and if M_1
and M_2 are the same except for their p_1−structures, α_1 and α_2, then

$$< M_2 >_p = \kappa^{\sigma(\alpha_2) - \sigma(\alpha_1)} < M_1 >_p$$

The reader may object that in the surgery formulas (iv) and (v)
we should really use p_1−surgery. But it turns out that there is no
problem: for surgery of index one and two, the p_1−structure on the
manifold obtained by surgery is determined, up to homotopy, by the
p_1−structure on the manifold we started with.

Note that by systematically choosing a p_1−structure with σ−in-
variant zero, we can obtain an invariant which does not depend on
the p_1−structure. This "renormalised" invariant would essentially
be the invariant θ_p of [BHMV1]. But it does not satisfy the index
two surgery formula.

Remark. There is in fact another way of resolving the framing
anomaly, by explicitly using the signature cocycle. See [Wa]. See
also Atiyah [A2] for an interpretation in terms of *2-framings*. The
invariant σ is closely related to Atiyah's σ.

2. The modules $V_p(\Sigma)$

We can extend our definition of 3−manifolds with structure in the obvious way to the case where the manifold has non-empty boundary. The boundary of a 3−manifold with structure is a *surface with structure*, i.e. a closed oriented surface equipped with a banded link and a p_1−structure. Here, a *banded link* l in a 2-manifold Σ is a finite collection of oriented intervals properly embedded in the interior of Σ.

In fact, we have a *cobordism category*, denoted by $C_2^{p_1}$ in [BHMV3], whose objects are closed oriented 2-manifolds with structure, and whose morphism set from an object Σ to an object Σ' is the set of isomorphism classes of 3−manifolds M with structure with boundary $-\Sigma \amalg \Sigma'$. In particular, if l (resp. l') denotes the banded link in Σ (resp. Σ'), then M contains a *banded tangle* with boundary $-l \amalg l'$.

Let Σ be a closed surface with structure. Denote by $\mathcal{V}_p(\Sigma)$ the k_p-module freely generated by the set $\mathcal{M}(\Sigma)$ of 3−manifolds with structure with boundary ∂M identified with Σ.

If M and M' are two elements of $\mathcal{M}(\Sigma)$, one can glue M and $-M'$ along their boundaries and the invariant $< M \underset{\Sigma}{\cup} -M' >_p$ is well defined in k_p. This defines a sesquilinear form $<\,,\,>_\Sigma$ on $\mathcal{V}_p(\Sigma)$. This form is hermitian, i.e. one has $< M', M >_\Sigma = \overline{< M, M' >_\Sigma}$. Hence its left kernel equals its right kernel. This kernel is called the *radical* of the form $<\,,\,>_\Sigma$.

2.1. Definition. *The k_p-module $V_p(\Sigma)$ is the quotient of $\mathcal{V}_p(\Sigma)$ by the radical of the form $<\,,\,>_\Sigma$.*

Thus, if $M_i \in \mathcal{M}(\Sigma)$ and $\lambda_i \in k_p$ $(i = 1, \ldots, n)$, then $\sum \lambda_i M_i$ is zero in $V_p(\Sigma)$ if and only if $\sum \lambda_i < M_i \underset{\Sigma}{\cup} -M' >_p$ is zero in k_p for all $M' \in \mathcal{M}(\Sigma)$.

Remark. This definition is in some sense analogous to the definition of singular homology, where one takes *all* cycles modulo *all* boundaries. In general, it is by no means clear that the modules constructed, as above, from a 3−manifold invariant, are finitely generated. However, this will be the case for the invariants $<\,>_p$.

Notation. The quotient map from $\mathcal{V}_p(\Sigma)$ to $V_p(\Sigma)$ will be denoted by Z_p. Hence, for every compact oriented 3-manifold M with structure, $Z_p(M)$ is an element of $V_p(\partial M)$.

2.2. Immediate Properties

1. The form $< , >_\Sigma$ induces a nondegenerate sesquilinear hermitian form on $V_p(\Sigma)$. This form will again be denoted by $< , >_\Sigma$. It induces a linear injection $V(-\Sigma) \to V(\Sigma)^\star$ (where $V(\Sigma)^\star$ is the dual of $V(\Sigma)$.)

2. The radical of the form $< , >_\emptyset$ is exactly the kernel of the map $< >_p: \mathcal{V}(\emptyset) \to k_p$. Therefore $< >_p$ induces a canonical isomorphism $V_p(\emptyset) \approx k_p$, and, if M is closed, then $Z_p(M) =< M >_p$, via this isomorphism. (This uses the multiplicativity of the invariant $< >_p$)

3. If W is a cobordism from a surface Σ to a surface Σ', the assignment $M \mapsto M \underset{\Sigma}{\cup} W$ defines a linear map $F_W : V_p(\Sigma) \to V_p(\Sigma')$. (This follows immediately from the definition of $V_p(\Sigma)$.)

Thus, we may view V_p as a functor from the category $C_2^{p_1}$ to the category of k_p-modules. In [BHMV3] such a functor is called a *quantisation* of the cobordism category $C_2^{p_1}$.

4. Let us fix a manifold with structure, M, with boundary Σ. If L is a banded link (tangle) in M, with boundary the given banded link in Σ, then the value of the pair (M, L) in $V(\Sigma)$ depends only on the image of L in the *skein module* (Jones-Kauffman module) $K(M, l)$. [5] (This follows immediately from the fact that the invariant $< >_p$ satisfies the Kauffman bracket relations.)

5. In the situation above, if moreover M is connected, the natural map

$$K(M, l) \to V_p(\Sigma)$$

is surjective. (This is not completely immediate. The proof relies on the surgery formulas (iv) and (v) of theorem 1.1, and the fact that every 3-manifold bounds.)

2.3. Example. If Σ is the torus $S^1 \times S^1$, equipped with some p_1-structure and the empty link, then we may take M to be a solid torus. The Jones-Kauffman module of the solid torus can be identified with

---

[5]By definition, $K(M, l)$ is the $\mathbf{Z}[A, A^{-1}]$-module freely generated by the set of (isotopy classes of) banded links L in M, meeting ∂M in l, quotiented by the Kauffman bracket relations as in property (iii) of theorem 1.1.

the polynomial ring $k_p[z]$, where z is represented geometrically by a standard unknotted untwisted band representing a generator of the first homology of the solid torus. It turns out [BHMV1] that the kernel of the canonical surjection $k_p[z] \to V_p(S^1 \times S^1)$ is an ideal, which is generated by a polynomial of degree n in the variable z, where $n = 1$ if $p = 1, 3, 4$, $n = 2$ if $p = 2, 5, 6$, and $n = [(p-1)/2]$ for $p \geq 5$. (Note that n is also the rank of $V_p(S^1 \times S^1)$.)

2.4. Relationship with the groups $SU(2)$ and $SO(3)$. Let us define a basis of $k_p[z]$ by setting $e_0 = 1$, $e_1 = z$ and $ze_j = e_{j+1} + e_{j-1}$. If we identify $k_p[z]$ in the usual way with the representation ring[6] $RSU(2)$ of the group $SU(2)$, the polynomial e_i corresponds to the irreducible representation of dimension $2i + 1$. Assume $p \geq 3$. Then the kernel of the map $k_p[z] \to V_p(S^1 \times S^1)$ is the ideal generated by e_n, if $p = 2n + 2$ is even, and generated by $e_n - e_{n-1}$, if $p = 2n + 1$ is odd [BHMV1]. In this way, one can describe $V_p(S^1 \times S^1)$ as a quotient of the representation ring $RSU(2)$. In the case where $p = 2n + 1$ is odd, it is also true that $V_p(S^1 \times S^1)$ is isomorphic to the quotient of $k_p[z](even)$ (the even polynomials in z) by the ideal generated by e_{2n}. Since $k_p[z](even)$ corresponds to the representations which lift to $SO(3)$, we can, in the $p-$odd case, present $V_p(S^1 \times S^1)$ as a quotient of the representation ring $RSO(3)$.

2.5. Definition. *Let V be a functor associated, as in definition 2.1, to an invariant $< >$ for closed 3−manifolds with structure. We say that V satisfies a Topological Quantum Field Theory (TQFT) if the following properties hold:*

(T1) (Finiteness) For all Σ, the module $V(\Sigma)$ is free of finite rank, and the canonical map $V(-\Sigma) \to V(\Sigma)^\star$ is an isomorphism.

(T2) (Multiplicativity) For all Σ_i, $i = 1, 2$, the canonical map

$$V(\Sigma_1) \otimes V(\Sigma_2) \to V(\Sigma_1 \amalg \Sigma_2)$$

is an isomorphism.

Remark. This definition of TQFT is essentially equivalent to the one by Atiyah and Segal (see [A1] [AHLS]), since most of the TQFT properties are obvious in our context.

---

[6]with coefficients in k_p

2.6. Main Theorem of [BHMV3]. *For all $p \geq 1$, the V_p-modules satisfy property (T1). If $p \geq 4$ is even, they also satisfy (T2). If $p \geq 3$ is odd, then (T2) holds provided the link in at least one of the Σ_i has an even number of components.*

Remark. In the p−even case, the V_p-modules satisfy all axioms of TQFT. In the p−odd case, we can consider the full subcategory $C_2^{p_1}(even)$ of $C_2^{p_1}$ whose objects are the surfaces with structure such that the banded link has an even number of components. Then it follows that in this case, the restriction of the functor V_p to the subcategory $C_2^{p_1}(even)$ is a TQFT. In view of Remark 2.4, we think that we have constructed Witten's TQFT [Wi] for $SU(2)$ ($p \geq 4$ even) and $SO(3)$ ($p \geq 3$ odd.)

Example. The multiplicativity property (T2) is by no means obvious. Here is an example. Let Σ be the sphere S^2 equipped with a banded link with an *odd* number, r, say, of components. Then $V_p(\Sigma)$ and $V_p(-\Sigma)$ are trivially zero (in fact, $\mathcal{M}(\Sigma)$ is empty!) If (T2) holds, then $V_p(-\Sigma \amalg \Sigma)$ must be the zero module. But $\mathcal{M}(-\Sigma \amalg \Sigma)$ contains the identity cobordism $\Sigma \times \mathbf{I}$, and there is no reason, in general, why $Z_p(\Sigma \times \mathbf{I})$ should be zero. Here is where the p−even and p−odd theories differ: If p is even, then $Z_p(\Sigma \times \mathbf{I})$ is indeed zero, but if $p = r + 2$, then $Z_p(\Sigma \times \mathbf{I})$ is a generator of $V_p(-\Sigma \amalg \Sigma) \approx k_p$.

3. The Verlinde formula

We shall not attempt to sketch the proof of theorem 2.6 in this talk. (An overview is given in section 1.D of [BHMV3].) However, we wish to indicate how one obtains the *Verlinde formula* for the ranks of the modules $V_p(\Sigma)$. For this, we need certain linear combinations of tangles which we shall now describe.

Let us denote by $\Sigma(n, m)$ the following surface with structure: We take the sphere S^2 viewed as the boundary of the cylinder $D^2 \times \mathbf{I}$ (with a standard p_1−structure), equipped with a standard banded link with n components in $D^2 \times 0$, and a standard banded link with m components in $D^2 \times 1$. The module $V_p(\Sigma(n, m))$ is generated by banded (n, m)−tangles in $D^2 \times \mathbf{I}$ (see remark 2.2.5.) One checks that the usual composition of tangles induces bilinear maps

$$V_p(\Sigma(n, m)) \times V_p(\Sigma(m, l)) \rightarrow V_p(\Sigma(n, l))$$

In particular, $V_p(\Sigma(n,n))$ is a k_p−algebra for all $n \geq 0$. [7] This algebra is a quotient of the Temperley-Lieb algebra on n strings, and it contains a central idempotent f_n, provided n is not too large: If $p = 2n + 2 \geq 4$ is even, then the f_i exist up to $i = n$, and if $p = 2n + 1 \geq 3$ is odd, then the f_i exist up to $i = 2n$. (These idempotents were first considered [8] by Jones [J1] and Wenzl [We]. See also Lickorish [Li1,Li2].) Let us think of banded (n,n)−tangles in $D^2 \times \mathbf{I}$ as *tangle diagrams* in a square $\mathbf{I} \times \mathbf{I}$, and note that modulo the Kauffman bracket relations, it is sufficient to consider diagrams without crossings. The important property of f_n is that if a banded tangle T is represented by such a diagram without crossings and without closed loops, and T is not the identity tangle, then

$$T.f_n = 0 = f_n.T$$

We like to express this by calling f_n the *augmentation idempotent* (with repect to the obvious augmentation homomorphism $V_p(\Sigma(n,n))$ $\rightarrow k_p$.)

We can now describe a basis of the modules $V_p(\Sigma)$. Let us assume for simplicity that the link in Σ is empty, and that Σ has genus at least 2 (the case of genus 1 was already considered in 2.3.) Let H be a handlebody with boundary Σ, and let $G \subset H$ be a banded trivalent graph [9] which is a deformation retract of H. We wish to define elements of $V_p(\Sigma)$ associated to colorings of the set of edges of G. If σ is a coloring of the edges of G by non-negative numbers, we "cable" the edges of G by idempotents f_i, given by the coloring. Thus, in the neighborhood of an edge e, we put the idempotent $f_{\sigma(e)}$. We think of the $f_{\sigma(e)}$ as "drawn" on the banded graph G. What do we do at a vertex? Assume that the edges meeting at that vertex are colored by i, j, k, then we connect the $i + j + k$ ends of the strings

---

[7]In fact, the collection of the modules $V_p(\Sigma(n,m))$ give an example of a k_p−linear category, whose set of objects is the set $\{0,1,2,\ldots\}$. It turns out that for the proof of theorem 2.6, it is quite helpful to consider k−linear categories, and to view them as generalised k−algebras, called k−*algebroids* in [BHMV3]. For example, there is an obvious notion of tensor product of modules over algebroids, and in this generalised sense, a tensor product formula similar to (T2) *always* holds!

[8]albeit in a somewhat different language

[9]i.e. an oriented surface (with boundary) together with a retraction onto a trivalent graph

coming from the edges by $\frac{1}{2}(i+j+k)$ little intervals near the vertex (the intervals are also "drawn" on G.) Of course, this is possible only if

(i) $$i+j+k \equiv 0 \ mod \ 2$$

Note that in this case there is, up to isotopy, precisely one way of doing this "without crossings". Now it follows immediately from the augmentation idempotent property of the f_i that the linear combination of banded links thus obtained can be non-zero only if the colors meeting at a vertex satisfy the triangle inequality

(ii) $$|\, i - j\, |\leq k \leq i+j$$

But even if this is the case, the result may be zero. Using the graphical computations in [MV], we find that the result is non-zero in $V_p(\Sigma)$ if and only if the colors meeting at a vertex satisfy

(iii) $\qquad i + j + k < \begin{cases} p-2 & \text{if } p \geq 4 \text{ is even} \\ 2p-2 & \text{if } p \geq 3 \text{ is odd} \end{cases}$

Let us denote by u_σ the element of $V_p(\Sigma)$ corresponding to the coloring σ of G. The theory developed in [BHMV3] allows one to describe completely the module $V_p(\Sigma)$ by cutting and pasting. It follows from this description that a basis of $V_p(\Sigma)$ is given by the elements u_σ such that σ satisfies conditions (i) - (iii) above at the vertices, and such that for all edges e, one has

$$\sigma(e) \in \begin{cases} \{0,1,2,\ldots n-1\} & \text{if } p = 2n + 2 \geq 4 \text{ is even} \\ \{0,2,4,\ldots 2n-2\} & \text{if } p = 2n + 1 \geq 3 \text{ is odd} \end{cases}$$

It is a well known exercise to count the number of such colorings, and the result is the following

3.1. Corollary (Verlinde's Formula.) *Let $p \geq 3$ and $g \geq 1$. Let Σ_g be a closed surface of genus g equipped with the empty link. Then*

$$rank \ V_p(\Sigma_g) = \left(\frac{p}{4}\right)^{g-1} \sum_{j=1}^{[(p-1)/2]} \left(\frac{1}{\sin\frac{2\pi j}{p}}\right)^{2g-2}$$

3.2. Remarks.

1. If $p = 2k + 4$, then this is Verlinde's formula [Ve] for the dimension of the space of conformal blocks of level k arising from the $SU(2)$ Wess Zumino Witten model of Conformal Field Theory. [10]

2. The case where the link in Σ is non-empty, can be treated similarly. More generally, one may define (and compute) $V-$modules for surfaces with colored structure (i.e. where the components of the link in Σ are colored also.) (See [BHMV3] for details.) The $V-$modules for surfaces with colored structure form part of a *modular functor* [AHLS] [Wa].

3. The basis of $V_p(\Sigma)$ constructed above is orthogonal with respect to the hermitian form $< , >_\Sigma$. For each homomorphism $k_p \to \mathbf{C}$, one may compute the signature of the induced form on $V_p(\Sigma) \otimes \mathbf{C}$. (A formula is given in [BHMV3].) In general, this induced form need not be positive definite.

4. A natural action of the Heisenberg group

Let Σ be an oriented surface (Σ need not be compact here.) Let us denote the (antisymmetric) intersection form on $H_1(\Sigma; \mathbf{Z})$ by $(x, y) \mapsto x \cdot y$.

4.1. Definition. The *Heisenberg group* of the surface Σ, denoted by $H(\Sigma)$, is the set $\mathbf{Z} \times H_1(\Sigma; \mathbf{Z})$ with the multiplication

$$(n, x)(m, y) = (n + m + x \cdot y, x + y)$$

We will denote the element $(1, 0)$ by u, and for $x \in H_1(\Sigma; \mathbf{Z})$, we write $[x] = (0, x)$. Thus u is central, and $[x][y] = u^{x \cdot y}[x + y]$. [11]

Let $\Gamma(\Sigma)$ denote the quotient of $H(\Sigma)$ by the normal subgroup generated by u^4 and the elements $[2x] = [x]^2$, where $x \in H_1(\Sigma; \mathbf{Z})$.

One has the following commutative diagram of short exact sequences:

---

[10] A good introduction into the algebro-geometric definition of this vector space is Atiyah's book [A3].

[11] In fact, $H(\Sigma)$ is the Heisenberg group of twice the intersection form on $H_1(\Sigma; \mathbf{Z})$.

$$0 \longrightarrow \mathbf{Z} \longrightarrow H(\Sigma) \dashrightarrow H_1(\Sigma; \mathbf{Z}) \longrightarrow 0$$

$$0 \longrightarrow \mathbf{Z}/4 \longrightarrow \Gamma(\Sigma) \longrightarrow H_1(\Sigma; \mathbf{Z}/2) \longrightarrow 0$$

Consider a surface with structure $\Sigma = (\Sigma, l)$ where l denotes the banded link in Σ. We have the following theorem:

4.2. Theorem. [BHMV3] *Assume $p \geq 2$. Then there is a natural isometric action of $\Gamma(\Sigma - l)$ on $V_{2p}(\Sigma)$, with the central element u acting by multiplication by $(-1)^{p+1} A^{p^2}$.*

Note that $A^{2p} = -1$ in k_{2p}, and $(-1)^{p+1} A^{p^2}$ is indeed a 4−th root of unity.

We shall sketch a proof of this result in section 5. First, let us apply it in the case where $p \equiv 0 \bmod 4$. Thus, let us write $p = 4k$.

Then u acts as -1, hence the action of $\Gamma(\Sigma - l)$ factors through an action of $\Gamma'(\Sigma - l) = \Gamma(\Sigma - l)/u^2$. For $a \in H_1(\Sigma - l; \mathbf{Z}/2)$, there is a well defined element $[a] \in \Gamma'(\Sigma - l)$, and we denote by τ_a the automorphism of $V_{8k}(\Sigma)$ given by the action of $[a]$. For a function $q : H_1(\Sigma - l; \mathbf{Z}/2) \to \mathbf{Z}/2$, we set

$$V_{8k}(\Sigma, q) = \{v \in V_{8k}(\Sigma) : \tau_a(v) = (-1)^{q(a)} v \text{ for all } a \in H_1(\Sigma - l; \mathbf{Z}/2)\}$$

Since $\tau_a^2 = id$, it is clear that

$$V_{8k}(\Sigma) = \bigoplus_q V_{8k}(\Sigma, q)$$

But since

$$\tau_{a+b} = (-1)^{a \cdot b} \tau_a \tau_b$$

the submodule $V_{8k}(\Sigma, q)$ can be non-zero only if the function q satisfies

$$q(a + b) = q(a) + q(b) + a \cdot b$$

i.e. the function q must be a *quadratic form* on $H_1(\Sigma - l, \mathbf{Z}/2)$ inducing the intersection form. It is well known [Jo] that such quadratic forms correspond bijectively to spin structures on $\Sigma - l$. Thus we have deduced the following:

4.3. Theorem. [BHMV3] *Let* $\Sigma = (\Sigma, l)$ *be a closed surface with structure. Let* $k \geq 1$. *Then there is a natural decomposition of* $V_{8k}(\Sigma)$ *as a direct sum of submodules* $V_{8k}(\Sigma, q)$ *canonically associated to spin structures* q *on* $\Sigma - l$.

Remark. Let $a \in H_1(\Sigma - l; \mathbf{Z}/2)$ be represented by a simple closed curve γ around a component of the banded link $l \subset \Sigma$. One can show that τ_a is multiplication by -1, hence $V_{8k}(\Sigma, q)$ is zero except if $q(a) = 1$. This means that the submodule associated to a spin structure on $\Sigma - l$ can be non-zero only if the the spin structure does *not* extend across any component of l.

The action of $\Gamma'(\Sigma - l)$ on $V_{8k}(\Sigma)$ can be described explicitly in terms of the basis elements described in section 3. This is used in [BHMV3] to compute the rank of the modules $V_{8k}(\Sigma_g)$. Here we only state the result which is as follows.

Recall that there are 2^{2g} spin structures on Σ_g, and that two spin structures are in the same orbit under the action of the diffeomorphism group of Σ_g if and only if they have the same *Arf invariant* in $\mathbf{Z}/2$ (see [Jo].) It follows that the rank of $V_{8k}(\Sigma_g, q)$ depends on q only through its Arf invariant. Explicit computation yields the following:

4.4. Corollary. (Spin Refinement of the Verlinde Formula)
Let $k \geq 1$. *Then*

$$rank \ V_{8k}(\Sigma_g, q) =$$
$$= \begin{cases} \frac{1}{4^g}\left(rank \ V_{8k}(\Sigma_g) + (2k)^{g-1}(2^g - 1)\right) & \text{if } Arf(q) = 0 \\ \frac{1}{4^g}\left(rank \ V_{8k}(\Sigma_g) - (2k)^{g-1}(2^g + 1)\right) & \text{if } Arf(q) = 1 \end{cases}$$

Example. Consider the case $k = 1$. If q has Arf invariant zero, then $V_8(\Sigma_g, q)$ is a free module of rank one. Otherwise, $V_8(\Sigma_g, q)$ is the zero module. In particular,

$$rank \ V_8(\Sigma_g) = 2^{g-1}(2^g + 1)$$

is the number of spin structures on Σ_g with Arf invariant zero.

Remarks.

1. As will be explained in [BM], the modules $V_{8k}(\Sigma, q)$ are related to the invariants on 3−manifolds with spin structure constructed (in-

dependently) by Turaev [Tu], Kirby and Melvin [KM] (using quantum groups), and Blanchet [B] (using the Kauffman bracket.)

2. If $p \equiv 2 \mod 4$, then the central element $u \in \Gamma(\Sigma - l)$ acts as the identity, and the action of $\Gamma(\Sigma - l)$ factors through an action of $H_1(\Sigma - l; \mathbf{Z}/2)$. In this case, one has a natural decomposition of $V_{2p}(\Sigma)$ as a direct sum of submodules $V_{2p}(\Sigma, h)$ canonically associated to mod 2 cohomology classes h on $\Sigma - l$, and one can compute the rank of these submodules. (See [BHMV3] for details.)

3. If p is odd, the situation is quite different. In this case, the element u acts by multiplication by a primitive 4−th root of unity. It turns out that, for odd p, one has a natural isomorphism

$$V_{2p}(\Sigma) \approx V_2'(\Sigma) \otimes V_p(\Sigma)$$

where $\Gamma(\Sigma - l)$ acts trivially on $V_p(\Sigma)$, and $V_2'(\Sigma)$ is an irreducible $\Gamma(\Sigma - l)$−module of rank 2^g (where g is the genus of Σ, and we assume that the link in Σ has an even number of components[12]) In fact, the modules $V_2'(\Sigma)$ define a functor belonging, in the sense of definition 2.1, to an invariant $< >_2'$ which is essentially $< >_2 / < >_1$. (See [BHMV3].)

5. Construction of the action

In this section, we explain how one constructs the action of $\Gamma(\Sigma - l)$ on $V_{2p}(\Sigma)$. We shall use the fact that a cobordism (with structure) from Σ to itself induces an endomorphism of $V_{2p}(\Sigma)$ (see Remark 2.2.3.) For this, we need a more geometric description of the group $\Gamma(\Sigma - l)$ in terms of banded links in $(\Sigma - l) \times \mathbf{I}$.

A geometric description of the Heisenberg group

As in definition 4.1, consider an oriented surface Σ. Let us denote by $\mathcal{L}^+(\Sigma)$ be the set of framed links (i.e., banded links with oriented cores) in $\Sigma \times \mathbf{I}$. Elements of $\mathcal{L}^+(\Sigma)$ are represented by oriented link diagrams on Σ. Putting one diagram above the other gives $\mathcal{L}^+(\Sigma)$ the structure of a monoid, with the empty link as identity element. Let $\mathcal{E}^+(\Sigma)$ be the quotient of $\mathbf{Z} \times \mathcal{L}^+(\Sigma)$ by isotopy and the following skein relations. We write an element $(n, E) \in \mathbf{Z} \times \mathcal{L}^+(\Sigma)$ as $u^n E$,

---

[12]otherwise, $V_{2p}(\Sigma) = 0$.

and define the notations

Then the relations are

(i) $\qquad u^p$ $= u^{p+\varepsilon}$ $\qquad$ (for $\varepsilon = \pm 1$)

(ii)

Here, the links are supposed to be identical except where depicted, and in (ii), the orientations are arbitrary.

Note that $\mathcal{E}^+(\Sigma)$ is again a monoid. In fact, it is not hard to check that $\mathcal{E}^+(\Sigma)$ is naturally isomorphic to the group $H(\Sigma)$. The isomorphism sends an element $u^n E$, where E is a framed link, to the element $u^{n+N(E)}[e] \in H(\Sigma)$, where $N(E)$ is the algebraic number of crossings of E, and e is the class of E in $H_1(\Sigma; \mathbf{Z})$.

Under this isomorphism, the quotient group $\Gamma(\Sigma - l)$ corresponds to the following quotient $\mathcal{E}(\Sigma - l)$ of $\mathcal{E}^+(\Sigma - l)$: It is the quotient by the relations $u^4 = 1$ and $u^n E = u^n E'$, if E and E' are framed links with the same underlying banded link (i.e., E and E' are the same up to changing some of the orientations of the cores of the bands.)

Sketch of proof of theorem 4.2

Here is how $\Gamma(\Sigma - l)$ acts on $V_{2p}(\Sigma)$. Assume $p \geq 2$. Let us denote by e_i the element of the skein module of the solid torus $D^2 \times S^1$ obtained from the idempotent f_i by identifying $D^2 \times 0$ and $D^2 \times 1$.[13] Set $b = (-1)^p e_{p-2}$. If E is a banded link in $(\Sigma - l) \times \mathbf{I}$, set

$$M(E) = (\Sigma \times \mathbf{I}, l \times \mathbf{I} \cup E(b))$$

---

[13] The e_i's thus defined coincide with the e_i's referred to in remark 2.4.

where the notation $E(b)$ means E with all components cabled by b. This is a linear combination of cobordisms with structure from Σ to itself, and it induces an endomorphism $\varphi(E)$ of $V_{2p}(\Sigma)$.

We must show that $\varphi(E)$ depends only on the value of E in $\mathcal{E}(\Sigma - l) \approx \Gamma(\Sigma - l)$, and that φ is the isometric action on $V_{2p}(\Sigma)$ whose existence was asserted in theorem 4.2. The proof relies on certain properties of the idempotent f_{p-2} which, for brevity, we won't state explicitly here. Nevertheless, let us try to sketch the main point of the proof.

First of all, if E is a framed link, we define $\varphi(E)$ as above, forgetting the orientation of the core of E. It is clear that $\varphi(E)$ is an isotopy invariant, and that φ is an action of $\mathcal{L}^+(\Sigma - l)$ on $V_{2p}(\Sigma)$.

Next, we show that this induces an action of $\mathcal{E}^+(\Sigma-l) \approx H(\Sigma-l)$. Consider the disjoint union of two copies of $D^2 \times \mathbf{I}$, each equipped with the idempotent f_{p-2}. We can embed this in several different ways into a 3−ball B. Here, the boundary ∂B is considered as a 2−sphere equipped with a banded link with $4(p-2)$ components, arranged in four groups of $p-2$ each. [14] We claim that for $\varepsilon = \pm 1$, the embeddings symbolized by the following two diagrams yield elements of $V_{2p}(\partial B)$ which are multiples of each other:

$$ \diagX_\varepsilon \; = \; C^\varepsilon \;)($$

The existence of such a relation follows from the fact that the submodule of $V_{2p}(\partial B)$ generated by linear combinations of banded links which "look like" f_{p-2} near each of the groups of $p-2$ components, has rank one. [15] A computation yields $C = -A^{p^2}$. Now if E has m components, then $E(b) = (-1)^{pm} E(e_{p-2})$. Hence the above implies

$$ \varphi\left(\diagX_\varepsilon \right) \; = \; (-1)^p(-A^{\varepsilon p^2}) \, \varphi\left( )(\right) $$

---

[14] In the language of surfaces with colored structure (see Remark 3.2.2), ∂B is the 2−sphere equipped with a 4−component banded link, with each component colored by $p-2$.

[15] In fact, this submodule is the V_{2p}−module associated to ∂B considered as a surface with colored structure, as in the preceding footnote.

since this move changes the number of components by ± 1. Another computation allows one to check that

$$\varphi\left(\right) \bigcirc \,) = \varphi\left(\right))$$

Thus, we have verified that, if we set $\varphi(u) = (-1)^{p+1} A^{p^2}$, then φ induces an action of $\mathcal{E}^+(\Sigma - l) \approx H(\Sigma - l)$ on $V_{2p}(\Sigma)$, as asserted.

The remaining assertions of the theorem can be checked quite easily using some more properties of the elements e_{p-2} (see [BHMV3].)

References.

[A1]   M. F. Atiyah, *Topological quantum field theories*, Publ. Math. IHES 68 (1989) 175-186.

[A2]   M. F. Atiyah, *On framings of 3−manifolds*, Topology, 29, No. 1, (1990) 1-7.

[A3]   M. F. Atiyah, *The Geometry and Physics of Knots*, Cambridge University Press (1990).

[AHLS]   M. Atiyah, N. Hitchin, R. Lawrence, and G. Segal, *Oxford seminar on Jones-Witten theory* (1988).

[B]   C. Blanchet, *Invariants of Three-manifolds with spin structure*, Comm. Math. Helv. 67 (1992) 406-427.

[BHMV1]   C. Blanchet, N. Habegger, G. Masbaum, P. Vogel, *Three-manifold invariants derived from the Kauffman bracket*, Topology 31 (1992) 685-699.

[BHMV2]   C. Blanchet, N. Habegger, G. Masbaum, P. Vogel, *Remarks on the Three-manifold Invariants θ_p*, NATO Advanced Research Workshop and Conference on Operator Algebras, Mathematical Physics, and Low-Dimensional Topology, Istanbul, July 1991 (to appear).

[BHMV3]   C. Blanchet, N. Habegger, G. Masbaum, P. Vogel, *Topological Quantum Field Theories Derived from the Kauffman Bracket*, Preprint May 1992 (in revised form January 1993)

[BM]   C. Blanchet, G. Masbaum, (in preparation).

[J1]   V. F. R. Jones, *Index of subfactors*, Invent. Math. 72 (1983), 1-25.

[J2]   V. F. R. Jones, *A polynomial invariant for links via von Neuman algebras,* Bull. Amer. Math Soc. **12** (1985), 103-111.

[Jo]   D. Johnson, *Spin structures and quadratic forms on surfaces,* J. London Math. Soc. (2) 22 (1980) 365-377.

[Ka]   L. H. Kauffman, *State Models and the Jones Polynomial,* Topology 26 (1987) 395-401.

[KM]   R. C. Kirby, P. Melvin, *The 3-manifold invariants of Witten and Reshetikhin-Turaev for* $sl(2, \mathbf{C})$, Inv. Math. 105, 473-545 (1991).

[Li1] W. B. R. Lickorish, *Three-manifold invariants and the Temperley-Lieb algebra* Math. Ann. 290, 657-670 (1991).

[Li2] W. B. R. Lickorish, *Calculations with the Temperley-Lieb algebra* Comm. Math. Helv. 67 (1992) 571-591.

[Li3] W. B. R. Lickorish, *Skeins and Handlebodies,* preprint January 1992

[MV]   G. Masbaum, P. Vogel, $3-valent$ *graphs and the Kauffman bracket*, preprint, Dec. 1991 (to appear in Pac. J. Math.)

[RT]   N. Yu. Reshetikhin, V. G. Turaev, *Invariants of 3-manifolds via link polynomials and quantum groups*, Inv. Math. 103, 547-597 (1991).

[Tu]   V. G. Turaev, *State sum models in low-dimensional topology,* Proc. ICM Kyoto 1990, vol I, 689-698.

[Ve]   E. Verlinde, *Fusion rules and modular transformations in 2d conformal field theory,* Nucl. Phys. B 300 (1988) 360-376.

[Wa]   K. Walker, *On Witten's 3-manifold Invariants,* (preprint).

[We]   H. Wenzl, *On sequences of projections,* C. R. Math. Rep. Acad. Sci. Canada IX (1987) 5-9.

[Wi]   E. Witten, *Quantum field theory and the Jones polynomial,* Comm. Math. Phys. 121 (1989), 351-399.

URA 212 du CNRS "Théories Géométriques", Université Paris VII, U.F.R. de Mathématiques, Tour 45-55, 5ème étage, 2 place Jussieu, 75251 PARIS Cedex 05, France
e-mail: masbaum@mathp7.jussieu.fr, vogel@mathp7.jussieu.fr

Contemporary Mathematics
Volume **164**, 1994

Knots that cannot be obtained from a trivial knot by twisting

KATURA MIYAZAKI AND AKIRA YASUHARA

ABSTRACT. A sufficient condition that a knot cannot be obtained from a trivial knot by twisting is expressed in terms of knot invariants. Consequently, for any odd number $m > 1$, the connected sum of m copies of the figure eight knot cannot be untied by twisting. It is also shown that the granny knot, the connected sum of two copies of the trefoil knot, cannot be untied by twisting.

1 Statement of Results

We shall work in the smooth category. Let K_0 be a knot in S^3, and D^2 a disk intersecting K_0 in its interior. Let $\omega = |lk(\partial D^2, K)|$. A $\frac{1}{n}$-Dehn surgery along ∂D^2 changes K_0 into a new knot K_n in S^3 (Figure 1). We say that K_n is obtained from K_0 by (n, ω)-*twisting* (or simply *twisting*) on D^2. Let $\mathcal{T}$ denote the set of knots which can be obtained from a trivial knot by some (n, ω)-twisting. (The twisting operation is a generalization of a crossing change. If the unknotting number of a knot is u, then there is a collection of u disks on which twistings untie the knot[1].) Yves Mathieu [6] studied the change of knots after twisting and showed that: if K_0 is a trivial knot and ∂D^2 links K_0 geometrically more than once, then all K_n

---

1991 Mathematics Subject Classification. Primary 57M25; Secondary 57R95.

This paper is in final form and no version of it will be submitted for publication elsewhere.

[1]In fact, every knot can be untied by twistings on *two* properly chosen disks (Ohyama [11]).

where $|n| > 1$ have Property P ([6, Theorem 4.1]). On the basis of this theorem he asked:

QUESTION. ([6, p-97].) Are all knots contained in the class $\mathcal{T}$? I.e. can we untie every knot by twisting?

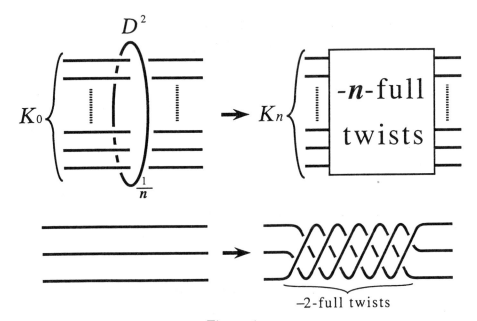

Figure 1

In previous papers, the authors independently answered the question in the negative by giving specific examples: $T(2, 75)\#k$ [16] and $T(2, 3)\#T(2, 13)$ [7] cannot be untied by twisting, where $T(p, q)$ is the (p, q) torus knot and k denotes any slice knot. The purpose of this paper is to (1) express a sufficient condition that a knot is not contained in $\mathcal{T}$ in terms of knot invariants, and (2) find a "smaller" knot not contained in $\mathcal{T}$. To state our results on (1) we use the following notation. The Arf invariant of a knot K is denoted by $\mathrm{Arf}(K)$; $\sigma_p(K)$ denotes the Tristram's p-signature [14]; $e_2(K)$ denotes the minimum number of generators of $H_1(X_K)$ where X_K is the 2-fold branched covering of S^3 along K.

Proposition 1.1 *If K is a knot such that $\mathrm{Arf}(K) = 1$, $\sigma_p(K) = 0$ for each prime number p, and $e_2(K) > 2$, then K is not contained in $\mathcal{T}$.*

Let K be an amphicheiral knot with $\mathrm{Arf}(K) = 1$ and $e_2(K) > 0$. The figure eight knot is an example. As K is amphicheiral, $\sigma_p(K) = 0$ for any prime p. Hence, by Proposition 1.1 $\#^m K$, the connected sum of m copies of K, is not contained in $\mathcal{T}$ where $m > 1$ is an arbitrary odd number. By [9] $\#^m K$ is concordant to a prime knot with the same Alexander invariant. Therefore, by applying Proposition 1.1 to such a prime knot, we obtain the following.

Corollary 1.2 *There are infinitely many prime knots not contained in* $\mathcal{T}$.

Then, what is the "smallest" knot not contained in $\mathcal{T}$? We can unknot by twisting all knots with 3, 4 or 5 crossings and the prime knots with 6 crossings. Here we prove that:

Proposition 1.3 *The granny knot* $3_1 \# 3_1$ *cannot be untied by twisting.*

REMARKS. (1) There are composite knots which *can* be untied by twisting: e.g. $3_1 \# \overline{3_1}, 3_1 \# 4_1, 4_1 \# 4_1$. See the Appendix for more examples.

(2) The granny knot satisfies none of the conditions given in Proposition 1.1.

The proofs of the propositions are outlined as follows. If (n, ω)-twisting unties a knot, it bounds a disk in a 4-manifold whose boundary is S^3. The second homology class represented by the disk can be computed from n and ω. There are several restrictions on the homology classes in a 4-manifold represented by embedded surfaces of given genus. Using these restrictions, we can preclude certain n and ω. The remaining n and ω will be precluded by using techniques of classical knot theory.

We conclude with some conventions and notation. All manifolds will be assumed to be oriented. For a manifold M, $-M$ denotes M with the opposite orientation. If M^4 is a closed 4-manifold, $\text{punc} M^4$ denotes M^4 with an open 4-ball deleted; the orientation of $\partial(\text{punc} M^4)$ is the one induced from $\text{punc} M^4$. For a manifold M and an integer n, nM indicates the connected sum of $|n|$ copies of εM where ε is the sign of n. For a knot K in S^3, $\overline{K}$ denotes the knot $-K$ in $-S^3$. Given a knot K, we write $\Delta_K(t)$ for the Alexander polynomial of K.

2 Proof of Proposition 1.1

Let K be a knot with $\text{Arf}(K) = 1$, $\sigma_p(K) = 0$ for any prime p, and $e_2(K) > 2$. Assume for a contradiction that K is obtained from a trivial knot by (n, ω)-twisting.

Lemma 2.1 ([Nakanishi].) *Suppose K is obtained from a trivial knot by* (n, ω)-twisting. If ω is even, then $e_2(K) \leq 2$.

Proof. Let D be a disk on which (n, ω)-twisting changes a trivial knot K_0 to K. Let X_K and X_{K_0} be the 2-fold coverings of S^3 branched along K and K_0, respectively. Note that $X_{K_0} \cong S^3$. Since ω is even, the preimage of ∂D in X_{K_0} is a 2-component link. It follows that X_K is obtained from M_{K_0} by Dehn surgeries along both components of the preimage of ∂D. Since $H_1(X_{K_0}) = 0$, we obtain $e_2(K) \leq 2$.   □

Since $e_2(K) > 2$, by Lemma 2.1 ω is odd. Then a contradiction will be deduced by using theorems of 4-dimensional topology.

Lemma 2.2 *Let K be a knot in S^3 obtained from a trivial knot by (n, ω)-twisting, and set $M = \mathrm{punc}(nCP^2)$. Consider K to be in ∂M. Then K bounds a disk in M representing $\omega(\gamma_1 + \cdots + \gamma_{|n|})$ in $H_2(M, \partial M)$, where $\gamma_1, \cdots, \gamma_{|n|}$ denote the standard generators of $H_2(M, \partial M)$.*

Proof. Let D be a disk on which the (n, ω)-twisting in the assumption of the lemma is performed. Namely $lk(\partial D, K) = \omega$ and a $\frac{1}{n}$-Dehn surgery along ∂D changes a trivial knot, say K_0, to K. Take $|n|$ parallel copies $D_1, \cdots, D_{|n|}$ of D (Figure 2). Note that ε-Dehn surgeries on the $|n|$ components of $\bigcup \partial D_i$ changes K_0 to K, where $\varepsilon = \mathrm{sgn}(n)$. Regard K_0 and $\bigcup D_i$ as being contained in the boundary of a 4-dimensional 0-handle h^0; the trivial knot K_0 bounds a properly embedded disk, Δ, in h^0. Then attach 2-handles $h_i^2, 1 \leq i \leq |n|$, to h^0 along ∂D_i with all framings ε. The resulting 4-manifold $h^0 \cup \bigcup h_i^2$ is $M = \mathrm{punc}(nCP^2)$. It is easy to verify that Δ represents the homology class stated in the lemma.   □

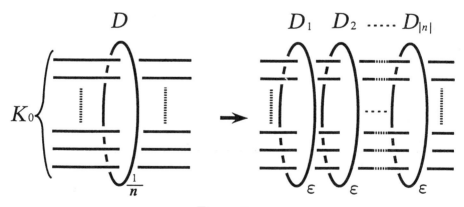

Figure 2

We use the following two theorems to preclude the case when ω is odd. The first theorem, a relative version of Rochlin's genus theorem, is due to Viro. It can be derived as a special case of the main result of Gilmer [3, Theorem 4.1]. The second is a well-known generalization of Rochlin's theorem on spin 4-manifolds. It is really Robertello's definition of the Arf invariant [12].

Theorem 2.3 (cf. [3, Remarks(a) on p-371], [15].) *Let M be an oriented, compact 4-manifold with boundary a 3-sphere, and K a knot in ∂M. Suppose that K bounds a surface of genus g in M representing an element ξ in $H_2(M, \partial M)$. If ξ is divisible by a prime odd number p, then we have:*

$$\left| \frac{2\,\xi \cdot \xi\, [\frac{p}{2}](p - [\frac{p}{2}])}{p^2} - \sigma(M) - \sigma_p(K) \right| \le \dim H_2(M; Z_p) + 2g,$$

where $[x]$ is the greatest integer not exceeding x.

Theorem 2.4 *Let M be an oriented, compact 4-manifold with boundary a 3-sphere, and K a knot in ∂M. If K bounds a disk in M representing a characteristic element ξ in $H_2(M, \partial M)$, then we have:*

$$\frac{\sigma(M) - \xi \cdot \xi}{8} \equiv \mathrm{Arf}(K) \bmod 2.$$

Let us return to the proof of Proposition 1.1. We may assume that our knot K is obtained from a trivial knot by (n, ω)-twisting, where ω is odd. By Lemma 2.2, $K(\subset \partial(\mathrm{punc}\ nCP^2))$ bounds a disk representing an element ξ such that $\xi \cdot \xi = n\omega^2$. If $\omega = 1$, then Theorem 2.4 implies $\mathrm{Arf}(K) = 0$, a contradiction. If $\omega > 1$, then there is a prime odd number, say p, dividing ω. Using Theorem 2.3 and the fact $\sigma_p(K) = 0$, we obtain:

$$\left| \frac{n\omega^2(p^2 - 1)}{2p^2} - n \right| \le |n|.$$

It follows that $\omega^2 \le \dfrac{4p^2}{p^2 - 1}$, so that $\omega^2 \le \dfrac{9}{2}$. Hence the odd number ω equals to 1. This is a contradiction, thus Proposition 1.1 is proved. □

3   Proof of Proposition 1.3

The proof is divided into three cases: $|n| > 1$ and $(n, w) \ne (\pm 2, 2)$; $(n, \omega) = (\pm 1, \text{ even})$; $(n, \omega) = (\pm 2, 2)$ or $(\pm 1, \text{ odd})$.

Claim 1. If $|n| > 1$ and $(n, \omega) \neq (\pm 2, 2)$, then (n, ω)-twisting cannot change a trivial knot to $3_1 \# 3_1$.

For the proof we recall the notion of congruence classes of knots (due to R. H. Fox [2]), an equivalence relation generated by certain twistings. A necessary condition for congruence is given by Fox [2] and Nakanishi-Suzuki [10] in terms of Alexander matrices and Alexander polynomials.

DEFINITIONS. (1) [2]. Let n, ω be non-negative integers. We say a knot K is *congruent* to a knot L *modulo* n, ω, and write $K \equiv L \pmod{n, \omega}$, if there is a sequence of knots $K = K_1, K_2, \cdots, K_l = L$ such that: for each i some (n_i, ω_i)-twisting changes K_i to K_{i+1} where $n_i \equiv 0 \pmod{n}$ and $\omega_i \equiv 0 \pmod{\omega}$.

(2) [10]. In (1) above, if the condition $\omega_i \equiv 0 \pmod{\omega}$ can be replaced by $\omega_i = \omega$ for all i, then we say K is ω-*congruent* to L *modulo* n, and write $K \overset{\omega}{\equiv} L \pmod{n}$.

Since a crossing change can be regarded as a $(1, 0)$- or $(-1, 0)$-twisting, all knots are congruent modulo $1, \omega$, for any ω. [10, Theorem 3] proved that there are infinitely many congruence classes modulo n, ω, where $n \geq 2$ and $(n, \omega) \neq (2, 1), (2, 2)$. Furthermore, their proof implies the following.

Proposition 3.1 *Let K be a knot whose i-th elementary ideal is $(t^2 - t + 1)$ for some i. Then a trivial knot is not congruent to K modulo n, ω, where $n \geq 2$ and $(n, \omega) \neq (2, 1), (2, 2)$.*

The second elementary ideal of $3_1 \# 3_1$ is $(t^2 - t + 1)$. Hence no (n, ω)-twisting trivializes $3_1 \# 3_1$ if $n \geq 2$ and $(n, \omega) \neq (\pm 2, 1), (\pm 2, 2)$. However, Lemma 3.2(2) below precludes the case $(n, \omega) = (\pm 2, 1)$, proving Claim 1.
□

Lemma 3.2 (1) *If $K \overset{\omega}{\equiv} L \pmod{n}$ where n is even and ω is odd, then $\Delta_K(-1) = \pm \Delta_L(-1)$.*

(2) *If n is even and ω is odd, (n, ω)-twisting cannot untie $3_1 \# 3_1$.*

Proof. (2) follows from (1). [10, Theorem 1] shows that if $K \overset{\omega}{\equiv} L \pmod{n}$, then $\Delta_K(t) \pm t^r \Delta_L(t)$ is a multiple of $(1 - t)p_n(t^{\omega})$ where r is some integer and $p_n(t) = (t^n - 1)(t - 1)^{-1}$. If n is even and ω is odd, then $p_n((-1)^{\omega}) = p_n(-1) = 0$. Hence (1) follows.
□

REMARK. Figure 3 shows that $3_1 \# 3_1$ is untied by doing $(2, 2)$-twistings on two disks, so $3_1 \# 3_1 \overset{2}{\equiv} 0 \pmod{2}$. Hence we need to use other techniques to exclude the remaining cases $n = \pm 1$ and $(n, \omega) = (\pm 2, 2)$.

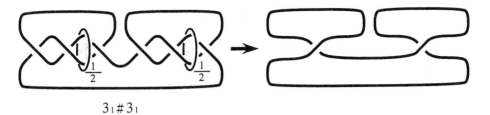

$3_1 \# 3_1$

Figure 3

Claim 2. Let ω be an even number and $\varepsilon = \pm 1$. Then (ε, ω)-twisting cannot change a trivial knot to $3_1 \# 3_1$.

Assume not for a contradiciton. Then by Lemma 2.2, $3_1 \# 3_1$ bounds a disk in $M =$punc(εCP^2) representing $\xi = \omega \gamma \in H_2(M, \partial M)$ where γ is the standard generator. Note that ξ is divisible by $p = 2$. Then, using Theorem 2.3 and the fact $\sigma_2(3_1 \# 3_1) = -4$, we obtain $|\varepsilon \omega^2/2 - \varepsilon + 4| \le 1$. It follows that $-8\varepsilon \le \omega^2 \le 4 - 8\varepsilon$, hence $8 \le \omega^2 \le 12$ or $-8 \le \omega^2 \le -4$. Since ω is even, this is absurd. □

Claim 3. (n, ω)-twisting cannot untie $3_1 \# 3_1$ if $(|n|, \omega) = (2, 2)$ or $(1, \text{ odd})$.

We shall use Theorem 3.3 below restricting the *characteristic* homology classes of a *closed* 4-manifold represented by an embedded sphere. Hence we need to find a punctured 4-manifold in which $3_1 \# 3_1$ bounds a disk representing a characteristic element (Lemma 3.4, Corollary 3.5 below).

Theorem 3.3 (K. Kikuchi [4].) *Let M be a closed, oriented 4-manifold such that (1) $H_1(M)$ has no 2-torsion and (2) $b_2^+(M) \le 3$ and $b_2^-(M) \le 3$. Let ξ be a characteristic element of $H_2(M)$. If ξ is represented by an embedded sphere, then $\xi \cdot \xi = \sigma(M)$.*

Lemma 3.4 *Let K be a knot in S^3 obtained from a trivial knot by $(\pm 2, 2)$-twisting. Set $M =$punc$(S^2 \times S^2)$ and regard $K \subset \partial M$. Then K bounds a disk in M representing $\pm 2\alpha + 2\beta \in H_2(M, \partial M)$, where α and β are the standard generators of $H_2(M, \partial M)$; i.e. $\alpha = [S^2 \times *], \beta = [* \times S^2]$, and $\alpha \cdot \beta = 1$.*

Proof. Let K_0 be a trivial knot contained in the boundary of a 0-handle h^0. Assume that a $(2, 2)$-twisting changes K_0 to a given knot K. Figure 4 shows that doing 0-surgeries along l_1 and l_2 have the same effect on K_0 as the $(2, 2)$-twisting. (Also see [1, Fig. 12 on p-506].) Attach two 2-handles h_1^2, h_2^2 to h^0 with framings 0 along $l_1 \cup l_2$. Then $h^0 \cup h_1^2 \cup h_2^2 \cong$ punc$(S^2 \times S^2) = M$ and $(\partial M, K_0) \cong (S^3, K)$. A properly embedded disk in h^0 bounded by K_0 represents the desired homology class $2\alpha + 2\beta$ of $H_2(M, \partial M)$, proving the lemma. If $(-2, 2)$-twisting changes K_0 to K, the similar arguments work. $\qquad\square$

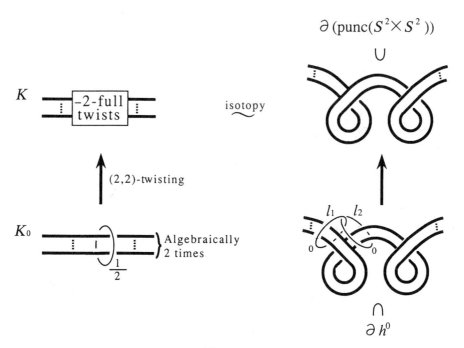

Figure 4

The above proof and Figure 3 imply the following.

Corollary 3.5 $3_1 \# 3_1$ *bounds a disk in* $\mathrm{punc}(2(S^2 \times S^2))$ *representing* $-2\alpha_1 + 2\beta_1 - 2\alpha_2 + 2\beta_2$, *where* α_1, β_1 *represent the standard generators of the first* $S^2 \times S^2$, *and* α_2, β_2 *the second* $S^2 \times S^2$.

Let us turn to Claim 3. First assume that $3_1 \# 3_1$ is obtained from a trivial knot by $(\pm 2, 2)$-twisting. $3_1 \# 3_1$ bounds a disk D_1 in $M_1 = \mathrm{punc}(S^2 \times S^2)$ such that $[D_1] = \pm 2\alpha + 2\beta$ (Lemma 3.4). On the other hand, by Corollary 3.5, $\overline{3_1 \# 3_1}$ bounds a disk D_2 in $M_2 = \mathrm{punc}(2(S^2 \times S^2))$ such that $[D_2] = 2(\alpha_1 + \beta_1 + \alpha_2 + \beta_2)$. Paste M_1 and M_2 along their boundaries so that $(\partial M_1, \partial D_1) = (-\partial M_2, -\partial D_2)$. Then the resulting 4-manifold $M = 3(S^2 \times S^2)$ contains the 2-sphere $S = D_1 \cup D_2$ such that $[S] = 2(\pm\alpha + \beta + \alpha_1 + \beta_1 + \alpha_2 + \beta_2)$, a characteristic class of $H_2(M)$. By Theorem 3.3 we have: 24 or 8 $= S \cdot S = \sigma(M) = 0$, a contradiction. (This also contradicts Kervaire-Milnor's theorem [5].) Then, assume $3_1 \# 3_1$ is obtained from a trivial knot by (ε, ω)-twisting where $\varepsilon = \pm 1$ and ω is odd. $3_1 \# 3_1$ bounds a disk D_3 in $M_3 = \mathrm{punc}(\varepsilon CP^2)$ such that

$[D_3] = \omega\gamma$ (Lemma 2.2). Paste M_2 and M_3 above so that $(\partial M_2, \partial D_2) = (-\partial M_3, -\partial D_3)$. Then $N = M_2 \cup M_3$ is $2(S^2 \times S^2)\#\varepsilon CP^2$ and the embedded sphere $S' = D_2 \cup D_3$ represents $2(\alpha_1 + \beta_1 + \alpha_2 + \beta_2) + \omega\gamma \in H_2(N)$. Since ω is odd, $[S']$ is characteristic. Then Theorem 3.3 implies that: $16 + \varepsilon\omega^2 = S' \cdot S' = \sigma(M) = \varepsilon$. For each $\varepsilon = \pm 1$, this leads to a contradiction. Claim 3 and hence Proposition 1.3 are proved.   □

Appendix

In [6], Mathieu also asked if there is a composite knot untied by twisting. Since then several families of composite knots are shown to give an affirmative answer [8], [13]. We shall list some examples; the last two are due to M. Teragaito.

EXAMPLE 1. Let K be a knot obtained as a band connected sum of a 2-component trivial link $l_1 \cup l_2$; $k\#\bar{k}$ is such a knot if k is a 2-bridge knot. Then K can be untied by twisting.

Proof. The knot K is obtained by connecting l_1 and l_2 by a band b (say) as in Figure 5(a). There is a disk D such that $\partial D = l_1, D \cap l_2 = \emptyset$ and D intersects the band b transversely. Figure 5(b) depicts K in a neighborhood of D. Take a simple loop c as in Figure 5(c). A 1-Dehn surgery along c will eliminate the intersections of D and b (Figure 5(d)). Hence K deforms to a trivial knot.   □

EXAMPLE 2. Let T_n be the n-half-twisted double of the unknot; in particular $T_2 \cong 4_1, T_{-2} \cong 3_1$. Figure 6 demonstrates that $T_n\#T_{-n}$ can be untied by $(-1, 1)$-twisting.

In Examples 1 and 2, each prime summand of the composite knots is a 2-bridge knot. However, this is not always the case.

EXAMPLE 3. Fix any integer $p > 1$. Let K_1 be the $(p, p+1)$ torus knot. Let K_2 be a satellite of a p-bridge knot with the pattern J in Figure 7. Although both K_i have bridge indices$\geq p$, [13] demonstrates $K_1\#K_2$ can be untied by some $(1, p + 2)$-twisting.

We conclude with a question raised in [13].

QUESTION. Is there a composite knot with more than two prime factors which can be trivialized by twisting?

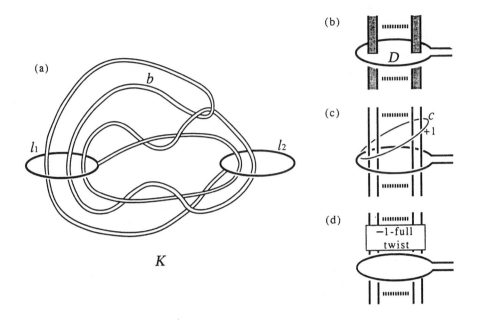

Figure 5

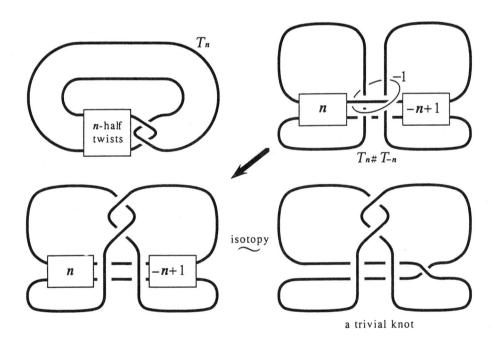

Figure 6

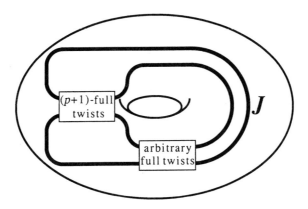

Figure 7

REFERENCES

1. T. D. Cochran and R. E. Gompf, Applications of Donaldson's theorems to classical knot concordance, homology 3-spheres and property P, Topology **27** (1988), 495–512.

2. R. H. Fox, Congruence classes of knots, Osaka Math. J. **10** (1958), 37–41.

3. P. M. Gilmer, Configurations of surfaces in 4-manifolds, Trans. Amer. Math. Soc. **264** (1981), 353–380.

4. K. Kikuchi, Representing positive homology classes of $CP^2 \# 2\overline{CP^2}$ and $CP^2 \# 3\overline{CP^2}$, Proc. Amer. Math. Soc. **117** (1993), 861–869.

5. M. Kervaire and J. Milnor, On 2-spheres in 4-manifolds, Proc. Nat. Acad. Sci. U. S. A. **47** (1961), 1651–1657.

6. Y. Mathieu, Unknotting, knotting by twists on disks and property (P) for knots in S^3, *Knots 90* (ed. by A. Kawauchi), Proc. 1990 Osaka Cof. on Knot Theory and Related Topics, de Gruyter, 1992, pp. 93–102.

7. K. Miyazaki, A solution to Mathieu's problem, *Proc. Kashikojima Conf. on Knot Theory and its Application*, 1991, pp. 199–202, (Japanese).

8. K. Motegi and T. Shibuya, Are knots obtained from a plain pattern prime?, Kobe J. Math. **9** (1992), 39–42.

9. Y. Nakanishi, Prime links, concordance and Alexander invariants, Math. Seminar Notes **8** (1980), 561–568.

10. Y. Nakanishi and S. Suzuki, On Fox's congruence classes of knots, Osaka J. Math. **24** (1987), 217–225.

11. Y. Ohyama, Twisting and unknotting operations, preprint.

12. R. Robertello, An arf invariant of knot cobordism, Comm. Pure and Appl. Math. **18** (1965), 543–555.

13. M. Teragaito, Composite knots trivialized by twisting, J. Knot Theory and its Ramifications **1** (1992), 467–470.

14. A. G. Tristram, Some cobordism invariants for links, Proc. Cambridge Philos. Soc. **66** (1969), 251–264.

15. O. Ya. Viro, Link types in codimension–2 with boundary, Uspehi Mat. Nauk **30** (1975), 231–232, (Russian).

16. A. Yasuhara, On slice knots in the complex projective plane, Rev. Mat. Univ. Complut. Madrid **5** (1992), 255–276.

Faculty of Engineering
Tokyo Denki University
2-2 Kanda-Nishikicho
Tokyo 101, Japan
E-mail: miyazaki@cck.dendai.ac.jp

Department of Mathematical Sciences
College of Sience and Engineering
Tokyo Denki University
Hatoyama-machi
Saitama 350-03, Japan
E-mail: yasu@r.dendai.ac.jp

Contemporary Mathematics
Volume **164**, 1994

The Arithmetic of Braids
and a Statement of Chisini

B. MOISHEZON

Abstract

In this paper we construct counter-examples to the so called "Chisini's Theorem" on braid monodromy factorizations corresponding to cuspidal curves. The construction is based on our recent results on the topology of generic polynomial maps.

1. Introduction

Let D be a closed disk in $\mathbf{C}^1$, $L \in D - \partial D$ a finite subset, $\#L = m$, v a point in ∂D. The braid group B_m can be defined as the group $B[D, L]$ of equivalence classes of diffeomorphisms of D which are identity maps on ∂D, map L to itself, and are equivalent if and only if they define the same automorphism of $\pi_1(D - L, v)$ (See [6]). Denote by B_m^+ the normal sub-semi-group of B_m generated by positive half-twists (that is, by all conjugates of a positive half-twist $X_1 \in B_m$).

Let S be an algebraic curve in $\mathbf{C}P^2$ of degree m. It can be considered as a complex-analytic analog of classical knots. The topology of the embedding $S \subset \mathbf{C}P^2$ is determined by the *braid monodromy* of S ([9, 4, 6]), which is described by a factorization of the "full twist" Δ_m^2 in the semi-group B_m^+ (in standard generators, $\Delta_m^2 = (X_1 \ldots X_{m-1})^m$). If S is a cuspidal curve (that is, the only singularities of S are nodes and cusps), then this factorization can be written as follows

$$\Delta_m^2 = \prod_k Q_k^{-1} X_1^{\rho_k} Q_k, \ \rho_k \in (1, 2, 3), \tag{1}$$

1991 *Mathematics Subject Classification.* Primary 14H30, 20F36.

This paper is in final form and no version of it will be submitted for publication elsewhere.

where X_1 is a positive half-twist in B_m.

Any factorization $\mathcal{E}$ of Δ_m^2 which looks like (1) we call a *cuspidal factorization*. We denote

$$
\begin{aligned}
\rho(\mathcal{E}) &= \text{number of degree 3 factors in } \mathcal{E}; \\
d(\mathcal{E}) &= \text{number of degree 2 factors in } \mathcal{E}; \\
\mu(\mathcal{E}) &= \text{number of degree 1 factors in } \mathcal{E}.
\end{aligned}
$$

In the case when a *cuspidal factorization* is a braid monodromy factorization of a cuspidal curve S, $\rho(\mathcal{E}), d(\mathcal{E}), \mu(\mathcal{E})$ are respectively the numbers of cusps, nodes and "simple branch points for a generic projection to $\mathbf{C}P^1$" of S.

The classical formula $\Delta_m^2 = (X_1 \dots X_{m-1})^m$ gives a *cuspidal factorization* with $\rho(\mathcal{E}) = 0, d(\mathcal{E}) = 0, \mu(\mathcal{E}) = m(m-1)$. We call this factorization the *basic factorization* of Δ_m^2. The *basic factorization* corresponds to the braid monodromy of a non-singular curve of degree m in $\mathbf{C}P^2$ [9].

Let

$$
h = g_1 g_2 \dots g_r \tag{2}
$$

be a factorization in B_m^+. The transformation which changes two neighboring factors in (2) as follows

$$
\begin{aligned}
g_i \cdot g_{i+1} &\mapsto (g_i g_{i+1} g_i^{-1}) \cdot g_i \text{ , or} \\
g_i \cdot g_{i+1} &\mapsto g_{i+1} \cdot (g_{i+1}^{-1} g_i g_{i+1}),
\end{aligned}
$$

is called a *Hurwitz move*. We say that two factorizations are *Hurwitz equivalent* if one can be obtained from the other by a finite sequence of Hurwitz moves. The braid monodromy factorization of Δ_m^2 is defined up to *Hurwitz equivalence* [4, 6].

Let $z \in B_m$ be such that $[z, h] = 1$. We denote by

$$
h_z = z^{-1} g_1 z \cdot z^{-1} g_2 z \cdot \dots z^{-1} g_r z
$$

and say that the factorized expression h_z is obtained from (2) by *simultaneous conjugation* (by z). Two factorizations are called *Hurwitz and conjugation equivalent* if one can be obtained from the other by a finite sequence of Hurwitz moves followed by a simultaneous conjugation. Let $g_i = \prod g_{i\alpha_i}$ be an expression of g_i ($\in B_m^+$) as a product of positive half-twists. We say that the factorization

$$
h = \prod g_{1\alpha_1} \cdot \prod g_{2\alpha_2} \cdot \dots \cdot \prod g_{r\alpha_r}
$$

is obtained from (2) by *a complete regeneration*.

For a given plane algebraic curve S any two factorizations of Δ_m^2 representing the braid monodromy of S are *Hurwitz and conjugation equivalent*.

Taking a non-singular curve C very close to S, we get a braid monodromy factorization for C from that of S using *complete regeneration*. The factorization of Δ_m^2 obtained from a braid monodromy factorization for S by *complete regeneration* must be *Hurwitz and conjugation equivalent* to the *basic factorization* (since both represent the braid monodromy of C).

DEFINITION 1. *We say that a factorization* $\Delta_m^2 = \prod_k g_k$ *in* B_m^+ *is*

1) geometric, if a complete regeneration of it is Hurwitz and conjugation equivalent to the basic factorization $\Delta_m^2 = \prod_{m-1 \; times}(X_1 \ldots X_{m-1})$;

2) analytic, if it corresponds to the braid monodromy of an algebraic curve.

It follows from above that any *analytic* factorization is *geometric*. The inverse is not true even for cuspidal factorizations, and one of the goals of our work is to show it.

Take any $a \in L$ and a simple path p_a in $D - (L - a)$ connecting $v(\in \partial D)$ with a. Let d be a very small disk in $D - (L - a)$ centered in a, $s = \partial d$ (taken with positive orientation). Assume that $p_a \cap d$ is a radius in d, and denote by $\tau' = p_a - p_a \cap d$. We denote by $e(p_a)$ the element of $\pi_1(D - L, v)$ represented by

$$\tau' \cup s \cup (-\tau').$$

We call such $e(p_a)$ a *geometric generator* of $\pi_1(D - L, v)$ corresponding to p_a.

We say that $\gamma_1, \ldots, \gamma_m$ is a *geometric basis* of $\pi_1(D - L, v)$, if there exists a set of simple paths $p_1, \ldots, p_m$ in D connecting v with the points of L and intersecting only at v, such that each $\gamma_j = e(p_j)$.

Let $\mathcal{C} = \prod_{j=1}^r g_j$ be a factorized expression in $B_m^+ \subset B_m = B[D, L]$, $\gamma_1, \ldots, \gamma_m$ be a *geometric base* of $\pi_1(D - L, v)$. Then $\mathcal{C}$ defines a "set of relations":

$$RC : (\gamma_i)g_j \cdot \gamma_i^{-1} \sim 1, \; 1 \le j \le r, \; 1 \le i \le m.$$

In many cases it is possible to replace the set $\{(\gamma_i)g_j \cdot \gamma_i^{-1} \sim 1\}$ by more convenient "sets of relations" which define the same quotient of $\pi_1(D-L, v)$. Such a more convenient set (defining the same quotient of $\pi_1(D-L, v)$) will still be denoted by RC and will be called *a full sets of relations defined by* $\mathcal{C}$.

DEFINITION 2. *Let $\mathcal{C}$, $R\mathcal{C}$ be as above. Denote by $N(\mathcal{C})$ the normal subgroup in $\pi_1(D - L, v)$ generated by the conjugates of the elements $(\gamma_i)g_j \cdot \gamma_i^{-1}$, $1 \leq j \leq r$, $1 \leq i \leq m$. Define*

$$G(\mathcal{C}) = \pi_1(D - L, v)/N(\mathcal{C}).$$

Let $\gamma_1, \ldots, \gamma_m$ be ordered in such a way that $\prod_{j=1}^m \gamma_j$ is represented by ∂D with the positive orientation. Denote by Π_0 the normal subroup of $\pi_1(D - L, v)$ generated by the conjugates of $\prod_{j=1}^m \gamma_j$. Define

$$\bar{G}(\mathcal{C}) = G(\mathcal{C})/\Pi_0.$$

$\bar{G}(\mathcal{C})$ will be called the Zariski-Van Kampen group of $\mathcal{C}$.

REMARK. One can check that $\forall\, i \in (1, \ldots, m)$ $(\gamma_i)\Delta_m^2$ is equal to γ_i conjugated by $\prod_{j=1}^m \gamma_j$. It implies that the image of $\prod_{j=1}^m \gamma_j$ in $G(\mathcal{C})$ lies in the center of $G(\mathcal{C})$, that is, Π_0 is a cyclic subgroup generated by this image.

It is easy to see that if two factorizations $\mathcal{C}_1, \mathcal{C}_2$ in B_m^+ are *Hurwitz and conjugation equivalent*, then there exists $\beta \in B[D, L] \subset Aut(\pi_1(D - L, v))$ such that

$$(N(\mathcal{C}_1))\beta = N(\mathcal{C}_2)$$

which implies that the groups $G(\mathcal{C}_1), G(\mathcal{C}_2)$ (as well as $\bar{G}(\mathcal{C}_1), \bar{G}(\mathcal{C}_2)$) are isomorphic and the corresponding isomorphism can be lifted to a "geometric" automorphism of $\pi_1(D - L, v)$ (that is , such that it sends geometric generators to geometric generators).

Let Z be a positive half-twist in $B_m = B[D, L]$, corresponding to a simple path z in D which connects points $a, b \in L$. We say that a pair (A, B) of two geometric generators of $\pi_1(D - L, v)$ is an *associated pair* for Z, if there exist two simple paths p_a, p_b in D such that $A = e(p_a)$, $B = e(p_b)$, $z \cap p_a = a$, $z \cap p_b = b$, and the interior of the triangle formed by z, p_a, p_b does not contain points of L. It is easy to see that for given Z an associated pair is well defined up to simultaneous conjugation. We choose the order in (A, B) so that $(B)Z = A$.

Consider a cuspidal factorization $\mathcal{E}$ of Δ_m^2 , that is, $\mathcal{E} = \prod_k Z_k^{\rho_k}$, where all $\rho_k \in (1, 2, 3)$, and each Z_k is a positive half-twist in B_m. For each k denote by (A_k, B_k) an associated pair for Z_k.

Let S_n be the symmetric group of degree n, $\bar{\Psi} : \bar{G}(\mathcal{E}) \to S_n$ and $\Psi :$
$G(\mathcal{E}) \to S_n)$ some homomorphisms. Denote by $\psi : \pi_1(D - L, v) \to S_n$ the
canonical lifting of $\bar{\Psi}$ (resp. of Ψ) to $\pi_1(D - L, v)$.

We say that $\bar{\Psi}$ (resp. Ψ) is *geometric* if it is surjective and has the
following properties:

1) $\forall$ geometric generators $\gamma \in \pi_1(D - L, v)$, $\psi(\gamma)$ is a transposition in S_n;

2) for each associated pair (A_k, B_k), the transpositions $\psi(A_k), \psi(B_k)$ have
exactly one common index when $\rho_k = 3$, and do not have common indices
when $\rho_k = 2$ (when $\rho_k = 1$ these transpositions are equal).

We say that a geometric $\bar{\Psi} : \bar{G}(\mathcal{E}) \to S_n$ (resp. $\Psi : G(\mathcal{E}) \to S_n$) is
solvable if $Ker\bar{\Psi}$ (resp. $Ker\Psi$) is a solvable group. It is clear that if $\mathcal{E}$ is
such that there exists a solvable geometric homomorphism $\bar{\Psi} : \bar{G}(\mathcal{E}) \to S_n$
with $n \geq 5$, then this homomorphism is unique in the following sense: any
other surjection $\bar{\Psi}' : \bar{G}(\mathcal{E}) \to S_{n'}$ with $n' \geq 5$ has $n' = n$ and coincides with
$\bar{\Psi}$ (up to an inner automorphism of S_n). The same is true for any solvable
geometric $\Psi : G(\mathcal{E}) \to S_n$ (with $n \geq 5$).

From the Remark above it follows that for any surjective $\Psi : G(\mathcal{E}) \to S_n$
with $n \geq 5$ there exists unique $\bar{\Psi} : \bar{G}(\mathcal{E}) \to S_n$ such that Ψ is the canonical
lifting of $\bar{\Psi}$.

If $\mathcal{C} = \Delta_m^2$ and corresponds to the braid monodromy of a cuspidal curve
$S \subset \mathbf{C}P^2$, then by the Zariski-Van Kampen Theorem [11, 10] we have for
the "generic" affine part S' of S

$$\pi_1(\mathbf{C}^2 - S') \simeq G(\mathcal{C}) \text{ and } \pi_1(\mathbf{C}P^2 - S) \simeq \bar{G}(\mathcal{C}).$$

In this case ($\mathcal{C}$ is *analytic*), for any geometric $\bar{\Psi} : \bar{G}(\mathcal{C}) \to S_n$ there exists a
non-singular algebraic surface X and a stable ramified covering $f : X \to \mathbf{C}P^2$
of degree n with the branch locus S and the monodromy homomorphism
$\pi_1(\mathbf{C}P^2 - S) \to S_n$ equal to $\bar{\Psi}$. The existence of geometric $\bar{\Psi} : \bar{G}(\mathcal{C}) \to S_n$
is a necessary and sufficient condition for the cuspidal curve S coresponding
to $\mathcal{C}$ to be the branch curve for some *stable* ramified cover.

The parameter space of degree m algebraic curves in $\mathbf{C}P^2$ has a finite
stratification such that for any two curves S_1, S_2 of the same stratum the
corresponding embeddings $S_1, S_2 \subset \mathbf{C}P^2$ are topologically equivalent (that
is, there exists an isotopy of $\mathbf{C}P^2$ maping S_1 to S_2). This fact implies that
for any m there exists only a finite number of *Hurwitz and conjugation
equivalence* classes of *analytic* cuspidal factorizations of Δ_m^2. For *geometric*

non-analytic cuspidal factorizations the situation is different. We have the following

THEOREM 1. $\exists$ *an infinite set* $\mathcal{N}_c$ *of positive integers such that for each* $m \in \mathcal{N}_c$ *the following is true:*

There exist positive integers μ_m, ρ_m, d_m *and infinitely many geometric cuspidal factorizations*

$$\Delta_m^2 = \tilde{\mathcal{E}}_{m,v}, \ v \in \mathbf{Z}$$

with

$$\mu(\tilde{\mathcal{E}}_{m,v}) = \mu_m, \ \rho(\tilde{\mathcal{E}}_{m,v}) = \rho_m, \ d(\tilde{\mathcal{E}}_{m,v}) = d_m$$

such that $\forall v_1, v_2$ *with* $v_1 \neq v_2$ *the factorizations* $\tilde{\mathcal{E}}_{m,v_1}, \tilde{\mathcal{E}}_{m,v_2}$ *are not Hurwitz and conjugation equivalent, and moreover, the corresponding groups* $G(\tilde{\mathcal{E}}_{m,k_1})$, $G(\tilde{\mathcal{E}}_{m,k_2})$ *are not isomorphic. Also the corresponding Zariski-Van Kampen groups* $\bar{G}(\tilde{\mathcal{E}}_{m,v_1}), \bar{G}(\tilde{\mathcal{E}}_{m,v_2})$ *are not isomorphic.*

In particular, for each $m \in \mathcal{N}_c$, *infinitely many of* $\tilde{\mathcal{E}}_{m,v}$ *are not analytic (that is, do not correspond to braid monodromies of cuspidal curves).*

In his articles [1, 2, 3] O. Chisini claimed that if a *geometric* cuspidal factorization $\Delta_m^2 = \mathcal{E}$ is such that

$$d(\mathcal{E}) + 2\rho(\mathcal{E}) \ < \ m(m+3)/2 \ - \ 3, \tag{3}$$

then $\mathcal{E}$ corresponds to the braid monodromy of a cuspidal curve of degree m (that is, it is *analytic* in our teminology).

Our construction of factorizations $\tilde{\mathcal{E}}_{m,v}$ is such that $54 \in \mathcal{N}_c$, $d_{54} = 756$, $\rho_{54} = 378$, so that for $m = 54$ the inequality

$$d_m + 2\rho_m \ < \ m(m+3)/2 \ - \ 3$$

is satisfied.

Thus, we have counter-examples to the "Chisini's Theorem".

Actually, we have also counter-examples with $m = 18$, but in them $d(\mathcal{E}) = 0$, which looks like a pathology. (Another "pathology" of our $m = 18$ counter examples is that the classical Veronese surface V_2, famous by "pathological" properties, is involved in their construction.)

REMARK. For any cuspidal factorization of $\Delta_m^2 = \mathcal{E}$ and a projection $: \mathbf{C}P^2- \to \mathbf{C}P^1$, it is possible to construct a cuspidal curve $\bar{S}$ and a topological embedding $\iota : \bar{S} \to \mathbf{C}P^2$ which is not complex-analytic, but ehaves like a complex-analytic one with respect to $\pi : \mathbf{C}P^2- \to \mathbf{C}P^1$ ($\iota : \bar{S} \to \mathbf{C}P^1$ becomes a ramified complex-analytic covering of degree m), o that the braid monodromy for $\iota(\bar{S})$ with respect to π will be well defined nd will be represented by $\Delta_m^2 = \mathcal{E}$. We denote $S(\mathcal{E}) = \iota(\bar{S})$ and call it a semi-algebraic curve" corresponding to $\mathcal{E}$. The standard proof of Zariski-'an Kampen Theorem works in the "semi-algebraic" case and gives

$$\pi_1(\mathbf{C}P^2 - S(\mathcal{E})) \simeq \bar{G}(\mathcal{E}).$$

Assume that there exists a geometric homomorphism $\bar{\Psi} : \bar{G}(\mathcal{E}) \to S_n$. 'hen, using the analogy with the case of *analytic* $\mathcal{E}$, one can construct a ifferentiable 4-manifold Y and a degree n $\mathcal{C}^\infty$ map $g : Y \to \mathbf{C}P^2$ which utside of $S(\mathcal{E})$ is a non-ramified covering of degree n with the monodromy omomorphism equal to $\bar{\Psi}$, and over a nbhd of $S(\mathcal{E})$ locally behaves like a omplex-analytic ramified covering branched at $S(\mathcal{E})$ and with local mon-dromy induced by $\bar{\Psi}$. We call such Y a "semi-algebraic surface", g a "semi-lgebraic covering" of $\mathbf{C}P^2$ and $S(\mathcal{E})s$ a "semi-algebraic stable branch curve".

In our construction of factorizations $\Delta_m^2 = \tilde{\mathcal{E}}_{m,v}$ we have for each $m \in \mathcal{N}_c$ positive integer n_m and geometric homomorphisms $\bar{\Psi}_{m,v} : \bar{G}(\tilde{\mathcal{E}}_{m,v}) \to S_{n_m}$. 'hus, we have 4-manifolds $Y_{m,v}$ and degree n_m "semi-algebraic coverings" $_{m,v} : Y_{m,v} \to \mathbf{C}P^2$ branched at $S(\tilde{\mathcal{E}}_{m,v})$ with monodromy homomorphisms qual to $\bar{\Psi}_{m,v} : \bar{G}(\tilde{\mathcal{E}}_{m,v}) \to S_{n_m}$. One can prove that these $Y_{m,v}$ are all simply-onnected. It seems also that all $Y_{m,v}$ with fixed m are homeomorphic to ach other. An interesting problem is, whether all such $Y_{m,v}$ have different iffeomorphic types and are "indecomposable". If it is so, then infinitely 1any of $Y_{m,v}$ with fixed m do not have a complex structure, and we will ave many new examples of simply-connected indecomposable 4-manifolds omeomorphic and not diffeomorphic to algebraic surfaces of general type.

The construction of $\tilde{\mathcal{E}}_{m,v}$ with given m (see Sectons 2 and 3 below) is based n a choice of a cuspidal geometric factorization $\Delta_{m_0}^2 = \mathcal{E}_{m_0}$ ($m_0 = m/3$) 'ith some special (non-trivial) properties (like the existence of a solvable eometric $\bar{\Psi} : \bar{G}(\mathcal{E}_{m_0}) \to S_n$ with $n \geq 5$ and infinite kernel) . In general we xpect that many stable branch curves will provide such $\mathcal{E}_{m_0}$'s. At the present moment not much is known about stable branch curves. Actually, only two classes of such curves can be used now. One of them : $\{S_{p,q},\ p,q \geq 2\}$ (branch curves of ramified coverings $f_{p,q} : CP^1 \times CP^1 \to CP^2$, defined by three generic bihomogeneous polynomials of bidegree (p,q)), was studied in [8] (to have infinite kernels, we have to consider only $\{S_{p,p},\ p \geq 2\}$). However, using that class we do not obtain $\mathcal{E}_{m,v}$'s satisfying the inequality (3), and thus, do not get counter-examples to "Chisini's Theorem". The second class $\{S(p),\ p \geq 3\}$ (branch curves of "generic polynomial maps of type p" $g_p : CP^2 \to CP^2$, defined by three generic homogeneous polynomials of degree p) can be treated similarly to the first one ([7, 5]). When $p = 3$, it provides counter-examples to "Chisini's Theorem" mentioned above.

2. Construction of Non-Equivalent Factorizations

Consider a large closed disk D in $\mathbf{C}^1$ centered at (0) and the set $K = \{a_j = e^{2\pi i j/l},\ j = 0,1,\dots,l-1\}$, forming a convex polygon Π_l. Denote by z_{jk} the diagonal in Π_l connecting a_j with a_k, and by Z_{jk} the positive half-twist in $B_l = B[D, K]$ defined by z_{jk}. Let Δ_l^2 be the "full twist" in B_l. The factorization

$$\Delta_l^2 = \prod_{j=1}^{l-1} \prod_{k=0}^{j-1} Z_{jk}^2 \qquad (4)$$

describes the braid monodromy of the "union of l lines in general position": $\cup_{j=0}^{l-1} L_j$ in $\mathbf{C}P^2$ [9, 6].

Consider small disks

$$d_j = \{z \in \mathbf{C}^1|\ |z - a_j| \leq r\ \}, D_j = \{z \in \mathbf{C}^1|\ |z - a_j| \leq 2r\ \}\ (r \ll 1)$$

and finite sets

$$K_j = \{a_{j\alpha} = a_j + re^{2\pi i \alpha/m_j},\ \alpha = 0,1,\dots,m_j - 1,\ j = 0,1,\dots,l-1\}.$$

Denote by $K' = \cup K_j$, $m = \sum_{j=1}^{l} m_j$. Let $a'_j = a_j(1 - 2r)$, Π'_l be the convex polygon formed by a'_j 's, and z'_{jk} the diagonal in Π'_l connecting a'_j with a'_k. Let $S_j = \partial D_j$, β_0 be a simple path in D_0 connecting a_{00} with a'_0 below the

?al axis, and such that $\beta_0 \cap d_0 = a_{00}$. Let $\beta_j = a_j \cdot \beta_0$ (β_j is obtained from
$'_0$ by rotation of $360°/l$ in D).

Let $\beta_{jk} = \beta_j \cup z'_{jk} \cup \beta_k^{-1}$, and $Z_{jk;00}$ the positive half-twist in $B_m = B[D, K']$
efined by β_{jk}.

Denote by $\bar{\rho}_j$ a diffeomorpnism of D_j which is the positive rotation on
$60°/m_j$ in d_j and is damping to Id_{S_j} in $D_j - d_j$. Let ρ_j be the corresponding
lement in $B[D_j, K_j] \subset B_m(= B[D, K'])$, so that $\Delta^2(j) = \rho_j^{m_j}$, and set

$$Z_{jk;\alpha_1,\alpha_2} = \rho_j^{-\alpha_1} \rho_k^{-\alpha_2} Z_{jk;00} \rho_k^{\alpha_2} \rho_j^{\alpha_1}.$$

teplacing each L_j in $\cup L_j$ by m_j lines $L_{j\alpha}$, $\alpha = 0, 1, \ldots, m_j - 1$, which are
lose to L_j and meet at a "far away" point p_j on L_j , we get the following
eometric factorization of $\Delta_m^2 (\in B[D, K'])$, describing the braid monodromy
$r \cup_{j=0}^{l} \cup_{\alpha=0}^{m_j-1} L_{j\alpha}$:

$$\Delta_m^2 = \Delta^2(0) \cdot \prod_{j=1}^{l-1}(\prod_{k=0}^{j-1} \prod_{\alpha_1=m_j-1}^{0} \prod_{\alpha_2=m_k-1}^{0} Z_{jk;\alpha_1,\alpha_2}^2)\Delta^2(j) \qquad (5)$$

'aking in (5) $l = 2$, $m_1 = m_2 = \lambda$ and denoting $Y_{\alpha_1,\alpha_2} = Z_{01;\alpha_1,\alpha_2}$, we obtain

$$\Delta_{2\lambda}^2 = \Delta^2(0) \cdot \prod_{\alpha_1=\lambda-1}^{0} \prod_{\alpha_2=\lambda-1}^{0} Y_{\alpha_1,\alpha_2}^2 \cdot \Delta^2(1)). \qquad (6)$$

LEMMA 1. *1) The following formula is true in $B_{2\lambda}$ and gives a geometric
ictorization:*

$$\Delta_{2\lambda}^2 = (\Delta^2(0))^2 \cdot (\prod_{\alpha=\lambda-1}^{0} Y_{\alpha,\alpha})^2 \cdot (\Delta^2(1))^2. \qquad (7)$$

) The equality

$$\prod_{\alpha_1=\lambda-1}^{0} \prod_{\alpha_2=\lambda-1}^{0} Y_{\alpha_1,\alpha_2}^2 = \Delta^2(0) \cdot (\prod_{\alpha=\lambda-1}^{0} Y_{\alpha,\alpha})^2 \cdot \Delta^2(1) \qquad (8)$$

(implied by (7) and (6)), after replacing each of $\Delta^2(0), \Delta^2(1)$ *by their* basic *factorizations, each* $Y^2_{\alpha_1,\alpha_2}$ *by* $Y_{\alpha_1,\alpha_2} \cdot Y_{\alpha_1,\alpha_2}$ *and* $(\prod^0_{\alpha=\lambda-1} Y_{\alpha,\alpha})^2$ *by*

$$\prod^0_{\alpha=\lambda-1} Y_{\alpha,\alpha} \cdot \prod^0_{\alpha=\lambda-1} Y_{\alpha,\alpha},$$

gives two Hurwitz equivalent *factorizations.*

PROOF. (An outline; the details will appear in [5]).

One can prove first (8) algebraically (using the induction), so that the *Hurwitz equivalence* will become evident. Then one gets (7) from (8) and (6).

One can use also geometrical arguments. Then it is more difficult to check the *Hurwitz equivalence* for "regenerated" (8). Let us explain the geometrical proof for (7) ((8) follows from (6) and (7)).

Consider the plane curve $E = \cup^{\lambda-1}_{\alpha=0} C(\alpha)$, where each $C(\alpha)$ is a conic in $\mathbf{C}P^2$ defined by the equation

$$5X^2_0 + (1 + r_\alpha)X^2_1 - (1 + \frac{4}{9}r_\alpha)X^2_2 = 0,$$

where r_α $(\alpha \in (0,\ldots,\lambda-1))$ are λ different small real numbers. It is clear that all $C(\alpha)$'s are passing through four points

$$(1:2:3), \ (1:-2:3), \ (1:2:-3), \ (1:-2:-3) \ \in \mathbf{C}P^2,$$

and are close to the "hyperbola" $\{5X^2_0 + X^2_1 - X^2_2 = 0\}$.

Computing the braid monodromy for E, one gets the formula (7). Q.E.D.

COROLLARY of LEMMA1. *Consider* $B[D_0, K_0] \times B[D_1, K_1]$ *as the subgroup of* $B_{2\lambda} = B[D, K_0 \cup K_1]$ *(it corresponds to the embedding* $D_0 \cup D_1 \subset D$*). Then* $(\prod^0_{\alpha=\lambda-1} Y_{\alpha,\alpha})^2$ *commutes with any element of* $B[D_0, K_0] \times B[D_1, K_1]$.

PROOF. Use the formula $(\prod^0_{\alpha=\lambda-1} Y_{\alpha,\alpha})^2 = \Delta^2_{2\lambda} \cdot (\Delta^2(0))^{-2} \cdot (\Delta^2(1))^{-2}$, implied by (7).

Q.E.D.

From now on we assume that all $m_j = m_0$, so that $m = m_0 l$.

LEMMA 2. *There exists the following* geometric *factorization of* Δ_m^2 :

$$\Delta_m^2 = \prod_{i=l-1}^{0} (\Delta^2(i))^l \cdot \prod_{j=1}^{l-1}\prod_{k=0}^{j-1}(\prod_{\alpha=m_0-1}^{0} Z_{jk;\alpha,\alpha})^2. \tag{9}$$

PROOF. Consider (5) with all $m_j = m_0$ and then, using (8) for $B_{2m_0} = B[D, K_j \cup K_k]$, replace each $\prod_{\alpha_1=m_0-1}^{0}\prod_{\alpha_2=m_0-1}^{0} Z_{jk;\alpha_1,\alpha_2}^2$ by $\Delta^2(j) \cdot (\prod_{\alpha=m_0-1}^{0} Z_{jk;\alpha,\alpha})^2 \cdot \Delta^2(k)$. By Lemma 1, 2), this replacement can be done via Hurwitz moves on "complete regenerations". Now (5) gives:

$$\Delta_m^2 = \Delta^2(0) \cdot \prod_{j=1}^{l-1}\prod_{k=0}^{j-1}(\Delta^2(j)(\prod_{\alpha=m_0-1}^{0} Z_{jk;\alpha,\alpha})^2\Delta^2(k)). \tag{10}$$

Let $\mathcal{X}_{jk} = \prod_{\alpha=m_0-1}^{0} Z_{jk;\alpha,\alpha}$. It is evident that

$$[\Delta^2(k'), \mathcal{X}_{jk}] = 1 \quad \forall j' \neq j, k.$$

From Corollary of Lemma 1 it follows that $\mathcal{X}_{jk}^2$ commutes with $\Delta^2(j), \Delta^2(k)$. So each $\Delta^2(i)$ commutes with any $\mathcal{X}_{jk}$. Using that, we obtain (9) from (10) collecting together all $\Delta^2(k)$'s. We can perform the last operation via Hurwitz moves.

Thus, we obtain (9) from (5) via Hurwitz moves on "complete regenerations". Since (5) is geometric, we get the same for (9). Q.E.D.

$\forall b \in B[D, K']$ denote

$$Z_{jk;\alpha,\alpha}(b) = b^{-1}Z_{jk;\alpha,\alpha}b$$

PROPOSITION 1. *For any collection*

$$\{b_{jk} \in B[D_j, K_j] \times B[D_k, K_k] \subset B_m = B[D, K'], \ k < j, \}$$

the formula :

$$\Delta_m^2 = \prod_{i=l-1}^{0} (\Delta^2(i))^l \cdot \prod_{j=1}^{l-1}\prod_{k=0}^{j-1}(\prod_{\alpha=m_0-1}^{0} Z_{jk;\alpha,\alpha}(b_{jk}))^2 \tag{11}$$

is true and gives a geometric factorization of Δ_m^2.

PROOF. Since $(\prod_{\alpha=m_0-1}^{0} Z_{jk;\alpha,\alpha}(b_{jk}))^2 = b_{jk}^{-1}(\prod_{\alpha=m_0-1}^{0} Z_{jk;\alpha,\alpha})^2 b_{jk}$, we get by Cor. of Lemma 1

$$(\prod_{\alpha=m_0-1}^{0} Z_{jk;\alpha,\alpha}(b_{jk}))^2 = (\prod_{\alpha=m_0-1}^{0} Z_{jk;\alpha,\alpha})^2,$$

which shows that (11) is implied by (9).

To see that (11) is *geometric*, we use the following

LEMMA 3. *Let $\mathcal{A}$ be a subgroup of B_m, A, C_1, C_2, C_3 factorized expressions in B_m^+, such that the factors of A generate $\mathcal{A}$ and*

$$[C_i, y] = 1, \forall y \in \mathcal{A}, \ i = 1, 2, 3.$$

(We use the same notation for a factorized expression and for the product of its factors.) *Then $\forall b \in \mathcal{A}$ the factorized expressions $AC_1C_2C_3$ and $AC_1C_{2;b}C_3$ are Hurwitz equivalent.*

PROOF. Let $A = g_1 \ldots g_r$. We can write any $b \in \mathcal{A}$ in the form

$$b = g_{i_1}^{\epsilon_1} \ldots g_{i_k}^{\epsilon_k}, \ \epsilon_j = \pm 1$$

with the minimal $\sum_{j=1}^{k} |\epsilon_j|$.

Let $b' = b \cdot g_{i_k}^{-\epsilon_k}$. Using induction on $\sum_{j=1}^{k} |\epsilon_j|$, we can assume tha $AC_1C_2C_3$ is Hurwitz equivalent to $AC_1C_{2;b'}C_3$. Denote the factorized ex pression $C_{2;b'}$ by $C_{2'}$. Clearly, $C_{2'}$ and C_2 are equal as elements of B_m, an so $\forall y \in \mathcal{A}$ $[C_{2'}, y] = 1$. Since $C_{2;b} = C_{2';g_{i_k}^{\epsilon_k}}$ (as factorized expressions), w just have to prove that $AC_1C_{2'}C_3$ is Hurwitz equivalent to $AC_1C_{2';g_{i_k}^{\epsilon_k}}C_3$.

In other words, because of induction, it is sufficient to prove the Lemm for $b = g_j^\epsilon$, $j = 1, \ldots, r$, $\epsilon = \pm 1$.

Let us use the symbol $\sim$ for Hurwitz equivalence, and $\equiv$ for *identity* two factorized expressions.

We can write $A \equiv A_1 g_j A_2 \sim A_1 A_{2;g_j^{-1}} g_j$. Let $A' \equiv A_1 A_{2;g_j^{-1}}$. We nee the following

CLAIM. $\forall k \in \mathbf{Z}$, an element g and a factorized expression H with $[H, g] = 1$,

$$H \cdot g \ \sim \ H_{g^k} \cdot g \ \sim g \cdot H_{g^k}.$$

Proof of the Claim. Use induction (on k), and the following observation:

$$H_{g^{k-1}} \cdot g \sim g \cdot H_{g^k} \sim H_{g^k} \cdot g_{H_{g^k}} \equiv H_{g^k} \cdot g$$

We use $g_{H_{g^k}} = g$.

Q.E.D. for the Claim.

Now we have, using the Claim :

$AC_1C_2C_3 \equiv A'g_jC_1C_2C_3 \sim A'C_1g_jC_2C_3 \sim A'C_1g_jC_{2;g_j^\epsilon}C_3 \sim$
$A'g_jC_1C_{2;g_j^\epsilon}C_3 \equiv AC_1C_{2;g_j^\epsilon}C_3$.

Q.E.D. for Lemma 3.

Since (9) is a geometric factorization, in order to finish the proof of Proposition 1, it is sufficient to show that *complete regenerations* of (9) and (11) are Hurwitz equivalent.

In order to get rid of the b_{jk}'s in a complete regeneration of (11), we apply $l(l-1)/2$ times Lemma 3 with $\mathcal{A} = \prod_{i=l-1}^0 B[D_i, K_i] \subset B[D, K']$ and A equal to a comlete regeneration of $\prod_{i=l-1}^0 (\Delta^2(i))^l$, where each $\Delta^2(i)$ is replaced by its *basic* factorization. Each time we use almost the same arguments, so to explain them, it is sufficent to consider a simple case, when only one of b_{jk}'s, say $b_{j'k'}$, is not the identity element. In this case, denote by C_2 the factorized expression $\prod_{\alpha=m_0-1}^0 Z_{j'k';\alpha,\alpha} \cdot \prod_{\alpha=m_0-1}^0 Z_{j'k';\alpha,\alpha}$, by C_1 (resp. C_3) the factorized expression standing between A and C_2 (respectively following C_2) in the complete regeneration of (11). From Cor. of Lemma 1 it follows that each C_i commutes with any element of $\mathcal{A}$. From Lemma 3 it follows that $AC_1C_{2;b_{j'k'}}C_3$ is Hurwitz equivalent to $AC_1C_2C_3$.

Q.E.D.

Consider the braid group B_{m_0} with standard generators $X_1, X_2, \ldots, X_{m_0-1}$ and isomorphisms $s_j : B_{m_0} \to B[D_j, K_j]$ defined by $s_j(X_\alpha)$ = positive half-twist in $B[D_j, K_j]$ corresponding to the arc $(a_{j,\alpha-1}, a_{j,\alpha})$ of ∂d_j.

Let as above $\mathcal{X}_{jk} = \prod_{\alpha=m_0-1}^0 Z_{jk;\alpha,\alpha}$. Denote by

$$\mathcal{X}_{jk}(b_{jk}) = \prod_{\alpha=m_0-1}^0 Z_{jk;\alpha,\alpha}(b_{jk}).$$

COROLLARY OF PROPOSITION 1. *For any geometric cuspidal factorization* $\Delta_{m_0}^2 = \mathcal{E}(m_0)$ *in* $B_{m_0}^+$, *we have a geometric cuspidal factorization in* B_m^+ *given by the formula*

$$\Delta_m^2 = \prod_{i=l-1 \; l\text{times}}^{0} s_i(\mathcal{E}(m_0)) \cdot \prod_{j=1}^{l-1}\prod_{k=0}^{j-1}(\mathcal{X}_{jk}(b_{jk}) \cdot \mathcal{X}_{jk}(b_{jk})). \tag{12}$$

PROOF. Follows immediately from Proposition 1.
Q.E.D.

Choose all $b_{j0} = 1$ and each $b_{jk} \in B[D_j, K_j] \times B[D_k, K_k]$, $0 < k < j \le l-1$, so that its component in $B[D_j, K_j]$ is equal to identity. In other words, we can consider b_{jk} as an element of $B[D_k, K_k](\in B[D, K'])$. In particular, there exists $c_{jk} \in B_{m_0}$ such that $b_{jk} = s_k(c_{jk})$.

Let $\underline{c} = \{c_{jk}\}$. Denote the right-hand side of (12) by $\mathcal{E}'(m_0, l, \underline{c})$, or simply by $\mathcal{E}'$. We want to relate the groups $G(\mathcal{E}(m_0))$ and $G(\mathcal{E}') = G(\mathcal{E}'(m_0, l, \underline{c}))$. Choose a geometric base of $\pi_1(D - K', v)$, $v \in \partial D \cap \mathbf{R}^+$, as follows:

Connect v with a'_0 by a simple path δ_0 going below real axis close to it, and such that $\delta_0 \cap (\cup_{j=o}^{l-1}D_j) = a'_0$ and $\delta_0 \cap \Pi'_l = a'_0$. Let $\delta_j = \delta_0 \cap z'_{0j}$ (connects v with a'_j), $\beta_j(\alpha) = (\beta_j)\rho_j^\alpha$ (connects a'_j with $a_{j\alpha}$), and $\tilde{\gamma}_{j\alpha} = \delta_j \cup \beta_j(\alpha)$ (connects v with $a_{j\alpha}$). Consider a geometric base $\{\gamma_{j\alpha}\}$ of $\pi_1(D - K', v)$ defined by $\gamma_{j\alpha} = e(\tilde{\gamma}_{j\alpha})$.

It is evident that for any $k < j$, $(\gamma_{j\alpha})Z_{jk;\alpha\alpha} = \gamma_{k\alpha}$, and so

$$(\gamma_{j\alpha})b_{jk} \cdot Z_{jk;\alpha\alpha}(b_{jk}) = (\gamma_{k\alpha})b_{jk}. \tag{13}$$

By our choce of b_{jk}'s (see above) we have $(\gamma_{j\alpha})b_{jk} = \gamma_{j\alpha}$. Therefore, (13) gives

$$(\gamma_{j\alpha})Z_{jk;\alpha\alpha}(b_{jk}) = (\gamma_{k\alpha})b_{jk}. \tag{14}$$

By the definition of $G(\mathcal{E}')$ we have, using (14),

$$
\begin{aligned}
G(\mathcal{E}') = \quad & \{ \; \{\gamma_{j\alpha}\}| \; \gamma_{j\alpha} \sim (\gamma_{k\alpha})b_{jk} \\
& \forall \; k, j, \; 0 \le k < j \le l-1, \alpha \in (0, \ldots, m_0 - 1), \\
& \text{and the relations of } G(s_i(\mathcal{E}(m_0))), \; i \in (0, \ldots, l-1) \\
& \text{written in generators } \gamma_{j\alpha} \; \}.
\end{aligned}
$$

Denote

$$s'_k = s_0 s_k^{-1} : B[D_k, K_k] \to B[D_0, K_0].$$

Our choice of the isomorphisms s_k, $k = 0, \ldots, l - 1$ is such that for any $b \in B[D_k, K_k]$ and any $\alpha_1 \in (0, \ldots, m_0 - 1)$ we have:
the word $(\gamma_{k\alpha_1})b$, after replacing in it each $\gamma_{k\alpha}$ by $\gamma_{0\alpha}$, is equal to $(\gamma_{0\alpha_1})s_k'(b)$.

Using

$$\gamma_{j\alpha} \sim (\gamma_{0\alpha})b_{j0} = \gamma_{0\alpha}$$

(recall that all $b_{j0} = 1$), we obtain a smaller set of generators $\{\gamma_{0\alpha}\}$. We can replace now each relation $\gamma_{j\alpha} \sim (\gamma_{k\alpha})b_{jk}$ by

$$\gamma_{0\alpha} \sim (\gamma_{0\alpha})s_k'(b_{jk})(= (\gamma_{0\alpha})s_0(c_{jk})) \tag{15}$$

(recall that $b_{jk} = s_k(c_{jk})$, and so $s_k'(b_{jk}) = s_k'(s_k(c_{jk})) = s_0(c_{jk})$).
From (15) and the presentation of $G(\mathcal{E}')$ written above, we obtain now

$$
\begin{aligned}
G(\mathcal{E}') = \ & \{ \, \{\gamma_{0\alpha}\} \mid \gamma_{0\alpha} \sim (\gamma_{0\alpha})s_0(c_{jk}) \\
& \forall \, k, j, \; 0 \le k < j \le l - 1, \alpha \in (0, \ldots, m_0 - 1), \\
& \text{and the relations of } G(s_0(\mathcal{E}(m_0))), \\
& \text{written in generators } \gamma_{0\alpha} \, \}.
\end{aligned}
\tag{16}
$$

Consider B_{m_0} as $B[D, L]$, $\#L = m_0$, and choose a geometric base $\gamma_0, \gamma_1, \ldots,$ γ_{m_0-1} in $\pi_1(D - L, v)$ such that $\forall i \in (1, \ldots, m_0 - 1$

$$
\begin{aligned}
(\gamma_{i-1})X_i \ &= \gamma_i; \\
(\gamma_i)X_i \ &= \gamma_i \gamma_{i-1} \gamma_i^{-1}; \\
(\gamma_j)X_i \ &= \gamma_j \; \forall j \neq i, i - 1.
\end{aligned}
$$

Define the embedding $\tau : \pi_1(D - L, v) \hookrightarrow \pi_1(D - K', v)$ by $\tau(\gamma_\alpha) = \gamma_{0\alpha}$. It is clear that $\forall \gamma \in \pi_1(D - L, v)$, $b \in B[D, L]$ we have

$$\tau((\gamma)b) = (\tau(\gamma))s_0(b). \tag{17}$$

Let $\sigma : \pi_1(D - L, v) \to G(\mathcal{E}(m_0))$ be the canonical surjection, and $N(\underline{c})$ be the normal subgroup in $G(\mathcal{E}(m_0))$ generated by all conjugates of

$$\sigma((\gamma_\alpha)c_{jk} \cdot \gamma_\alpha^{-1}).$$

Call the elements of $G(\mathcal{E}(m_0))/N(\underline{c})$ *geometric generators* if they are images of geometric generators of $G(\mathcal{E}(m_0))$ (which by definition are images of geometric generators of $\pi_1(D - L, v)$.

PROPOSITION 2. *The group $G(\mathcal{E}') = G(\mathcal{E}'(m_0, l, \underline{c}))$ is isomorphic to the quotient $G(\mathcal{E}(m_0))/N(\underline{c})$. Moreover, under this isomorphism geometric generators of $G(\mathcal{E}'(m_0, l, \underline{b}))$ correspond to geometric generators of $G(\mathcal{E}(m_0))/N(\underline{c})$.*

PROOF. We get the isomorphism sending $\gamma_{0\alpha}$ to γ_α and using (16) and (17).

Q.E.D.

DEFINITION 3. *We say that a geometric cuspidal factorization $\Delta_{m_0}^2 = \mathcal{E}(m_0)$ in $B_{m_0}^+$ has Property (*) if the following is true:*

0) There exists a geometric solvable surjection $\Psi : G(\mathcal{E}(m_0)) \to S_n$ with $n \geq 5$, and a half-twist Z in B_{m_0} with an associated pair $(\gamma_{1'}, \gamma_1)$ in the corresponding $\pi_1(D - L, v)$ such that for the images $\Gamma_1, \Gamma_{1'}$ of $\gamma_1, \gamma_{1'}$ in $G(\mathcal{E}(m_0))$ and $A = \Gamma_{1'} \cdot \Gamma_1^{-1}$, we have:

(a) $\Psi(\gamma_1) = \Psi(\gamma_{1'})$, that is, $A \in \text{Ker } \Psi$;

(b) the images of $\Gamma_1 A \Gamma_1^{-1}$ and A^{-1} in $Ab(\text{Ker } \Psi)$ are equal.

1) Denote for a positive integer k by $N(k; Z)$ the subgroup of $Ab(\text{Ker } \Psi)$ generated by the S_n-orbit of the image of A^k in $Ab(\text{Ker } \Psi)$ (S_n acts naturally on $Ab(\text{Ker } \Psi)$). Then there exists an infinite sequence $\{k_i\}$ such that in the sequence $\{Ab(\text{Ker } \Psi)/N(k_i; Z)\}$ any two groups are not isomorphic to each other.

Assume now that $l = 3$, that is, $m = 3m_0$. Recall that in our construction all $b_{0j} = 1$, that is, all $c_{0j} = 1$. In the case $l = 3$, we could have only c_{12} non-trivial, that is, we can take $\underline{c} = c_{12}$.

PROPOSITION 3. *Let $\Delta_{m_0}^2 = \mathcal{E}(m_0)$ be a geometric cuspidal factorization in $B_{m_0}^+$ having Property (*) with the corresponding half-twist Z and the infinite sequence $\{k_i\}$. Denote by $\mathcal{E}_{m,k}$ the geometric cuspidal factorization $\mathcal{E}'(m_0, 3, Z^k)$ of Δ_m^2. (It is given by (12) with*

$$l = 3, \quad b_{01} = b_{02} = 1, \quad b_{12} = s_1^{-1}(Z^k) \;).$$

Then

1) $\forall k \in \mathbf{Z}$,

$\mu(\mathcal{E}_{m,k}) = 9\mu(\mathcal{E}(m_0)) + 6m_0;$

$\rho(\mathcal{E}_{m,k}) = 9\rho(\mathcal{E}(m_0)); d(\mathcal{E}_{m,k}) = 9d(\mathcal{E}(m_0)).$

(So these numbers are independent of k.)

2) In the sequence $\{\mathcal{E}_{m,k_i}\}$ any two factorizations $\mathcal{E}_{m,k_1}$, $\mathcal{E}_{m,k_2}$ with $k_1 \neq k_2$ re not Hurwitz and conjugation equivalent. Moreover, the corresponding roups $G(\mathcal{E}_{m,k_1}), G(\mathcal{E}_{m,k_2})$ are not isomorphic.

PROOF. 1) immediately follows from the formula (12).

Let N_k be the normal subgroup in $G(\mathcal{E}(m_0))$ generated by all conjugates f all

$$\sigma((\gamma_\alpha)Z^k \cdot \gamma_\alpha^{-1}), \tag{18}$$

vhere $\{\gamma_\alpha\}$ is a geometric base of $\pi_1(D - L, v)$.

By Proposition 2, $G(\mathcal{E}_{m,k})$ is isomorphic to $G(\mathcal{E}(m_0))/N_k$. We can extend $\gamma_{1'}, \gamma_1)$ to a geometric base

$$\gamma_{1'}, \gamma_1, \gamma_2, \ldots, \gamma_{m_0-1}$$

f $\pi_1(D - L, v)$ such that

$$(\gamma_j)Z = \gamma_j \quad \forall j \in (2, \ldots, m_0 - 1).$$

Jsing this base and (18), we see that N_k is generated by all the conjugates f

$$\sigma((\gamma_1)Z^k \cdot \gamma_1^{-1}), \sigma((\gamma_{1'})Z^k \cdot \gamma_{1'}^{-1}). \tag{19}$$

Denote $\Pi_1 = \gamma_{1'}\gamma_1$ and

$$\nu_k = \begin{cases} \sigma(\Pi_1^{k'}\gamma_1\Pi_1^{-k'}\gamma_1^{-1}) & \text{for} \quad k = 2k' \\ \sigma(\Pi_1^{k'+1}\gamma_1^{-1}\Pi_1^{-k'}\gamma_1^{-1}) & \text{for} \quad k = 2k' + 1. \end{cases}$$

Jsing

$$(\gamma_1)Z = \gamma_{1'}, \quad (\gamma_{1'})Z = \gamma_{1'}\gamma_1\gamma_{1'}^{-1}$$

nd (19), it is easy to check that N_k is generated by all the conjugates of ν_k. Since $A = \gamma_{1'}\gamma_1^{-1} = \Pi_1\gamma_1^{-2}$, we have the following formula for ν_k:

$$\nu_k = \begin{cases} \sigma((A\gamma_1^2)^{k'}(\gamma_1^2 \cdot \gamma_1^{-1}A\gamma_1)^{-k'}) & \text{for} \quad k = 2k' \\ \sigma((A\gamma_1^2)^{k'}A(\gamma_1^2 \cdot \gamma_1^{-1}A\gamma_1)^{-k'}) & \text{for} \quad k = 2k' + 1. \end{cases} \tag{20}$$

Denote by $\tilde{A}$ (resp. $\tilde{\Gamma}$, $\tilde{\nu}_k$, $\tilde{N}_k$) the image of A (resp. the images of γ_1, ν_k, $\sqrt{}_k$) in $G(\mathcal{E}(m_0))/(Ker\ \Psi)'\ (\supset Ab(Ker\ \Psi))$. Since $\tilde{\Gamma}\tilde{A}\tilde{\Gamma}^{-1} = \tilde{A}^{-1}$ (Def. 3), nd so $\tilde{A}$ commutes with $\tilde{\Gamma}^2$, we obtain from (20) that for any k

$$\tilde{\nu}_k = \tilde{A}^k. \tag{21}$$

It is clear that the image in $Ab(Ker\ \Psi)$ of the set of all conjugates of ν_k is the same as the S_n-orbit of $\tilde{\nu}_k = \tilde{A}^k$.

(21) implies that $\tilde{N}_k$ is generated by the S_n-orbit of $\tilde{A}^k$, that is, $\tilde{N}_k$ is the same as the subgroup $N(k, Z)$ of Def. 3.

Let

$$\Psi_k : G(\mathcal{E}(m_0))/N_k \to S_n$$

be the surjection canonically corresponding to Ψ. We see that

$$Ab(Ker\ \Psi_k)(= Ab(Ker\ \Psi)/\tilde{N}_k)) \simeq Ab(Ker\ \Psi)/N(k, Z). \qquad (22)$$

Now we can prove 2). Consider any two $\mathcal{E}_{m,k_1}$, $\mathcal{E}_{m,k_2}$ with $k_1 \neq k_2$. Assume that they are Hurwitz and conjugation equivalent. Then the corresponding groups $G(\mathcal{E}_{m,k_1}), G(\mathcal{E}_{m,k_2})$ are isomorphic.

Consider the homomorphisms Ψ_{k_i}, $i = 1, 2$. Since $Ker\ \Psi$ is solvable, we get that $Ker\ \Psi_{k_i}$, $i = 1, 2$, are solvable groups. Because $n \geq 5$, we have that the groups $Ker\ \Psi_{k_i}$, $i = 1, 2$, also are isomorphic, which implies isomorphism of the groups $Ab(Ker\ \Psi_{k_i})$, $i = 1, 2$. From (22) it follows now that the quotients $Ab(Ker\ \Psi)/N(k_i, Z)$, $i = 1, 2$, are isomorphic. This contradicts to Property (*).

Q.E.D.

3. Generic Polynomial Maps in Complex Dimention Two

The Proposition 3 shows that in order to prove Theorem 1 we need a cuspidal geometric factorization (of some $\Delta^2_{m_0}$) having Property (*). At the present moment we know only two classes of such factorizations (see Introduction). Because we want to get counter-examples to "Chisini's Theorem", we consider the second class which was studied in our recent work on the topology of generic polynomial maps in complex dimension two [7, 5]. We have to give a short description of it.

DEFINITION 4. *We say that*

$$g_p : \mathbf{C}P^2 \to \mathbf{C}P^2$$

is a generic polynomial map of type p if it is defined by three "generic" homogeneous polynomials of degree p.

Let $S(p) \subset \mathbf{C}P^2 (= Im(g_p))$ be the branch curve of such g_p. Denote by V_p the Veronese surface of order p, that is, the projective embedding of $\mathbf{C}P^2$ defined by the monomials of degree p. The map g_p is the same as a generic projection of V_p to $\mathbf{C}P^2$. The topology of g_p is essentially defined by the topological type of the complement $\mathbf{C}P^2 - S(p)$, in particular, by the homotopy type of it. The space $\mathbf{C}P^2 - S(p)$ is homotopically equivalent to a 2-dimensional complex, so the main ingredients of its homotopy type are $\pi_1(\mathbf{C}P^2 - S(p))$ and $\pi_2(\mathbf{C}P^2 - S(p))$ as a $\pi_1(\mathbf{C}P^2 - S(p))$-module.

Let $n = p^2 = \deg g_p$, S_n be the symmetric group of order n. Denote by $S'(p)$ the affine part of $S(p)$ corresponding to a generic choice of "a line at ∞" in $\mathbf{C}P^2$, and by

$$\Psi : \pi_1(\mathbf{C}^2 - S'(p)) \to S_n, \ \overline{\Psi} : \pi_1(\mathbf{C}P^2 - S(p)) \to S_n$$

the canonical monodromy surjections corresponding to g_p. Let $m_0 = \deg S(p) = 3p(p-1)$, and $\Delta^2_{m_0} = \mathcal{E}_p$ be a cuspidal factorization corresponding to the braid monodromy of $S'(p)$, so that

$$\pi_1(\mathbf{C}^2 - S'(p)) = G(\mathcal{E}_p), \pi_1(\mathbf{C}P^2 - S(p)) = \bar{G}(\mathcal{E}_p).$$

Ψ, $\bar{\Psi}$ can be considered as geometric homomorphisms

$$\Psi : G(\mathcal{E}_p) \to S_n, \ \bar{\Psi} : \bar{G}(\mathcal{E}_p) \to S_n$$

In [7, 5] we use the technique developed in [8] to compute explicitly the groups $G(\mathcal{E}_p)$, $\bar{G}(\mathcal{E}_p)$ in the case $p \geq 3$. The case $p = 1$ is trivial, while in the case $p = 2$ the groups $\pi_1(\mathbf{C}^2 - S'(p))$, $\pi_1(\mathbf{C}P^2 - S(2))$ are too big for explicit description.

In the case $p \geq 3$ it follows from our explicit description that the groups $\pi_1(\mathbf{C}P^2 - S(p))(= \bar{G}(\mathcal{E}_p))$ are "small" in the sense that they have solvable sugroups of finite index. For such "small" fundamental groups the computation of second homotopy groups as π_1−modules already makes sense. Thus, the interesting next problem in the topology of generic polynomial maps would be to describe explicitly the π_1−modules $\pi_2(\mathbf{C}P^2 - S(p))$.

Our main result in [7, 5] is the following theorem:

THEOREM 2. Let $\overline{\mathcal{H}}_p = Ker \, \overline{\Psi}$, $\mathcal{H}_p = Ker \, \Psi$. Let $m' = \frac{m_0}{2} (m_0$ is always even).

Then the structure of $\overline{\mathcal{H}}_p$ (resp. $\mathcal{H}_p$) can be described as follows:

1) $\overline{\mathcal{H}}_p'$ (the commutant of $\overline{\mathcal{H}}_p$) is trivial for p even and isomorphic to $\mathbf{Z}/2$ for p odd. The same is true for the commutant $\mathcal{H}_p'$.

2) $\overline{\mathcal{H}}_p' \subset center(\bar{G}(\mathcal{E}_p))$ (resp. $\mathcal{H}_p' \subset center(G(\mathcal{E}_p))$).

3) There is a subgroup $\overline{\mathcal{H}}_{p,0}$ in $\overline{\mathcal{H}}_p$ with $\overline{\mathcal{H}}_p' \subset \overline{\mathcal{H}}_{p,0} \subset \overline{\mathcal{H}}_p$ (resp. $\mathcal{H}_{p,0}$ in $\mathcal{H}_p$ with $\mathcal{H}_p' \subset \mathcal{H}_{p,0} \subset \mathcal{H}_p$) such that $\overline{\mathcal{H}}_p/\overline{\mathcal{H}}_{p,0} \simeq \mathbf{Z}/m'$, and

$$\overline{\mathcal{H}}_{p,0}/\overline{\mathcal{H}}_p' \simeq \begin{cases} (\mathbf{Z} \oplus \mathbf{Z}/3)^{n-1} & \text{if } p \equiv 0 \bmod 3 \; ; \\ \mathbf{Z}^{n-1} & \text{if } p \not\equiv 0 \bmod 3 \end{cases}$$

(resp. $\mathcal{H}_p/\mathcal{H}_{p,0} \simeq \mathbf{Z}$ and

$$\mathcal{H}_{p,0}/\mathcal{H}_p' \simeq \begin{cases} (\mathbf{Z} \oplus \mathbf{Z}/3)^{n-1} & \text{if } p \equiv 0 \bmod 3 \; ; \\ \mathbf{Z}^{n-1} & \text{if } p \not\equiv 0 \bmod 3 \end{cases}$$

Actually,

$$\mathcal{H}_{p,0} = Ker \{\mathcal{H}_p \rightarrow Ab(G(\mathcal{E}_p))(= H_1(\mathbf{C}^2 - S'(p)))\}$$

("degree zero" part of $\mathcal{H}_p$), and $\overline{\mathcal{H}}_p$ is the image of $\mathcal{H}_{p,0}$ in $\overline{\mathcal{H}}_p$.

In particular, we see that for $p \geq 3$ the geometric homomorphisms Ψ, $\bar{\Psi}$ are *solvable* (with $n \geq 9$). To apply Proposition 3 with $\mathcal{E}(m_0) = \mathcal{E}_p$, we have to establish that $\mathcal{E}_p \, \forall \, p \geq 3$ has Property (*). This fact follows from more explicit description of $G(\mathcal{E}_p)$ contained in Theorem 3 of [7] (it is analogous to Theorem 3 of [8]).

We need some terminology.

For any A, B we denote by $A_B = B^{-1}AB$. Consider the braid group B_n, $n \geq 4$, with standard generators $X_1, \ldots, X_{n-1}$. Denote by N_t the normal subgroup of it normally generated by

$$[X_1, X_{2;X_1X_3}]$$

and by $\tilde{B}_n$ the quotient B_n/N_t (see Section 1.1 of [8]). Let $\tilde{P}_n = Ker(\tilde{B}_n - S_n)$ and

$$\tilde{P}_{n,0} = Ker(\tilde{P}_n \rightarrow Ab(\tilde{B}_n)(= \mathbf{Z}))$$

("degree zero elements" in $\tilde{P}_n$).

Denote by $x_1, \ldots, x_{n-1}$ the images of $X_1, \ldots, X_{n-1}$ in $\tilde{B}_n$.

DEFINITION 5 (comp. Def. 8 and Rmk 8 from [8]). $G_0(n)$ is the $\tilde{B}_n$-group presented as follows:

Generators: $u_1, \ldots, u_{n-1}, \tau$.

Relations:

$$[u_i, u_j] = \begin{cases} 1 & \text{if } |i - j| \geq 2 \\ \tau & \text{if } |i - j| = 1, \end{cases}$$

$$\tau^2 = 1, \quad \tau_{u_i} = \tau \ \forall \ i = 1, \ldots, n - 1.$$

The $\tilde{B}_n$-action on $G_0(n)$ is given by the formulas:

$$(u_{k-1})x_k = u_k u_{k-1}; \quad (u_k)x_k = u_k^{-1}\tau; \quad (u_{k+1})x_k = u_k u_{k+1};$$

and $(u_j)x_k = u_j \ \forall \ j \neq k - 1, k, k + 1$.

It is clear that $G_0(n)' \simeq \mathbf{Z}/2$ and $Ab(G_0(n)) \simeq \mathbf{Z}^{n-1}$. The conjugation defines on $\tilde{P}_{n,0}$ and $\tilde{P}_n$ the structure of $\tilde{B}_n$-groups. We proved in ([8], Th.1) that $\tilde{P}_n/\tilde{P}_{n,0} \simeq \mathbf{Z}$ and $\tilde{P}_{n,0}$ is isomorphic as $\tilde{B}_n$-group to $G_0(n)$. Under this isomorphism τ corresponds to $c = [x_1, x_2]$ and u_1 to $\xi_1 = x_{1;x_2^{-1}}^2 \cdot x_2^2$. We denote by $\xi_1, \ldots, \xi_{n-1}$ the elements of $\tilde{P}_{n,0}$ corresponding to $u_1, \ldots, u_{n-1}$.

Let $\lambda(p) = p(p-1)/2$, and assume as above that $n = p^2$, $p \geq 3$. Using the $\tilde{B}_n$-action on $G_0(n)$, we define canonically the corresponding semi-direct product $G_0(n) \underline{\times} \tilde{B}_n$.

Denote by $\underline{N}(p)$ the normal subgroup of $G_0(n) \underline{\times} \tilde{B}_n$ normally generated by the elements

$$\bar{n}_1 = u_1^p \xi_1^{-2p+3} c^{\lambda(p)-1}, \quad \bar{n}_2 = c\tau, \quad \bar{n}_3 = \begin{cases} c & \text{if } p \text{ is even} \\ 1 & \text{if } p \text{ is odd} \end{cases}.$$

Define $\underline{G}_p = G_0(n) \underline{\times} \tilde{B}_n/\underline{N}(p)$.

Using the canonical surjection $\tilde{B}_n \to S_n$ and sending $G_0(n) \to \text{Id} \in S_n$, we define the surjection $\underline{\phi} : \underline{G}_p = G_0(n) \underline{\times} \tilde{B}_n \to S_n$. Because $\underline{\phi}(\underline{N}(p)) = \text{Id}$, $\underline{\phi}$ defines canonically the surjection $\Phi : \underline{G}_p \to S_n$.

Let $\underline{H}_p = Ker\Phi = G_0(n) \underline{\times} \tilde{P}_n/\underline{N}(p)$, $\underline{H}_{p,0}$ the subgroup of $\underline{H}_p$ generated by the images of $G_0(n)$ and $\tilde{P}_{n,0}$ ("degree zero elements" in $\underline{H}_p$).

Denote by U_j the images of u_j in $\underline{G}_p$.

THEOREM 3 ([7, Thm 3]). *There exists an isomorphism*

$$s : \underline{G}_p \to \pi_1(\mathbf{C}^2 - S'(p)) \ (= G(\mathcal{E}_p))$$

under which Φ *corresponds to* Ψ, *that is,* $s(\underline{H}_p) = \mathcal{H}_p$ *and* $s(\underline{H}_{p,0}) = \mathcal{H}_{p,0}$. *Moreover,* s *can be chosen so that* $s(U_1) = A_1$, *where* A_1 *can be written in the form* $\Gamma_{1'} \cdot \Gamma_1^{-1}$ *with* $\Gamma_{1'}, \Gamma_1$ *being the images of a* X_1 - *associated pair* $(\gamma_{1'}, \gamma_1)$ *from* $\pi_1(D - L, v)$, *and* $s(x_1) = \Gamma_1$.

Clearly, $A_1 \in Ker \ \Psi$. Denote by $\tilde{U}_1$ and $\tilde{x}_1$ the images of U_1 and x_1 in $\underline{G}_p/(Ker \ \Phi)'$ $(\supset Ab(Ker \ \Phi))$. From Def. 5 it follows that

$$U_{1;x_1^{-1}} = U_1^{-1}\tau, \quad \tau \in (Ker \ \Phi)',$$

and so $\tilde{U}_{1;\tilde{x}_1^{-1}} = \tilde{U}_1^{-1}$, which implies that the images of $\Gamma_1 A \Gamma_1^{-1}$ and A^{-1} in $Ab(Ker \ \Psi)$ are equal.

Denote for a positive integer k by $\underline{N}(k; X_1)$ the subgroup of $Ab(Ker \ \Phi)$ generated by the S_n-orbit of the image of U_1^k in $Ab(Ker \ \Phi)$ (S_n acts naturally on $Ab(Ker \ \Phi)$). Let $W(k, p) = Ab(Ker \ \Phi)/\underline{N}(k; X_1)$.

It is clear, that to prove that $\mathcal{E}_p$ has the Property (*) (of Def. 3), we only have to check now that any two groups $W(k_1, p), W(k_2, p)$ with $k_1 \neq k_2$ are not isomorphic to each other. This fact follows from the following elementary

STATEMENT.

$$W(k, p) \simeq \begin{cases} (\mathbf{Z}/((2p - 3)k))^{n-1} & \text{if } k \not\equiv 0 \ mod \ 3 \text{ or } p \not\equiv 0 \ mod \ 3 \\ (\mathbf{Z}/((2p_1 - 1)k) \oplus \mathbf{Z}/3)^{n-1} & \text{if } k \equiv 0 \ mod \ 3 \text{ and } p = 3p_1. \end{cases}$$

In particular,

$$\#(W(k, p)) = ((2p - 3)k)^{n-1}.$$

So we proved the following

LEMMA 4. *The factorization $\Delta^2_{m_0} = \mathcal{E}_p$ (corresponding to the braid mon-tromy of the cuspidal curve $S'(p)$, $p \geq 3$) has Property (*) with $\{k_i\} =$ all positive integers).*

PROOF of THEOREM 1 (see Introduction). Denote by $\tilde{\mathcal{E}}_{m,k}$ the geomet-c cuspidal factorization $\mathcal{E}'(m_0, 3, Z^k)$ of Δ^2_m ($m = 3m_0$), corresponding to $(m_0) = \mathcal{E}_p$ ($\tilde{\mathcal{E}}_{m,k}$ is given by (12) with

$$l = 3, \ \mathcal{E}(m_0) = \mathcal{E}_p, \ b_{01} = b_{02} = 1, \ b_{12} = s_1^{-1}(Z^k)).$$

:om Proposition 3 and Lemma 4 it follows now that any two factorizations $_{n,k_1}$, $\tilde{\mathcal{E}}_{m,k_2}$ with $k_1 \neq k_2$ are not Hurwitz and conjugation equivalent, and oreover, the corresponding groups $G(\tilde{\mathcal{E}}_{m,k_1}), G(\tilde{\mathcal{E}}_{m,k_2})$ are not isomorphic.

Similarly one can prove that the groups $\bar{G}(\tilde{\mathcal{E}}_{m,k_1}), \bar{G}(\tilde{\mathcal{E}}_{m,k_2})$ are also not omorphic.

Q.E.D.

OUNTER-EXAMPLES TO "CHISINI'S THEOREM". Denote by $m_0(p)$ the gree of $S(p)$, and let $\rho(p)$ (resp. $d(p)$) be the number of cusps (resp. of des) for $S(p)$ (that is, $\rho(p) = \rho(\mathcal{E}_p)$, $d(p) = d(\mathcal{E}_p)$).

One can check that

$$
\begin{aligned}
m_0(p) &= 3p(p-1) \\
\rho(p) &= 3(p-1)(4p-5) \\
d(p) &= 3(p-1)((p-1)(3p^2-14)+2)/2.
\end{aligned}
$$

t $m(p) = 3m_0(p)$. Using Proposition 3, 1), we get

$$
\begin{aligned}
\rho(\tilde{\mathcal{E}}_{m,k}) = 9\rho(p) &= 27(p-1)(4p-5) \\
d(\tilde{\mathcal{E}}_{m,k}) = 9d(p) &= 27(p-1)((p-1)(3p^2-14)+2)/2.
\end{aligned}
$$

d $m(p) = 9p(p-1)$.

Consider the case $p = 3$. We have $m(3) = 54$ and

$$\rho(\tilde{\mathcal{E}}_{54,k}) = 378, \quad d(\tilde{\mathcal{E}}_{54,k}) = 756,$$

the inequality (3) is satisfied ($1512 < 1536$).

From Theorem 1 it follows that infinitely many of $\tilde{\mathcal{E}}_{54,k}$ are not *analytic* (that is, do not correspond to the braid monodromies of cuspidal curves). They are counter-examples to "Chisini's Theorem".

Aknowledgments. The present work came as a result of very stimulating discussions with F. Catanese (started in 1987). I would like to express my gratitude to him.

References

[1] Chisini O., *Il teorema d'esistenza delle trecce algebriche, Nota I*, Rend. Acc. Lincei, ser.8, **XVII** (1954), 143-149.

[2] Chisini O., *Il teorema d'esistenza delle trecce algebriche, Nota II*, Rend. Acc. Lincei, ser.8, **XVII** (1954), 307-311.

[3] Chisini O., *Il teorema d'esistenza delle trecce algebriche, Nota III*, Rend. Acc. Lincei, ser.8, **XVIII** (1955), 8-13.

[4] Moishezon B., *Algebraic surfaces and the arithmetic of braids, I*, Progress in Math. **36** (1983), 199-269.

[5] Moishezon B., *Algebraic surfaces and the arithmetic of braids, III* (to appear).

[6] Moishezon B., Teicher M.,*Braid group technique in complex geometry, I* , Contemp. Math. **78** (1988), 425-555.

[7] Moishezon B., *Topology of generic polynomial maps in complex dimension two*, Preprint.

[8] Moishezon B., *On cuspidal branch curves*, J. Algebraic Geometry **2**, n. 2 (April 1993), 309-384.

[9] Moishezon B., *Stable branch curves and braid monodromies*, Lect. Notes in Math. **862** (1981), 107-192.

10] Van Kampen E.R., *On the fundamental group of an algebraic curve*, Am. J. Math. **55** (1933), 255-260.

11] Zariski O., *Algebraic Surfaces (Chapter VIII)*, Springer, Berlin-Heidelberg-New York, 1971.

DEPARTMENT OF MATHEMATICS, COLUMBIA UNIVERSITY, NEW YORK, √ Y , 10027.

Current address: Department of Mathematics, Columbia University, New ʹork, N Y, 10027.

E-mail address: bm@shire.math.columbia.edu

Contemporary Mathematics
Volume **164**, 1994

On tunnel number and connected sum

of knots and links

Kanji Morimoto

Dedicated to Mutsuko

Let L be a link (or knot) in the 3-sphere S^3, i.e. L is a disjoint union of finitely many tame simple closed curves in S^3. Let $t(L)$ be the tunnel number of L, here the tunnel number of L is the minimum number of mutually disjoint arcs properly embedded in the exterior of L in S^3 whose complementary space is a handlebody. Let $r(L)$ be the minimum number of elements of the fundamental group $\pi_1(S^3 - L)$ which generate the group. Then by definition, we have the inequality $r(L) \leq t(L) + 1$.

The relation between tunnel number and connected sum of knots and links has been studied by several people. Norwood and Jones proved in [N] and [J] that every two generator knot is prime. By the above inequality this implies that every tunnel number one knot is prime (cf. [S] and [MS]). On the other hand, Norwood showed in [N] that the link which is a connected sum of a trefoil knot and a Hopf link has tunnel number one. However, Jones proved in [J] that every composite two generator link has a Hopf link summand.

Motivated by the above, we can ask the questions : determine $t(K_1)$ and $t(K_2)$ when $t(K_1 \# K_2) = 2$, and which composite links have tunnel number one ? In this article we report the answers to the above questions.

1. Composite tunnel number two knots

1991 Mathematical Subject Classification : 57M25

The final version of this paper will be submitted for publication elsewhere

We say that a knot K in S^3 admits a (g, b)-decomposition for some integers $g(\geq 0)$ and $b(> 0)$ if there is a genus g Heegaard splitting (V_1, V_2) of S^3 such that $V_i \cap K$ is a b-string trivial arc system in V_i $(i = 1, 2)$. We note here that if a knot K admits a (g, b)-decomposition, then the tunnel number of the knot is at most $g + b - 1$. Our first result is :

Theorem 1. (Theorem of [M1])   *Let K_1 and K_2 be non-trivial knots in S^3, and suppose $t(K_1 \# K_2) = 2$. Then :*

(1)   *if neither K_1 nor K_2 are 2-bridge knots, then $t(K_1) = t(K_2) = 1$ and at least one of K_1 and K_2 admits a $(1, 1)$-decomposition, or*

(2)   *if one of K_1 and K_2, say K_1, is a 2-bridge knot, then $t(K_2)$ is at most two and K_2 is prime.*

Corollary 2.   *Every tunnel number two knot has at most two connected sum summands.*

Proof of Corollary 2.   Suppose $t(K_1 \# K_2 \# K_3) = 2$ for some non-trivial knots K_1, K_2 and K_3. Put $K_4 = K_2 \# K_3$. Then $t(K_1 \# K_4) = 2$. Then by Theorem 1, both K_1 and K_4 are prime because tunnel number one knots are prime. This contradicts that K_4 is not prime, and hence completes the proof of the corollary. ∎

An elementary observation yields the inequality $t(K_1 \# K_2) \leq t(K_1) + t(K_2) + 1$ for any two knots K_1 and K_2. As a consequence of Theorem 1, we have :

Corollary 3. *Let K_1 and K_2 be tunnel number one knots. Then $t(K_1 \# K_2) = 2$ if and only if at least one of K_1 and K_2 admits a $(1, 1)$-decomposition.*

Proof of Corollary 3.   Suppose $K_1 \# K_2$ has tunnel number two. Then by Theorem 1 and since 2-bridge knots admit $(1,1)$-decompositions, at least one of K_1 and K_2 admits a $(1,1)$-decomposition.

Conversely, suppose at least one of K_1 and K_2 admits a $(1,1)$-decomposition. Then by the definition of a $(1,1)$-decomposition, we can construct a genus three Heegaard splitting (V_1, V_2) of S^3 such that the knot $K_1 \# K_2$ is a core of a handle of V_1. Hence $K_1 \# K_2$ has tunnel number two. This completes the proof of the corollary. ∎

By the definitions of tunnel number and (1,1)-decomposition, we see that the 2-fold branched covering space of S^3 along a tunnel number one knot has a Heegaard splitting of genus three, and that if a knot admits a (1,1)-decomposition, then the 2-fold branched covering space of S^3 along the knot has a Heegaard spliiting of genus two. Hence to find a tunnel number one knot which admits no (1,1)-decomposition, it is sufficient to find a tunnel number one knot such that the 2-fold branched covering space of S^3 along the knot has no Heegaard splitting of genus two. Moriah and Rubinstein showed in [MR] that there are infinitely many such tunnel number one knots, and hence they showed in [MR] that there are infinitely many pairs of knots K_1 and K_2 such that $t(K_1) = t(K_2) = 1$ and $t(K_1 \# K_2) = 3$.

Now, by Theorem 1 we can ask whether the estimate of Theorem 1 is the best possible or not. The answer is yes. In fact, the next theorem shows that there are infinitely many tunnel number two knots K such that the tunnel number of $K \# K'$ is two again for any 2-bridge knot K'.

Theorem 4. (Theorem 3 of [M2])   *Let n be an integer (> 1) and K_n a knot illustrated in Figure 1. Then we have :*

   (1) : *$t(K_n) = 2$,*

   (2) : *$t(K_n \# K) = 2$ for any 2-bridge knot K, and*

   (3) : *K_n and $K_{n'}$ are mutually different types if n and n' are mutually different integers (> 1).*

Remark 5.   Let K be a tunnel number two knot. If $t(K \# K') = 2$ for some 2-bridge knot K', then $t(K \# K') = 2$ for any 2-bridge knot K'.

2. Composite tunnel number one links

For a tunnel number one link L, we call an arc properly embedded in the exterior of L in S^3 whose complementary space is a handlebody an unknotting tunnel for L. If a composite link has tunnel number one, then since tunnel number one knots are prime and 3-component links have tunnel number at least two, it is a connected sum of a knot and a 2-component link. The next theorem completely determines composite tunnel number one links and the unknotting tunnels.

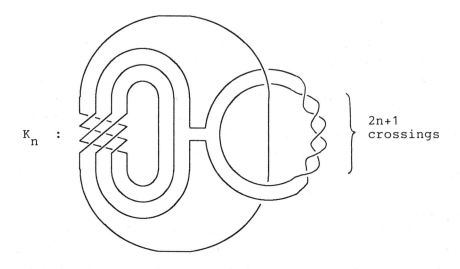

K_n :

$2n+1$ crossings

Figure 1

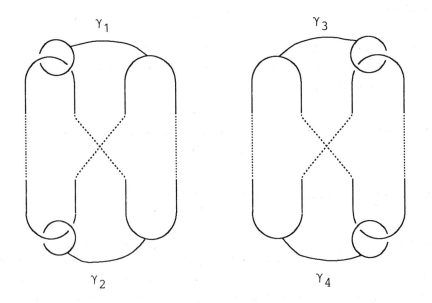

γ_1

γ_3

γ_2

γ_4

Figure 2

Theorem 6. (Theorem of [M3])   *Let K be a non-trivial knot in S^3 and L a 2-component link in S^3. Then $K \# L$ has tunnel number one if and only if K is a 2-bridge knot and L is a Hopf link. Moreover, any unknotting tunnel for $K \# L$ is isotopic in the exterior of $K \# L$ in S^3 to one of the four arcs γ_1, γ_2, γ_3 and γ_4 indicated in Figure 2, which are obtained from the upper or the lower tunnels for the 2-bridge knot K.*

References

[J]. Jones A, C., *Composite two-generator links have a Hopf link summand*, preprint.

[MR]. Moriah, Y. and Rubinstein, A talk in workshop on geometric topology at Haifa 1992.

[M1]. Morimoto, K., *On the additivity of tunnel number of knots*, to appear in Topology Appl..

[M2]. ─────────, *There are knots whose tunnel numbers go down under connected sum*, preprint.

[M3]. ─────────, *On composite tunnel number one links*, preprint.

[MS]. Morimoto, K. and Sakuma, M., *On unknotting tunnels for knots*, Math. Ann. **289** (1991), 143–167.

[No]. Norwood, F. H., *Every two generetor knot is prime*, Proc. A.M.S. **86** (1982), 143–147.

[S]. Scharlemann, M., *Tunnel number one knots satisfy the Poenaru conjecture*, Topology Appl. **18** (1984), 235–258.

Department of Mathematics

Takushoku University

Tatemachi, Hachioji

Tokyo 193, Japan

e-mail address : morimoto@la.takushoku-u.ac.jp

Contemporary Mathematics
Volume **164**, 1994

Essential Laminations in 3-Manifolds Obtained by Surgery on 2-Bridge Knots

Ramin Naimi

Loosely speaking, a lamination in a manifold is a foliation of a closed subset of that manifold. In [N1] we prove:

Theorem 1 *Surgery on a 2-bridge torus knot $T_{2,q}$, with coefficient $\in (-\infty, q-2)$ yields a manifold which admits essential laminations.*

Theorem 2 *Nontrivial surgery on a non-torus 2-bridge knot yields a manifold which admits essential laminations.*

This gives as a corollary, for example, a new and simpler proof that property P is true for 2-bridge knots ([T] gives a proof of this which is over 100 pages); by a theorem of Gabai and Oertel [GO] (mentioned below), existence of an essential lamination in M implies $\pi_1(M)$ is infinite . So nontrivial surgery on a non-torus 2-bridge knot never yields a homotopy sphere. And for torus knots, one gets Seifert manifolds, which are not homotopy spheres either; see [M].

In trying to understand 3-manifolds (with the hope of eventually classifying them, as with 2-manifolds), one approach that has turned out to be fruitful is to study objects of codimension one in them, more specifically, incompressible surfaces, Reebless foliations (i.e. foliations without Reeb components), and essential laminations.

For a 3-manifold M containing an incompressible surface (with some extra hypotheses), Waldhausen proved theorems such as: M has infinite fundamental group, the universal cover of M is $\mathbf{R}^3$, homotopic homeomorphisms of M are isotopic, and $\pi_1(M)$ determines M up to homeomorphism. Similar theorems for manifolds containing Reebless foliations were proven by Haefliger, Novikov, Palmeira, Rosenberg, and others.

The essential lamination was developed in the 1980's as a generalization of the incompressible surface and the Reebless foliation, which themselves qualify as essential laminations. In fact they are just extreme cases of essential laminations: at one end we have surfaces, which are compact and properly embedded,

The detailed version of this paper has been submitted for publication elsewhere.

1991 *Mathematics Subject Classification.* Primary 57M99; Secondary 57R30.

and at the other end we have foliations, which have empty complement and can have non-compact leaves. A "typical" lamination is in general somewhere in between; it has non-empty (and usually dense) complement as in the case of surfaces, but can have non-compact leaves as in foliations.

This seems to suggest that essential laminations might be more abundant than incompressible surfaces and Reebless foliations. And in fact, they are more abundant than incompressible surfaces: in contrast to Theorems 1 and 2, Hatcher and Thurston [HT] show that for each 2-bridge knot only finitely many surgeries yield manifolds which contain incompressible surfaces. So far, however, we do not know of any manifolds which admit essential laminations but not foliations. Essential laminations do however seem to be easier to find and work with, since they are carried by branched surfaces, which are finite objects.

The study of essential laminations in 3-manifolds can be divided into two sub-disciplines:

  1. *Which 3-manifolds admit essential laminations?*
  2. *What can we say about such manifolds?*

Towards answering the second question, analogues of some of Waldhausen's theorems have been proven ([GO]) for closed manifolds admitting essential laminations: they are irreducible, have infinite fundamental group, and are covered by $\mathbf{R}^3$. Other questions such as whether homotopic homeomorphisms are isotopic are being worked on (with partial results by Gabai and Kazez [GK]).

Theorems 1 and 2 above, together with [N2] completeley answer the first question for 2-bridge knots. Theorem 2 was also proven, using different constructions, by Delman [D1], and later generalized to most Montesinos knots [D2]. And Roberts [Rb], together with Delman [DR] answer the first question for alternating knots.

Hatcher [H] already constructed essential laminations for "most" 2-bridge knots; to get Theorem 2, we construct essential laminations for the rest of the 2-bridge knots. The main method of [N1] used to find new essential branched surfaces and laminations can be summarized as follows: we start with two different surgery descriptions of a manifold, where for one of the surgery descriptions we have an essential lamination, but not for the other. Then we "pull back" this lamination through some Kirby Calculus to the second surgery description, and try to generalize it to other knots. This is quite general and could potentially be fruitful for many other knots and links as well.

Eisenbud, Hirsch, Jankins, and Neumann [EHN], [JN1], [JN2], proved Theorem 1 using different methods. (Actually, they showed existence of foliations transverse to Seifert fibers, which can easily be shown to be essential.) The construction given here is used to get Theorem 2. Hatcher [H] constructed a branched surface for $T_{2,q}$ which carried laminations (with a projective transverse structure) for surgery coefficients $\in (-\infty, 0]$. The proof of Theorem 1 consists of showing that this branched surface in fact carries laminations (also with a projective transverse structure) for surgery coefficients $\in (-\infty, q-2)$. Brittenham

[B], Claus [C], and [JN2] show there exist no essential laminations for surgery coefficients $\in [q-1, +\infty)$. [JN2] conjectured non-existence for $[q-2, q-1)$, which was proven by the author in [N2].

[N1] is a rewriting of the author's Ph.D. thesis (Caltech, 1992).

I would like to thank my advisor, Dave Gabai, for all his time, encouragement, enthusiasm, and advise (which still continue).

References

[B]   M. Brittenham, Essential laminations in Seifert-fibered spaces, Topology 32(1993) 61-85.

[C]   W. Claus, Essential laminations in closed Seifert-fibered spaces, Ph.D. Thesis, University of Texas at Austin, (1991).

[D1]   C. Delman: Essential laminations in knot complements, Ph.D. Thesis, 1992, Cornell University.

[D2]   C. Delman: personal communication.

[DR]   C. Delman, R. Roberts: personal communication.

[EHN]   D. Eisenbud, U. Hirsch, and W. Neumann: Transverse foliations of Seifert Bundles and self homeomorphism of the circle, Comment. Math. Helv. 56 (1981), 638-660.

[GK]   D. Gabai, W. Kazez: In preparation.

[GO]   D. Gabai, U. Oertel: Essential laminations in 3-manifolds, Ann. of Math., 130 (1989), 41-73.

[H]   A. Hatcher: Some examples of essential laminations in 3-manifolds, Ann. Inst. Fourier, Grenoble, 42 (1992) 313-325.

[HT]   A. Hatcher, W. Thurston: Incompressible surfaces in 2-bridge knot complements, Inventiones Math. 79 (1985), 225-246.

[JN1]   M. Jankins, W. Neumann: Homomorphisms of Fuchsian groups to $PSL(2, \mathbf{R})$, Comment. Math. Helv., 60 (1985) 480-495.

[JN2]   —: Rotation numbers of products of circle homeomorphisms, Math. Annalen 271 (1985), 381-400.

[M]   L. Moser: Elementary surgery along a torus knot, Pacific Journal of Math., 38(1971), No. 3.

[N1]   R. Naimi: Constructing essential laminations in 2-bridge knot surgered
        3-manifolds, preprint.

[N2]   —: Foliations transverse to fibers of Seifert manifolds, Comment. Math.
        Helv., to appear.

[Rb]   R. Roberts: Constructing taut foliations, Ph.D. Thesis, 1992, Cornell
        University.

[T]   M. Takahasi: Mem. of AMS 29 (1981), No. 239.

Contemporary Mathematics
Volume **164**, 1994

A dynamical system approach
to free actions on ℝ-trees :
a survey with complements[*][†]

Frédéric Paulin

Abstract : This paper is a survey of the dynamical system approach
of D. Gaboriau, G. Levitt and the author to E. Rips' theorem about free
group actions on ℝ-trees, with lots of complements as : a simplified proof
of S. Adams' theorem on the number of ends of orbits of finite systems
of isometries of ℝ, a study of the quasi-isometry types of the orbits, a
canonical decomposition of any geometric ℝ-tree endowed with a free
action of a finitely generated group. In particular, we develop R. Skora's
notion of a *graph of* ℝ-*trees* analogous to the Bass-Serre notion of a graph
of groups.

The theory of one-dimensional dynamical systems studies the itera-
tions of one, or a group of, or a pseudogroup of selfmap(s) of a subset of
the line. We are interested in the isometric case with some finiteness assumptions. That is we are considering *systems of isometries* $X = (D, \varphi_j)$,
where D is a disjoint union of closed intervals, and $\{\varphi_j\}$ is a finite set of
partial isometries between closed intervals in D.

These systems arise for instance in transversely measured codimension
1 foliations on compact manifolds, by taking transversals and first return
maps (see for instance [Hae1, Hae2, Sa]] where intervals are open instead
of closed). As E. Rips, who named them Makanin Combinatorial Objects,
has showed it (see [GLP1]), they also arise (as $X(K)$) in the study of

---

[*]AMS codes : 58 F 03, 20 E 06, 20 E 08.

[†]This paper is in final form and no version of it will be submitted for publication
elsewhere

actions of finitely generated groups G on $\mathbb{R}$-trees T, by taking a finite
subtree K of T, taking the restrictions to K of generators of G, and
cutting K at branch points to turn it into intervals. Recall that an $\mathbb{R}$-tree
is an arcwise connected metric space in which every arc is isometric to an
interval of $\mathbb{R}$. See for instance [Sha1, Sha2, Mor1] for historical remarks,
references and motivations on $\mathbb{R}$-trees.

The systems X may be interpreted (see [GLP2]) as generating systems
of finitely generated *pseudogroups* $\mathcal{P}$ on D. That is, $\mathcal{P}$ is the smallest
set of partial isometries between closed subintervals of D, stable under
composition, inverse, restrictions and finite gluings, containing the φ_j's
and the identity. In this category, there is a notion of equivalence, which
basically says that changing a complete transversal of a measured foliation
yields equivalent pseudogroups.

As for every good dynamical object, there is a notion of *orbit* for the
systems X. Two points x, y in D are in the same orbit if there is a word
in the φ_j's and their inverses, defined at x with image y. Every orbit is
the set of vertices of a graph, called the *Cayley graph of the orbit*, where
there is an edge labelled j between x, y whenever $y = \varphi_j(x)$. As in the
Cayley group case, the quasi-isometry types of the Cayley graphs of the
orbits of a finitely generated closed pseudogroup do not depend on the
generating system (see [GLP2] for a proof of this non trivial fact). Even if
the Cayley graph of an orbit may be non isomorphic to the Cayley graph
of any group (that is the graph may be non homogeneous), it satisfies
(at least when non singular) a *quasi-homogeneous* behavior (see section 1
Proposition 1.4).

The orbits of X satisfy the following Hausdorff property (see [GLP2]),
a stronger version of the "segment closed" property introduced in [Rim1] :
there exists $\delta > 0$ such that if p, q are isometric embeddings $[0, \eta] \to D$
with $0 < \eta < \delta$ such that q_t and p_t are in the same orbit of X for all but
countably many t, then there is a word w in the φ_j's and their inverses
such that w is defined on p_t with image q_t for every $t \in [0, \delta]$.

The following result answers the question about the quasi-isometry
type of the orbits.

Proposition 0.1 *The orbits of X have polynomial growth. Except finitely
many of them, they are (uniformly) quasi-isometric to a tree or some $\mathbb{R}^p$.*

There is a dynamical decomposition of the systems X into dynamically
simple pieces (see [Ima, MS1, AL] for the case of some singular measured
foliations, [GLP1] for a direct proof in our context). The system X splits
up canonically into finitely many pieces. On each piece, either every orbit
is finite or every orbit is dense. Furthermore, there is a classification of the
pieces with dense orbits into the homogeneous case, the interval exchange

case, the exotic case (also called the Levitt case by E. Rips, since G. Levitt was the first one to exibit these systems of isometries), see [GLP1].

One may also ask for more precise asymptotic properties of the orbits, at least for the generic case (either for the Baire's category or Lebesgue's measure). D. Gaboriau [Gab] shows that, except finitely many of them, the orbits have no more than two ends, improving definitely S. Adams' result [Ada], whose proof, that almost every (for the Lebesgue measure) orbit has $0, 1$ or 2 ends, we give in Section 1. D. Gaboriau [Gab] also proves that, in the exotic case, the union of orbits with one end contains a dense G_δ and the orbits with *two* ends are uncountable.

There is a group $G(X)$ (see [GLP1]) and an ℝ-tree $T(X)$ with an action of $G(X)$ (see [GLP2]) canonically associated to X. If two systems generate equivalent pseudogroups, then their associated groups are isomorphic and their associated ℝ-trees are equivariantly isometric (see [GLP2]).

The group $G(X)$ and the tree $T(X)$ may be viewed geometrically as follows. There is a compact foliated 2-complex $\Sigma(X)$ canonically associated to X, obtained by gluing a foliated band on D for every φ_j between the domain and the image. The group $G(X)$ is obtained from $\pi_1\Sigma(X)$ by killing all loops contained in leaves. When D is connected, $G(X)$ has the following presentation : the generators are the φ_j's, and the relations are the words in the φ_j's and their inverses having a fixed point inside D (see Subsection 2.2 for the general case). Let $\tilde{\Sigma}(X) \to \Sigma(X)$ be the covering with transformation group $G(X)$. The tree $T(X)$ is the space of leaves made Hausdorff of the lifted foliation on $\tilde{\Sigma}(X)$.

If $X_1, \cdots, X_k$ are the minimal components of X, then the group $G(X)$ canonically splits as a free product $G(X_1) * \cdots * G(X_k) * F$ where F is a finitely generated free group (see [GLP1]). We show that a corresponding decomposition holds for the associated tree (see Section 2). Recall R. Skora [Sko2] notion's of a *graph of* ℝ-*trees* : a graph Γ is given, and for every vertex v (resp. edge e) is given an action of a group G_v (resp. G_e) on an ℝ-tree T_v, and an isometric embedding $T_e \to T_{v(e)}$ equivariant with respect to a given monomorphism $G_e \to G_{v(e)}$. There exists a *fundamental* ℝ-*tree* of a graph of ℝ-trees, satisfying an obvious universal property (see [Sko2] for a topological proof à la G.P. Scott-C.T.C. Wall [SW], and Theorem 2.12 for a proof more in J.-P. Serre's spirit [Ser]).

Proposition 0.2 *If $X_1, \cdots, X_k$ are the minimal components of X, then the ℝ-tree $T(X)$ canonically splits as the fundamental ℝ-tree of a graph of ℝ-trees with vertex ℝ-trees $T(X_1), \cdots, T(X_k)$ with the action of $G(X_1), \cdots, G(X_k)$ and (finitely many) finite segments with the trivial group action, and with edge ℝ-trees reduced to points.*

The main result of [GLP1] (see also [BF1]) is to give a proof of Rips' theorem, saying that a finitely generated group acting freely on an $\mathbb{R}$-tree is a free product of free abelian groups and surface groups.

The action of $G(X)$ on $T(X)$ is free (see [GLP2]) provided X satisfies the following "without reflection condition" : there is no x such that $x + t$ is in the orbit of $x - t$ for $t > 0$ small. Note that there are examples ([Lev4][GLP2]) with $G(X) \simeq \mathbb{Z}/2\mathbb{Z}$; this condition rules them out. To be without reflection implies that every leaf in $\tilde{\Sigma}(X)$ is closed, and the Hausdorff property mentioned above implies that the leaf space of $\tilde{\Sigma}(X)$ is in fact Hausdorff. This condition is satisfied if $X = X(K)$ is obtained from a free action on an $\mathbb{R}$-tree as above (see [GLP2]). In this case, the corresponding free actions on $T(X(K)))$ are approximations of the original action for K big enough (see [GLP2]).

Using these approximations and the result of [GL] for the free case, we get the following result (see Theorem 2.17), originally proved by R. Jiang [Jia].

Corollary 0.3 *(R. Jiang) The number of orbits of branch points of a free action of a finitely generated group on an $\mathbb{R}$-tree is uniformly bounded.*

I thank G. Levitt and D. Gaboriau for useful remarks.

1 Asymptotic properties of orbits

We recall first notations and definitions from [GLP1] that will be used subsequently without further notices.

A *multi-interval D* is a union of finitely many disjoint closed intervals of $\mathbb{R}$. Components of D may be *degenerate intervals*, i.e. consist of only one point.

Definition 1.1 *A system of isometries is a pair $X = (D, \{\varphi_j\}_{j=1,\cdots,k})$, where D is a multi-interval and each $\varphi_j : A_j \to B_j$ (called a generator) is an isometry between closed (possibly degenerate) subintervals of D.*

The intervals A_j, B_j are called *bases*. A generator $\varphi_j : A_j \to B_j$ is a *singleton* if A_j is degenerate.

An *X-word* is a word in the $\varphi_j^{\pm 1}$'s. It is a partial isometry of D, whose domain (defined in the obvious maximal way) is a closed interval (possibly degenerate or empty). If φ_j is not a singleton, define $\overset{\circ}{\varphi}_j : \overset{\circ}{A}_j \to \overset{\circ}{B}_j$ as the restriction of φ_j to the interior of A_j. An $\overset{\circ}{X}$-word is a word in the $\overset{\circ}{\varphi}_j^{\pm 1}$'s. Its domain is a (possibly empty) open interval.

Two points x, y in D *belong to the same X-orbit* (resp. $\overset{\circ}{X}$-orbit) if there exists an X-word (resp. $\overset{\circ}{X}$-word) sending one to the other. Note that the orbits are countable and that an orbit of $\overset{\circ}{X}$ is contained in an orbit of X with equality except perhaps for a finite number of them. An orbit of X or $\overset{\circ}{X}$ is *singular* if it consists of an endpoint of D, or if it meets some ∂A_j or ∂B_j, and is *regular* otherwise.

We can associate a sign $\pm$ (called its *derivative*) to every partial isometry of $\mathbb{R}$ whose domain has nonempty interior: it is precisely the value of the derivative of the associated global isometry of $\mathbb{R}$. By convention, singletons are neither positive nor negative. A *reflection* is a negative $\overset{\circ}{\varphi}$-word having a fixed point, the *center* of the reflection. The orbit of $x \in \overset{\circ}{D}$ by $\overset{\circ}{X}$ is *one-sided* if x is the center of a reflection, and *two-sided* otherwise.

Every orbit of X may be viewed as the set of vertices of a graph, called the *Cayley graph of the orbit* : there is an oriented edge labelled j (resp. $-j$) form x to y whenever $y = \varphi_j(x)$ (resp. $y = \varphi_j^{-1}(x)$, the inverse of an edge being the obvious one.

Recall that the geometric realization of a graph is endowed with the maximal distance such that the edges are isometric to $[0,1]$. A (λ, μ)-quasi-isometry between metric spaces E, E' is a map $f : E \to E'$ such that

$$\forall x, y \in E, \quad \frac{1}{\lambda} d(x,y) - \mu \le d(f(x), f(y)) \le \lambda d(x,y) + \mu$$

and there is some $C \ge 0$ such that $d(x', f(E)) \le C$ for all x' in E'.

The following result shows that the problem of the quasi-isometric type of orbits is intrinsic, that is does not depend on the generating set of our dynamical system.

Theorem 1.2 *([GLP2], Proposition 3.5) Let X, X' be systems of isometries on D having the same orbits. Then for every x in D, the Cayley graphs of the X-orbit and X'-orbit of x are quasi-isometric, with constants λ, μ independant of x.* $\square$

Let us recall the decomposition into minimal components of X, which is the basic decomposition step in [GLP1] to prove E. Rips' theorem, and which is also needed in order to understand the quasi-isometry types of orbits.

Theorem 1.3 *([Ima][AL],[GLP1] Theorem 3.1) Let X be a system of isometries on D, and X' the system obtained by removing all singletons. Let $E \subset D$ be the union of all finite singular $\overset{\circ}{X'}$-orbits. Then $D \setminus E$ is a disjoint union of open $\overset{\circ}{X'}$-invariant sets $U_1, \ldots, U_p$, where each U_i admits one of the following descriptions :*

- *family of finite orbits* : U_i consists of intervals of equal length meeting only finite $\overset{\circ}{X'}$-orbits ; the family may be *untwisted* (every $\overset{\circ}{X'}$-orbit contained in U_i is two-sided and meets each interval exactly once) or *twisted* (there is a one-sided orbit meeting each interval once, while all other orbits are two-sided and meet each interval twice).

- *minimal component* : every $\overset{\circ}{X'}$-orbit contained in U_i is dense in U_i.
  □

The U_i are not necessarily connected, and p is bounded in terms of the number of generators of X and components of D. Define a system of isometries X_i on the multi-interval D_i which is the closure of U_i, the generators of X_i being the restrictions of the non singleton generators of X to every closure of component of U_i. The orbits of X_i are either all finite or all infinite and dense in D_i (see [GLP1]). We say that X is *pure* if X equals some X_i of the second type.

An orbit of X is of one of the following exclusive types :

- a finite one (either singular or regular)

- an infinite singular one, containing the bases of a singleton of X whose bases are not in the same orbit of any X_i

- an infinite orbit of X_i.

In the second case, the Cayley graph of the orbit is obtained by taking finitely many orbits of X_i's (at least two), and joining them by finitely many edges (corresponding to singletons). By sending the orbits of X_i's to points, this graph maps onto a finite connected graph having at most m edges, where m is the number of singletons of X, hence at most $m + 1$ vertices.

Hence, if one wants to understand the quasi-isometry type of orbits, one has only to look at the pure cases. Indeed, finite orbits are (uniformly) quasi-isometric to points, since there is only a finite number of families of finite orbits and all orbits in a family (except at most one) have isometric Cayley graphs. An orbit of the second kind is quasi-isometric to the join of at most $m+1$ orbits of the pure components, where the constants of the quasi-isometry are bounded in terms of Λ where Λ is the maximum over all pairs x, y of endpoints of bases of the minimum length of an X_i-word sending x to y, if there is any (0 otherwise).

Proposition 1.4 *Except finitely many of them, the orbits are quasi-homogeneous.*

This means that for every two-sided regular orbit $\mathcal{O}_x$ and n in $\mathbb{N}$, there exists N in $\mathbb{N}$ such that for every y in $\mathcal{O}_x$, there exists y' in $\mathcal{O}_x$ with $d(y, y') \leq N$ such that $B(x, n)$ and $B(y', n)$ are isomorphic (in an edge-label preserving way), where $B(z, k)$ is the ball of radius k and center z in $\mathcal{O}_x$.

Proof. Note that (see for instance [GLP2] Proposition 3.7 or [Gus]), there are only finitely many one-sided orbits. Since x is regular and has a two-sided orbit, there is a closed segment I containing x in its interior such that

- every X-word of length less than $n + 1$ that is defined at x, is defined on I,

- every X-word of length less than $n + 1$ that is defined at some point of I, is defined at x,

- if an X-word of length less than $2n + 1$ has a fixed point inside I, then it has also x as fixed point.

Indeed, take $I = [x - \frac{\epsilon}{2}, x + \frac{\epsilon}{2}]$ where ϵ is the minimum of $d(x, wy)$ where w is an X-word of length $\leq 2n$, y is x or some base endpoint, and $wy \neq x$. According to a remark of E. Rips (see an easy corollary of Theorem 1.3 in [GLP1]), there is an N in $\mathbb{N}$ such that the union of images of I by X-words of lengths less than N contains the minimal component of X to which x belongs. Since the whole orbit of x is contained in this minimal component, the result follows. $\square$

The next step in [GLP1] in order to analyse the pure systems of isometries is the following. It will also be our second step in order to analyse the quasi-isometry types of orbits.

First a definition ([GLP1] section 4). A pure system of isometries X on D is called *homogeneous* if there exists a dense subgroup P in Isom($\mathbb{R}$) and a non empty open interval J in D such that $x, y \in J$ belong to the same $\overset{\circ}{X}$-orbit if and only if they belong to the same P-orbit. The group P (named the *group of periods*) is finitely generated, since it is a subgroup of the finitely generated subgroup of Isom($\mathbb{R}$) generated by the global isometries of $\mathbb{R}$ whose restrictions are the $\overset{\circ}{\varphi}_j$'s. Up to changing the embedding of components of D in $\mathbb{R}$, which will not change the isometry types of orbits (it changes only the labels of the vertices), two points in $\overset{\circ}{D}$ belong to the same $\overset{\circ}{X}$-orbit if and only if they belong to the same P-orbit.

Theorem 1.5 *Let X be pure and non homogeneous, then every orbit is quasi-isometric to a tree, with constants independant of the orbit.*

Proof. First recall the following deep result of D. Gaboriau [Gab], which is the best way of getting independant generators (for instance, it does not increase the number of generators), see [Lev1] théorème 5, and [Rim2] for previous results. The generators of X are called *independant* (see [GLP1] section 5) if no nontrivial reduced word in the $\overset{\circ}{\varphi_j}^{\pm 1}$ with non empty domain is the restriction of the identity.

Theorem 1.6 *(D. Gaboriau) If X is pure and non homogeneous, one may replace each generator φ_j by its restriction to a closed (possibly empty) subinterval of A_j, so as to get a system Y with independent generators and having the same orbits as X.* $\square$

Theorem 1.5 follows from Theorem 1.2, the fact that the orbits of Y and $\overset{\circ}{Y}$ only differs on singular ones, and the fact that if a $\overset{\circ}{Y}$-word is a reflection, its square is a restriction of the identity. Hence the Cayley graphs of the regular orbits of Y are trees since Y has independant generators. For the singular orbits, as D. Gaboriau pointed it out, there is a bounded subset of the Cayley graph whose complement is a union of trees. Indeed, any reduced cycle must contain an endpoint of a base. $\square$

Proposition 1.7 *If X is pure homogeneous, then every orbit of X is quasi-isometric to $\mathbb{Z}^{p-1}$ where p is the rank of the group of periods, hence has one end if $p \geq 3$ and two ends if $p = 2$.*

Proof. If P is purely translation, generated as a free abelian group by the translations of lengths $t_i, i = 1 \cdots p$, then every P-orbit in $\mathbb{R}$ is quasi-isometric to $\mathbb{Z}^p$. For every x in $\overset{D}{circ}$, there exists a system of generators $t_1, \cdots, t_p$ of P such that $x \pm t_1 \in D$. Up to changing the embedding of components of D in $\mathbb{R}$, we may assume that the orbit of x is isometric to $O_x = \{(n_1, \cdots, n_p) \,/\, n_1 t_1 + \cdots + n_p t_p \in D\} \subset \mathbb{Z}^d$ with the usual square metric of $\mathbb{Z}^d$. Since $\mathbb{R} = \bigcup_{n \in \mathbb{Z}} D + n t_1$ and since D is compact, the projection map $O_x \to \mathbb{Z}^{p-1}$ defined by $(n_1, n_2, \cdots, n_p) \mapsto (n_2, \cdots, n_p)$ is onto and preimages of points are finite, hence is a quasi-isometry.

The general statement follows from the fact that P has a finite index subgroup which consists of translations. $\square$

D. Gaboriau [Gab] has proved that only finitely many X-orbit may have more than two ends. The proof of the following weaker result is a simplification of arguments of S. Adams [Ada].

Theorem 1.8 *Almost every (for the Lebesgue measure on D) X-orbit has no more than two ends.*

Proof. According to Proposition 1.7, the proof of Theorem 1.5 and the analysis above it (finite orbits have no ends, and singular orbits are finitely many), we may suppose that X is pure non homogeneous with independant generators.

Let k be the number of generators of X, which is bounded in terms of the number of generators of the original system of isometries. Let T_k be the $2k$-homogeneous tree, with base vertex t_0, where every oriented edge with origin any given vertex is labelled (in a one-to-one way) by the $\pm j$'s. Let $\mathcal{O}_x$ be the orbit of $x \in D$ with base point at x. Since every regular orbit is a tree, there is a unique label-preserving embedding of any regular $\mathcal{O}_x$ into T_k sending x to t_0. If x, y are in the same orbit, there is an isomorphism $\alpha(x, y) : T_k \to T_k$, called shift of base point, sending $x \in \mathcal{O}_y$ to $x \in \mathcal{O}_x$ and $\mathcal{O}_y$ one-to-one onto $\mathcal{O}_x$ in a label preserving way, satisfying the following cocycle relations :

$$\alpha(x, x) = id, \quad \alpha(x, y)\alpha(y, z) = \alpha(x, z)$$

Let $B(x, n)$ be the ball of radius n and center x in $\mathcal{O}_x$. Recall that the *growth* of $\mathcal{O}_x$ is defined (in I. Babenko's term) as

$$\limsup_{n \to \infty} \frac{\log \operatorname{Card} B(x, n)}{\log n}$$

if this is finite (polynomial growth) and

$$\limsup_{n \to \infty} \frac{\log \log \operatorname{Card} B(x, n)}{\log n}$$

otherwise.

Lemma 1.9 *(well-known) The growth of $\mathcal{O}_x$ is polynomial, bounded by p the rank of the group of periods.*

Proof. The X-orbit of x is contained in the orbit of x by the subgroup of $\operatorname{Isom}(\mathbb{R})$ generated by the isometries whose restrictions are the generators of X. A finitely generated subgroup of $\mathbb{R}$ has a free abelian subgroup with finite index, hence has polynomial growth. $\square$

If T is a locally finite tree, let ∂T be the space of ends of T. It is a totally disconnected compact metrizable space, empty if and only if T is finite, and gives a compactification $T \cup \partial T$ of T. Note that $\partial \mathcal{O}_x$ is canonically identified with a subset of ∂T_k for every regular $\mathcal{O}_x$. If Y is topological space, let $\mathcal{M}(Y)$ denote the space of probability measures (positive, borelian, of total mass 1), endowed with the weak topology. Every isomorphism of a locally finite tree T extends to an homeomorphism of ∂T, hence acts on $\mathcal{M}(\partial T)$.

For every regular orbit $\mathcal{O}_x$, let $\mathcal{T}_x \subset \mathcal{O}_x$ be the convex hull of $\partial \mathcal{O}_x$, called the *trunk* of the orbit. It is the (maybe empty) subset of points belonging to a geodesic between two ends, obtained by removing recursively all terminal edges. Notice that $\partial T_x \subset \partial \mathcal{O}_x$ with equality unless T_x is empty and $\mathcal{O}_x$ has only one point. Let $\Omega \subset D$ be the measurable set of points belonging to the trunks of their orbits.

Lemma 1.10 *There is a measurable map $\mu : \Omega \to \mathcal{M}(\partial T_n)$ with $x \mapsto \mu_x$ such that*

- *the support of μ_x is contained in ∂T_x,*

- *if x, y are in the trunk of the same orbit, $\alpha(x, y)_* \mu_y = \mu_x$.*

Proof. For every $x \in \Omega$ and $y \in T_x$, let δ_y be the unit Dirac mass at y. Consider the sequence of probability measures on $T_n \cup \partial T_n$ defined by

$$\frac{1}{\operatorname{Card} B(x,n) \cap T_x} \sum_{y \in B(x,n) \cap T_x} \delta_y$$

Since $\mathcal{O}_x$ hence T_x has polynomial growth, this sequence converges (as n goes to ∞ outside a subset of $\mathbb{N}$ of density 0) to an element μ_x of $\mathcal{M}(\partial T_n)$, satisfying the conditions. $\square$

For every oriented edge e of T_x, let $\bar{e}$ be the edge e with the opposite direction and $\partial_e T_x$ be the subset of ends arriving at the endpoint of e (see Figure 1 (a)). In particular, $\{\partial_{\bar{e}} T_x, \partial_e T_x\}$ is a partition of ∂T_x. For every x in Ω, consider

$$k_x(e) = \mu_x(\partial_{\bar{e}} T_x) - \mu_x(\partial_e T_x).$$

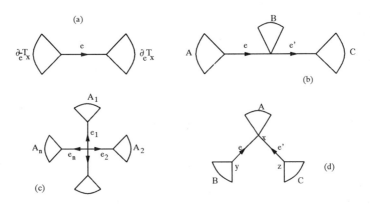

Figure 1 : Flow on the orbits.

Lemma 1.11 *The map k_x from the set of oriented edges of T_x to $\mathbb{R}$ satisfies the following properties :*

1. *$|k_x|$ is bounded by 1 ;*

2. *if e, e' are oriented edges with the extremity of e being the origin of e', then $k_x(e) \le k_x(e')$ with equality only if $\mu(\partial_{\bar{e}} \mathcal{T}_x) + \mu(\partial_{e'} \mathcal{T}_x) = 1$;*

3. *for every vertex t of $\mathcal{T}_x$, there is an oriented edge e with origin t such that $k_x(e) \ge 0$. If the valence of t is at least 3, then there is such an e with $k_x(e) > 0$.*

Proof. (2) With the notations of Figure 1 (b), one has $k_x(e) = \mu_x(A) - \mu_x(B) - \mu_x(C) \le \mu_x(A) + \mu_x(B) - \mu_x(C) = k_x(e')$.

(3) Otherwise, with the notations of Figure 1 (c), one would have $\mu_x(A_i) > \sum_{j \ne i} \mu_x(A_j)$. Hence a summation over i implies $1 > n - 1$, a contradiction since $n \ge 2$ (in a trunk, every valence is at least 2). The second assertion follows similarly. $\square$

For every x in Ω and y in $\mathcal{T}_x$, order the oriented edges with origin y so that $k_x(e_1) \ge \cdots \ge k_x(e_{n_y})$, using the order on the labels in the equality case, where n_y is the valence of y in $\mathcal{T}_x$. According to Lemma 1.11 (3), one has $k_x(e_1) \ge 0$ with strict inequality if $n_y \ge 3$. Then define $f_x : \mathcal{T}_x \to \mathcal{T}_x$ by $f_x(y)$ being the endpoint of e_1. Recall that if an oriented edge e has origin α and endpoint ω, we set $e = \overline{\alpha, \omega}$. Define

$$
\begin{aligned}
s(x) &= k_x(\, \overline{t_0, f_x(t_0)}\,) \\
r(x) &= \begin{cases} 0 & \text{if } t_0 \notin f_x(\mathcal{T}_x) \\ k_x(\, \overline{u_0, f_x(u_0)}\,) & \text{if } f_x(u_0) = t_0 \end{cases}
\end{aligned}
$$

Lemma 1.12   *1. s, r are measurable maps from Ω to $[0, 1]$,*

2. *$r \le s$*

3.
$$
\int_\Omega r = \int_\Omega s
$$

4. *Let Ω' be the subset of $x \in \Omega$ such the valence of x in $\mathcal{T}_x$ is at least three. Then $r(x) < s(x)$ for $x \in \Omega'$.*

Proof. (1) The only non trivial fact is that r is well defined. If $f_x(y) = f_x(z) = x$, then with the notations of Figure 1 (d), one has $k_x(e) = \mu_x(B) - \mu_x(A) - \mu_x(C) \ge 0$ and $k_x(e') = \mu_x(C) - \mu_x(A) - \mu_x(B) \ge 0$. Hence $\mu_x(A) = 0$ and $k_x(e) = k_x(e')$.

(2) follows by Lemma 1.11 (2).

(3) Let $g_i = \varphi_i$ and $g_{i+k} = \varphi_i^{-1}$ for $i = 1 \cdots k$. Let $F : \Omega \to \Omega$ given by $F(x) = f_x(t_0)$. Define $\Omega_i = \{x \in \Omega \, / \, F(x) \ne g_1(t_0), \cdots, F(x) \ne g_{i-1}(t_0), F(x) = g_i(t_0)\}$. Since $F(x)$ is adjacent to t_0, $\{\Omega_i\}_{1 \le i \le 2k}$ is a partition of Ω into measurable subsets. Furthermore, the $g_i^{-1}(\Omega_i)$'s are pairwise disjoints, $r(x) = s(g_i(x))$ for x in $g_i^{-1}(\Omega_i)$ according to Lemma

1.10 (2), and $r(x) = 0$ on $\Omega - \bigcup g_i^{-1}(\Omega_i)$. Let us check for instance the second assertion. Since $x \in g_i^{-1}(\Omega_i)$, one has $f_{g_i x}(t_0) = g_i(t_0)$ hence $f_x(u_0) = t_0$) with $u_0 = g_i^{-1}(t_0)$. Furthermore,

$$s(g_i(x)) = \mu_{g_i(x)}(\partial \overline{t_0, f_{g_i(x)}(t_0)} \mathcal{T}_{g_i(x)}) - \mu_{g_i(x)}(\partial \overline{t_0, f_{g_i(x)}(t_0)} \mathcal{T}_{g_i(x)})$$

$$= \mu_x(g_i(\partial \overline{g_i(u_0, f_x(u_0))} g_i^{-1}(\mathcal{T}_x)) - \mu_x(g_i(\partial_{g_i(\overline{u_0, f_x(u_0)})} g_i^{-1}(\mathcal{T}_x)) = r(x) .$$

The Lebesgue measure is invariant by isometries, hence

$$\int_\Omega r = \int_{\bigcup g_i^{-1}(\Omega_i)} r(x)\, dx = \sum_i \int_{g_i^{-1}(\Omega_i)} s(g_i(x))\, dx = \int_\Omega s .$$

(4) follows from the equality case in Lemma 1.11 (2) and the fact that if an oriented edge e with origin x satisfies $\mu(\partial_e \mathcal{T}_x) = 0$ (and is the first one for the order on the edges), then $f_x(x)$ is the endpoint of e. □

Assume by contradiction that the subset S of $x \in D$ whose orbits have at least three ends has non zero Lebesgue measure. We claim that Ω' has non zero Lebesgue measure. Recall that a countable union of measure zero sets has measure zero, and that the isometries of $\mathbb{R}$ preserve the Lebesgue measure. Since S is the union of some singular orbits and of the $w(\Omega')$'s for all $\overset{\circ}{X}$-words w, the claim follows. But then Lemma 1.12 (3) gives a contradiction. □

2   The group and $\mathbb{R}$-tree associated to a finitely generated closed pseudogroup

2.1   The group $G(X)$

We recall here notations and definitions from [GLP1] and [GLP2] that will be used subsequently without further notices.

Let X be a system of isometries on a multi-interval D. We define (see [GLP1], section 1) a *foliated 2-complex* $(\Sigma(X), \mathcal{F})$ (or simply Σ) associated to X. Start with the disjoint union of D (foliated by points) and strips $A_j \times [0,1]$ (foliated by $\{*\} \times [0,1]$). We get Σ by glueing the $A_j \times [0,1]$ to D, identifying each $(t,0) \in A_j \times \{0\}$ with $t \in A_j \subset D$ and each $(t,1) \in A_j \times \{1\}$ with $\varphi_j(t) \in B_j \subset D$. We will identify D with its image in Σ.

The foliation $\mathcal{F}$ is the decomposition of Σ into the leaves. A leaf is an equivalence class for the equivalence relation $\sim$ generated by $x \sim y$ if there is a $j = 1, \cdots, k$ with x, y corresponding to two points in the same leaf $\{*\} \times [0,1]$ of $A_j \times [0,1]$. The leaves are simplicial 0 or 1-complexes, where the vertices are the intersections of the leaves with D, and where the edges correspond to the leaves $\{*\} \times [0,1]$ of the $A_j \times [0,1]$. Two points of D are in the same leaf of $\mathcal{F}$ if and only if they are in the same X-orbit. For instance, if a point in D belongs to no base, then its leaf consists of itself.

This suspension process is well known for interval exchanges. See J. Morgan's notes [Mor2] for the first appearance under the above generality, and [AL][Lev3] for suspensions as measured foliations with Morse singularities on manifolds.

In what follows, we assume that X *is connected*, i.e. Σ is connected. The 2-complex Σ has the homotopy type of a finite graph, so that its fundamental group (for any choice of base point) $\pi_1(\Sigma)$ is a finitely generated free group. We will denote by $\overline{\mathcal{L}}$ the normal subgroup of $\pi_1(\Sigma)$ normally generated by the free homotopy classes of loops contained in leaves of $\mathcal{F}$. We will denote by $\mathcal{L}$ the normal subgroup of $\pi_1(\Sigma)$ normally generated by the free homotopy classes of loops contained in leaves of $\mathcal{F}$ and disjoint from each $\partial A_j \times (0,1)$.

Definition 2.1 *If X is a connected system of isometries, we define*

$$G(X) = \pi_1(\Sigma)/\overline{\mathcal{L}}.$$

In the case where D is connected, there is an easy presentation of the group $G(X)$ associated to a system of isometries X. The generators are the elements φ_j, and the relations are X-words having a fixed point. See [Lev1] for instance.

We define an action of a group on a metric space to be a left isometric action. A *branch point* in an ℝ-tree T is a point x such that $T \setminus \{x\}$ has at least 3 components. A *finite* ℝ-*tree* is a compact ℝ-tree which is the convex hull of finitely many points, hence has only finitely many branch points.

Let G be a finitely generated group acting on an ℝ-tree T, and let us fix a system of generators $\{\gamma_1, \cdots, \gamma_p\}$ for G, stable by taking inverses. Let K be a finite subtree of T. We define a system of isometries $X = X(K)$ as follows (see [GLP1], section 2).

Let $I_1, \cdots, I_n$ be the segments of K that are the closures of the connected components of K minus its branch points. Consider them as embedded disjointly in ℝ, and let D be their union. The system X has two types of generators.

For every branch point b of K, let $x_{i_1}, \cdots, x_{i_s}$ be the endpoints of the segments $I_{i_1}, \cdots, I_{i_s}$ corresponding to b. Consider the following finite set of singletons on D: $\Phi_0 = \{x_{i_1}^b \mapsto x_{i_2}^b, \cdots, x_{i_1}^b \mapsto x_{i_s}^b \ / \ b$ branch point of $K\}$. We could have taken all the possible pairs, but those one suffice.

Now the elements $\gamma_1, \cdots, \gamma_p$ of G define partial isometries of K (defined on a maybe empty closed finite subtree of K) $g_i : K \cap \gamma_i^{-1}(K) \to \gamma_i(K) \cap$

K with $g_i(t) = \gamma_i(t)$. The partial isometries g_i's of K induce partial isometries of D in the natural way. That is, for every $1 \leq i,j \leq n$ and $1 \leq k \leq p$, g_k defines, by maximal restriction, an isometry φ_{ijk} between a closed interval of I_i onto a closed interval of I_j. Set

$$\Phi = \Phi_0 \cup \{\varphi_{ijk}\}_{1 \leq i,j \leq n,\, 1 \leq k \leq p}$$

and $X = (D, \Phi)$.

2.2 Rips' combinatorial presentation of $G(X)$

We suppose that X is stable by taking inverses and does not contain a restriction of the identity. A *full* (resp. *half*) *reflection* is a negative φ_i with $A_i = B_i$ (resp. $\overset{\circ}{A}_i \cap \overset{\circ}{B}_i = \emptyset$, $A_i \cap B_i \neq \emptyset$). Let $\mathcal{G}_X$ (or simply $\mathcal{G}$) be the partially oriented labelled graph having as vertices the components of D and as edges the generators φ_i's of X, from the component I_1 of D towards the component I_2 if $A_i \subset I_1$ and $B_i \subset I_2$, the inverse of the edge φ_i being φ_i^{-1}. If X contains a full reflection (equal to its inverse), then the corresponding edge is non oriented.

Assume $\mathcal{G}$ is connected (this is clearly equivalent to assuming that X is connected) and fix I_0 a base vertex. Let $\pi_1(\mathcal{G}, I_0)$ be the fundamental group of the geometric realization of $\mathcal{G}$ based at I_0 (as usual, we identify the geometric realization of an edge and of its inverse). Let F_X (or simply F) be the quotient of $\pi_1(\mathcal{G}, I_0)$ by the normal subgroup normally generated by the square of the loops corresponding to full reflections. Hence F is a free product of a finitely generated free group and finitely many copies of $\mathbb{Z}/2\mathbb{Z}$. And F is free if X has no full reflection. Any element of F is represented by a unique allowable reduced word $\overline{\varphi_{i_1}} \cdots \overline{\varphi_{i_n}}$, where $(\varphi_{i_1}, \cdots, \varphi_{i_n})$ is an oriented path of edges in $\mathcal{G}$, allowable meaning that the path starts and ends at I_0, reduced meaning that $\varphi_{i_p} \neq (\varphi_{i_{p+1}})^{-1}$ for $p = 1, \cdots, n-1$. The identity is represented by the empty word, the inverse of $\overline{\varphi_{i_1}} \cdots \overline{\varphi_{i_n}}$ is $\varphi_{i_n}^{-1} \cdots \varphi_{i_1}^{-1}$ and the product in F corresponds to the concatenation with reduction of the words.

Define N_X (or simply N) to be the normal subgroup of F normally generated by the allowable reduced words $r = w_r \overline{\varphi_{i_1}} \cdots \overline{\varphi_{i_n}} w_r^{-1}$ (called *relations*) such that $n \geq 1$, w_r is a word associated to a path starting at I_0, and there exists $x \in D$ with x belonging to the domain of $\varphi_{i_p} \circ \cdots \circ \varphi_{i_n}$ for $p = 1, \cdots, n$ and

$$\varphi_{i_1} \circ \cdots \circ \varphi_{i_n}(x) = x.$$

The word $\overline{\varphi_{i_1}} \cdots \overline{\varphi_{i_n}}$ will be called the *core* of r. Note that if φ_i has a fixed point, in particular if it is a full or half reflection, then $\overline{\varphi_i}$ is the core of a relation. Note that any cyclic conjugate of the core of a relation is itself the core of a relation. The inverse of a relation is also a relation. A relation is said to be minimal if there is no proper subword of the core,

$\overline{\varphi_{i_p}} \cdots \overline{\varphi_{i_q}}$ with $1 \leq p < q \leq n$, such that if $y = \varphi_{i_{q+1}} \circ \cdots \circ \varphi_{i_n}(x)$ ($y = x$ if $q = n$) then $\varphi_{i_p} \circ \cdots \circ \varphi_{i_q}(y) = y$. Clearly N is normally generated by the minimal reduced relations.

The following fact is obvious.

Proposition 2.2 $G(X)$ *is isomorphic to* F/N. □

There is another presentation of the group $G(X)$, slightly less canonical (see also [Rim2]). It allows us to repair the fact that a multi-interval may be disconnected, for the price of getting a finite ℝ-tree.

For simplicity, we suppose that X has no full nor half reflection. Consider a maximal tree $T_{\max}$ in $\mathcal{G}$ containing the base point, and $T^1_{\max}$ the set of its edges. Then F is isomorphic to the free group, freely generated by the elements of X that are neither edges of $T_{\max}$ nor their inverses. (Here we use the convention that $\{a, a^{-1}, b, b^{-1}\}$ freely generates a free group of rank 2.) The isomorphism is given by the "forgetful" map which associates, to an allowable reduced word, the word where we have deleted the elements of $T^1_{\max}$ and their inverses. Note that N is normally generated by the relations r with w_r a word in $T^1_{\max}$ and their inverses.

Consider the quotient $K_{T_{\max}}$ (or simply K) of D by the equivalence relation generated by the relation $x \in D \sim y \in D$ if and only if there exists $\varphi \in T_{\max}$ with x in the domain of φ and $\varphi(x) = y$. Since $T^1_{\max}$ is finite, since the identifications are induced by isometries with connected domain, and since $T_{\max}$ is a tree, it is obvious that K is a finite ℝ-tree. (See [GL] for generalizations of this gluing process).

The elements of $X' = X - (T^1_{\max} \cup (T^1_{\max})^{-1})$ define partial isometries of K (defined on closed segments of K), that will be denoted by the same symbols. The group $G(X)$ is isomorphic to the group with the following presentation. The set of generators is X' (we overline them to distinguish them from the corresponding partial isometry of K). The relations are the words $\overline{\varphi_{i_1}} \cdots \overline{\varphi_{i_n}}$ such that $\varphi_{i_p} \in X'$ and there exists $x \in K$, with x belonging to the domain of the partial isometry $\varphi_{i_p} \circ \cdots \circ \varphi_{i_n}$ of K for $p = 1, \cdots, n$, such that $\varphi_{i_1} \circ \cdots \circ \varphi_{i_n}(x) = x$. With a meaning similar as the one above, we will consider only reduced and minimal relations.

2.3   The ℝ-tree $T(X)$

We recall here notations and definitions from [GLP2] that will be used subsequently without further notices.

We define a metric space $T(X)$ as follows (see for instance [Lev3]). Let $\tilde{\Sigma}(X)$ be the covering of Σ defined by $\overline{\mathcal{L}}$, and $\tilde{\mathcal{F}}$ be the measured foliation lifting $\mathcal{F}$. Any path γ in $\tilde{\Sigma}(X)$ piecewise transverse to $\tilde{\mathcal{F}}$ has a *length*, defined as the total mass of the measure induced on γ by the transverse measure of $\tilde{\mathcal{F}}$. Given x, y in $\tilde{\Sigma}(X)$, let $d_{\tilde{\mathcal{F}}}(x, y)$ be the lower bound of the lengths of all piecewise transverse paths from x to y. This defines a pseudodistance on $\tilde{\Sigma}(X)$ (two points x, y in the same leaf have $d_{\tilde{\mathcal{F}}}(x, y) = 0$, but the converse may be false).

The metric space $T(X)$ is obtained by identifying two points in $\tilde{\Sigma}(X)$ at pseudodistance 0 from each other. Since $\tilde{\mathcal{F}}$ is invariant by $G(X)$, there is a natural (isometric) action of $G(X)$ on $T(X)$. The natural projection $\pi : \tilde{\Sigma}(X) \to T(X)$ is equivariant and continuous.

Using ideas of [GS] Theorem 5.20, [Pau3] Proposition 4.6, G. Levitt has proved the following result.

Proposition 2.3 *([Lev3] Corollary III.5) For any system of isometries X, $T(X)$ is an $\mathbb{R}$-tree.* $\square$

In the case X is without reflection, this result is proved in [GLP2] Theorem 4.3, with the additional fact that the space of leaves is already Hausdorff, that is $d_{\tilde{\mathcal{F}}}(x, y) = 0$ if and only if x, y are in the same leaf. This last result is highly non trivial, since it uses in an essential way one of the main results of [GLP2], whose statement follows.

Theorem 2.4 *([GLP2] Theorem 2.6) If X is a system of isometries, if p, q are isometric embeddings $[a, b] \to \mathbb{R}$ with $a < b$ such that q_t and p_t are in the same X-orbit for all but countably many t's, then there are finitely many X-words $\omega_1, \cdots, \omega_n$ such that for every $t \in [a, b]$ there is an $i \in \{1, \cdots, n\}$ with p_t belonging to the domain of ω_i and $\omega_i(p_t) = q_t$.* $\square$

2.4   Generated pseudogroups and equivalences

It is proved in [GLP2] that the objects $G(X)$ and $T(X)$ are intrinsic for the dynamical system generated by X, for instance depends only on the orbits. More precisely, see [GLP2] section 4, X generates a pseudogroup, in the following sense.

A *closed pseudogroup* (or simply pseudogroup) on D is a set $\mathcal{P}$ of partial isometries between closed subintervals of D, containing the identity of every component of D, and stable by composition (defined on the maximal possible way), by taking inverses, by restrictions to closed subintervals (including points) and by finite extensions (if two elements coincide on the (non empty) intersection of their domains, and define a partial isometry on the union of the domains, then this isometry is also in $\mathcal{P}$). The pseudogroup generated by X is the smallest pseudogroup containing X, that is the set of restrictions of finite extensions of X-words. A pseudogroup is finitely generated if it is generated by some system of isometries.

Proposition 2.5 *In a finitely generated closed pseudogroup, every non singleton element is a restriction of a unique element with a maximal domain of definition.*

Proof. Let $\phi : A \to B \in \mathcal{P}$ and let $\{\phi_\alpha : A_\alpha \to B_\alpha\}_\alpha$ be the equivalence class of ϕ for the equivalence relation generated by $\phi_\alpha \sim \phi_\beta$ if and only if $A_\alpha \cap A_\beta \neq 0$ and ϕ_α, ϕ_β define on $A_\alpha \cup A_\beta$ a partial isometry of D. Since A has non empty interior, there is an isometry ψ from $\bigcup_\alpha A_\alpha$ to $\bigcup_\alpha B_\alpha$, whose restriction to A_α coincides with ϕ_α. According to Theorem 2.4, $\bigcup_\alpha A_\alpha$ is a closed interval and ψ belongs to $\mathcal{P}$, so the result follows. $\square$

In this category, there is a notion of equivalence : an equivalence between two pseudogroups $\mathcal{P}, \mathcal{P}'$ on multi-intervals D, D' is a finite set $\mathcal{Q} = \{\varphi_i : C_i \to D_i\}_{i=1\cdots m}$ of partial isometries from closed intervals in D onto closed intervals in D', with the C_i's (resp. B_i's) covering D (resp. D'), and $\mathcal{Q} \circ \mathcal{P} \circ \mathcal{Q}^{-1} \subset \mathcal{P}'$ and conversely. In particular (see [GLP2] Proposition 3.3), if two systems of isometries on D have the same orbits, then their generated pseudogroups are equivalent.

E. Rips has introduced examples of equivalences (that he called "elementary moves"). The only two we have needed in [GLP1] were the following ones. We recall them since we have slightly generalized them to get rid of the assumption of pureness (in [GLP1], Proposition 3.5 and section 7).

Splitting Let X be a system of isometries. Let x be an interior point of a component I of D. We first split the bases A_j (resp. B_j) containing x in their interior, and split B_j (resp. A_j) at the corresponding point $\varphi_j(x)$ (resp. $\varphi_j^{-1}(x)$). We replace φ_j by two isometries, the restrictions of φ_j to the closures of the two components of $A_j - \{x\}$ (resp. $A_j - \{\varphi_j^{-1}(x)\}$). We then replace I by two disjoint intervals I_1, I_2, isometric to the closures of the components of $I - \{x\}$, embedded anyhow in $\mathbb{R}$ disjointly from the other components of D, so that x is replaced by two points x_1, x_2. Let D' be the new multi-interval thus obtained. Each isometry in X defines a partial isometry of D' in the obvious way, by transfering the bases that were above I either to I_1 or I_2 accordingly. If $\{x\}$ was the base of a singleton, then transfer it arbitrarily above either x_1 or x_2. The new system of isometries X' on D' is then obtained by taking the partial isometries such defined, and adding a singleton taking x_1 to x_2.

It is easy to see that X and X' have equivalent generated closed pseudogroups.

Pruning Let X be a system of isometries. Suppose there is a point x in D, which is not a base of a singleton, which belongs to one and only one base A_j or B_j of only one isometry of X. Let I be the maximal open interval of D containing x, such that points of I belong to only one base. Then $A_j - I$ (resp. $B_j - I$) consists of $0, 1$ or 2 nondegenerate closed intervals. Suppose that every point of $\varphi_j(I)$ (resp. $\varphi_j^{-1}(I)$) is covered at least twice, i.e. belongs to a base of at least one isometry of X besides $\{\varphi_j\}$. Define a new system of isometries X' on the multi-interval

$D' = D - I$ in the obvious way: the φ_i's with $i \neq j$ are unchanged, while φ_j is replaced by $0, 1$ or 2 isometries defined on the components of $A_j - I$ (resp. $A_j - \varphi_j(I)$) by taking the restrictions of φ_j.

It is easy to see that X and X' have equivalent generated closed pseudogroup.

The main result with respect to equivalences is the following, which uses Theorem 2.4.

Proposition 2.6 *([GLP2] Propositon 3.10 and 4.1) If X, X' are systems of isometries with equivalent generated closed pseudogroups, then $G(X), G(X')$ are isomorphic and $T(X), T(X')$ are isometric by an isometry which commutes with the actions of $G(X), G(X')$.* $\square$

Ideas in the proof in [GLP2] of Proposition 2.6 may be used to give a conceptual computation of the group associated to a homogeneous system (see [GLP1], Proposition 4.2).

Proposition 2.7 *If the system of isometries X is pure and homogeneous, then its associated group $G(X)$ is a free abelian group isomorphic to the group of periods P, in the case P is purely translation ; otherwise, $G(X)$ is trivial.*

Proof. Let $\mathcal{P}$ be the closed pseudogroup generated by P, defined on the whole $\mathbb{R}$ (the definition is the same as for multi-intervals). We associate a group to $\mathcal{P}$ (resp. P), in the following way. We define the foliated 2-complex Σ associated to $\mathcal{P}$ (resp. P) as in Section 2.1, by gluing strips to $\mathbb{R}$, one for every element of $\mathcal{P}$ (resp. P). To Σ is given the obvious structure of a (in general non locally finite) connected CW-complex. The fundamental group $\pi_1(\Sigma)$ is a free group (in general not finitely generated). If $\mathcal{L}$ is the normal subgroup of $\pi_1(\Sigma)$ normally generated by free homotopy classes of loops contained in leaves, then define $G(\mathcal{P})$ (resp. $G(P)$) to be $\pi_1(\Sigma)/\overline{\mathcal{L}}$. By definition of the topology, $G(\mathcal{P})$ (resp. $G(P)$) is the direct limit of the groups $G(X_\alpha)$ where X_α is any finite set of elements of $\mathcal{P}$ (resp. P). From the proof of Lemma 4.8 in [GLP2], it follows that $G(\mathcal{P})$ is isomorphic to $G(P)$ (here we know that $\mathcal{P}$ is generated by P, so we do not need to use Theorem 2.4).

Define an equivalence between $\mathcal{P}(X)$ and $\mathcal{P}$ to be the set of (restrictions to components of the domain of the) isometries $\varphi_i = \gamma^i_{|D} : A_i = D \to B_i = \gamma^i(D)$, for $i \in \mathbb{Z}$, where γ is some translation of P with small enough translation length, so that the $\gamma^n(D)$'s cover $\mathbb{R}$. The conditions in the definition of an equivalence are easily checked. We claim that the groups associated to X and to $\mathcal{P}$ are isomorphic. Indeed, the only place where we essentially used the finiteness in the proof of 2.6 was for the splittings along endpoints of the elements of the equivalence. But here, we have the local finiteness of those endpoints, which is sufficient.

Now, it is clear that the group associated to $\mathcal{P}$ is P in the purely translation case, and is trivial otherwise. Indeed, the group associated to $\mathcal{P}$ obviously has a presentation with generators any set of generators of P and relations the words in the generators of P such that the corresponding isometry of $\mathbb{R}$ has a fixed point. In the case with symmetry, note that the symmetries are generating P, hence the result. □

2.5 Decompositions of $G(X)$ and $T(X)$ into minimal components

We are going to consider the decompositions of the group $G(X)$ and of the $\mathbb{R}$-tree $T(X)$ induced by the decomposition of X into minimal pieces. The following result is more or less contained in [GLP1] Proposition 3.5.

Proposition 2.8 *(E. Rips) Let $X_1, \cdots, X_m$ be connected systems of isometries on multi-intervals $D_1, \cdots, D_m$, satisfying $\overset{\circ}{D}_p \cap \overset{\circ}{D}_q = \emptyset$ for $1 \leq p < q \leq m$. Let X be a connected system of isometries on the multi-interval $D = \bigcup_{p=1}^{m} D_p$ with $X = X_0 \cup (\bigcup_{p=1}^{m} X_p)$ where X_0 is a finite set of singletons.*

Then $G(X)$ is isomorphic to the free product of $G(X_1), \cdots, G(X_m)$ and F where F is a finitely generated free group.

Proof. As an illustration, let us give a proof using the combinatorial presentation of $G(X)$ given in Subsection 2.2.

Since the splittings preserve the group (see Proposition 2.6), we may suppose that $D_p \cap D_q = \emptyset$, for every $1 \leq p < q \leq m$, and that if $X_0 = \{\varphi : \{a_i\} \rightarrow \{b_i\}/ i = 1 \cdots k\}$, then a_i, b_i are not in the interior of $D = \bigcup_{p=1}^{m} D_p$. We may also suppose that X has no full reflection (by a splitting at the fixed point of every full reflection).

We identify F_X with the free group freely generated by the elements of X that are neither edges of $T_{\max}$ nor their inverses.

Since the elements of X_0 have one-point domains, if r is a minimal relation, then the sum of the number of occurences in the word r of an element φ of $X_0 - T^1_{\max}$ and the number of occurences of its inverse is at most one. Indeed, if φ sends x to y, and $\overline{\varphi} w \overline{\varphi}$ (resp. $\overline{\varphi} w \varphi^{-1}$) is a subword of a relation r (necessarily in the core of the relation), then $\overline{\varphi} w$ (resp. w) is the core of a relation, since the associated isometry sends y to y (resp. x to x), hence r is not minimal. Up to cyclic conjugation, we may suppose φ appears only at the beginning of the core of relations. But assume φ or its inverse does occur in a relation, say with $w_r = \varphi w'$. Remove φ, φ^{-1} from the set of generators, remove the relation r from the set of relations, and replace every occurence of φ (resp. φ^{-1}) in any other relation r' by w'^{-1}

(resp. w'). We then get a presentation of $G(X)$ with one generator less. By finitely many such operations, we get a presentation of $G(X)$ as generated by $S_1 = X - (T^1_{\max} \cup (T^1_{\max})^{-1} \cup X_0)$ and by $S_2 \subset (X_0 - (T^1_{\max} \cup (T^1_{\max})^{-1}))$ with relations only between the elements of S_1. This proves that $G(X)$ is of the required form, the additional free factor being the free group on S_2. $\square$

Let us look now at $T(X)$. Recall that if $f : G \to G'$ is a group morphism, then a map $g : Y \to Y'$ between sets endowed with actions of G, G' respectively is called *equivariant* if $g(\gamma x) = f(\gamma)g(x)$ for every x in Y and γ in G. A *segment* in an $\mathbb{R}$-tree T is an isometric embedding of an interval.

Definition 2.9 *(J. Morgan, J.-P. Otal) Let G be a group and T, T' be $\mathbb{R}$-trees endowed with actions of G. A morphism from T to T' is a map $f : T \to T'$ such that*

  1. *f is equivariant and continuous*

  2. *for every non degenerate segment $[x, y]$ in T, there is a non degenerate segment $[x, z] \subset [x, y]$ such that f restricted to $[x, z]$ is an isometry.*

It is proved in [MO] that the second assertion is equivalent to the requirement that every segment of T may be decomposed into finitely many subsegment, each of which is mapped isometrically into T', that the composition of two morphisms is a morphism, that a morphism is distance non-increasing and that a bijective morphism is an isometry.

The following notions are due to R. Skora [Sko2] (see also [BF2][CL] [GL] [Lev4]). Note that in an $\mathbb{R}$-tree, two subtrees are either disjoint (hence there is a unique connecting arc), or they meet in a common subtree. One may thus define two notions of graph of $\mathbb{R}$-trees, the one in [Lev4] corresponding to the first case, and the one corresponding to the second case, that follows. If e is an edge of a graph, $v(e)$ is its endpoint and $\bar{e}$ its opposite edge. If Z is a graph, its set of vertices (resp. edges) is denoted by $V(Z)$ (resp. $E(Z)$).

Definition 2.10 *A graph of $\mathbb{R}$-trees $(\Gamma, \mathcal{T})$ consists of*

  • *a non empty connected graph Γ,*

  • *a non empty $\mathbb{R}$-tree T_v with an action of a group G_v, for every vertex v of Γ,*

  • *a non empty $\mathbb{R}$-tree T_e with an action of a group G_e, for every edge e of Γ,*

- an isomorphic identification of G_e and $G_{\bar{e}}$ and an isometric identification of T_e and $T_{\bar{e}}$ which is equivariant with respect to the previous identification, for every edge e,

- a monomorphism from G_e into $G_{v(e)}$ and an isometric embedding of T_e onto a <u>closed</u> subtree of $T_{v(e)}$, for every edge e, which is equivariant with respect to the above embedding of G_e into $G_{v(e)}$.

Underlying a graph of ℝ-trees $(\Gamma, \mathcal{T})$, there is a graph of groups $(\Gamma, \mathcal{G})$, in the Bass-Serre's sense (see [Ser]). Solving an universal problem, there is an ℝ-tree $T(\Gamma, \mathcal{T})$ with an action of the fundamental group of the underlying graph of groups $\pi_1(\Gamma, \mathcal{G})$. Recall that the G_v's inject into this fundamental group (see [Ser] page 64).

Definition 2.11 *A fundamental* ℝ*-tree of a graph of* ℝ*-trees* $(\Gamma, \mathcal{T})$ *is an* ℝ*-tree* T

1. *endowed with an action of* $\pi_1(\Gamma, \mathcal{G})$,

2. *endowed with a* G_v-*equivariant isometric embedding* $T_v \to T$, *for every vertex* v *of* Γ,

3. *such that the compositions* $T_{\bar{e}} \to T_e \to T_{v(e)} \to T$ *and* $T_{\bar{e}} \to T_{v(\bar{e})} \to T$, *coincide, for every edge* e *of* Γ,

and such that for every T' *satisfying* (1),(2),(3), *there is a* $\pi_1(\Gamma, \mathcal{G})$-*equivariant morphism* $\phi : T \to T'$ *such that the compositions* $T_v \to T \to T'$ *and* $T_v \to T'$, *coincide, for every edge* e.

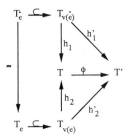

Figure 2 : Universal property of the fundamental ℝ-tree.

An *amalgamated product* $T_{v(\bar{e})} *_{T_e} T_{v(e)}$ or an *HNN extension* $T_{v(e)}*_{T_e}$ is obtained when the underlying graph consists of a segment or a loop respectively. Recall that an action of a group on an ℝ-tree is called *minimal* if there is no proper invariant subtree, and *transitive* if there is one orbit which is dense, which is equivalent, since the action is by isometries, to every orbit is dense.

Theorem 2.12 *A fundamental $\mathbb{R}$-tree of a graph of $\mathbb{R}$-trees exists and is unique up to equivariant isometry. Furthermore, if the vertex $\mathbb{R}$-trees are minimal (resp. transitive), then the fundamental $\mathbb{R}$-tree is minimal (resp. transitive). If the vertex groups act freely on the vertex $\mathbb{R}$-trees, if the underlying graph is a tree, if the edge $\mathbb{R}$-trees consist of a point, then the fundamental group acts freely on the fundamental tree.*

Proof. The existence (and uniqueness which follows by abstract nonsense) is due to R. Skora [Sko2], who uses topological methods à la Scott-Wall [SW]. Here, we give a more combinatorial proof, mimicking the proof of [GL] Theorem I.1, based on ideas of E. Rips. The idea is the following: consider the Bass-Serre tree of the underlying graph of groups. The fundamental $\mathbb{R}$-tree is obtained by taking copies of vertex $\mathbb{R}$-trees and gluing them along edge $\mathbb{R}$-trees, with the combinatoric given by the Bass-Serre tree. More precisely, let π_1 be the fundamental group of the underlying graph of groups. Recall from [Ser] the following facts.

Let $g \mapsto g^e$ be the monomorphism $G_e \to G_{v(e)}$. Let $\mathcal{G}$ be a maximal subtree of Γ. Let Π be the free product of the groups G_v and the free group with basis $E(\Gamma)$, with t_e the generator corresponding to $e \in E(\Gamma)$. Then π_1 is the quotient of Π by the subgroup normally generated by the relations

$$t_{\bar{e}} = t_e^{-1}, \quad t_e g^e t_e^{-1} = g^{\bar{e}} \quad \text{if } e \in E(\Gamma), g \in G_e$$

and

$$t_e = 1 \quad \text{if } e \in E(\mathcal{G}).$$

Up to isomorphism, this group does not depend on the choice of the maximal subtree (see [Ser] page 63).

Let A be an orientation of $E(\Gamma) - E(\mathcal{G})$, that is a choice of one of the two edges $e, \bar{e}$ for every e in $E(\Gamma) - E(\mathcal{G})$. Let $b(e) = 0$ if $e \in A$ and $b(e) = 1$ otherwise, so that $b(\bar{e}) = 1 - b(e)$. The Bass-Serre tree is the graph $T_{\mathcal{BS}}$ defined by

$$V(T_{\mathcal{BS}}) = \bigvee_{v \in V(\Gamma)} \pi_1/G_v$$

$$E(T_{\mathcal{BS}}) = \bigvee_{e \in E(\Gamma)} \pi_1/G_e$$

$$\overline{gG_e} = gG_{\bar{e}}, \quad v(gG_e) = gt_e^{1-b(e)} G_{v(e)}.$$

Then $T_{\mathcal{BS}}$ is a tree, with an action of π_1 induced by the left action of π_1 on its left classes, and does not depend, up to equivariant isomorphism, on the choice of $\mathcal{G}, A$ (see [Ser] page 71).

For every γ in π_1 and v, w in $V(\Gamma)$, we claim that there is a *path* $(g_0, e_1, \cdots, g_{n-1}, e_n, g_n)$, where $e_1, \cdots, e_n$ are consecutive edges in $E(\Gamma)$

(i.e. $v(e_i) = v(\overline{e_{i+1}})$), starting at v (i.e. $v(\overline{e_1}) = v$) and ending at w (i.e. $v(e_n) = w$), and where $g_0 \in G_v$ and $g_i \in G_{v(e_i)}$, such that $\gamma = g_0 t_{e_1} g_1 \cdots t_{e_n} g_n$ in π_1. Such a path is said to *represent* γ from v to w. For instance, (g_0) represents any $g_0 \in G_v$ from v to v. To prove this claim, note the following fact. Assume there is an edge fG_e from gG_v to hG_w in the Bass-Serre tree. Then $g = ft_e^{-b(e)}g'$ with $g' \in G_v$ and $h = ft_e^{1-b(e)}h'$ with $h' \in G_w$, hence $g^{-1}h = (g')^{-1}t_e h'$. Applying this fact inductively to every edge of the geodesic between G_v and γG_w in the Bass-Serre tree yields the claim. A path is *geodesic* if $g_i \notin \mathrm{Im}(G_{e_i} \to G_{v(e_i)})$ for every i such that $e_{i+1} = \overline{e_i}$. Since otherwise we could reduce its length, every element of π_1 is represented by a geodesic path.

Let F (as Forest) be the disjoint union of the vertex ℝ-trees. Let $z \mapsto z^e$ be the equivariant isometric embedding $T_e \to T_{v(e)}$. Let

$$T = (F \times \pi_1)/\sim$$

where $\sim$ is the equivalence relation generated by

$$(x, gh) \sim (hx, g) \quad \text{if } x \in T_v, g \in \pi_1 \text{ and } h \in G_v$$

and

$$(x^e, g) \sim (x^{\overline{e}}, g) \quad \text{if } x \in T_e \text{ and } g \in \pi_1.$$

The set T is endowed with a left action of π_1 induced by $g(x, h) = (x, gh)$.

Let d_{T_v} be the distance in T_v. Let δ be the pseudodistance on $F \times \pi_1$ defined by

$$\delta((x, g), (y, h)) = \inf d_{T_v}(x, g_0 x_1^{\overline{e_1}}) + d_{T_{v(e_1)}}(x_1^{e_1}, g_1 x_2^{\overline{e_2}}) + \cdots$$

$$+ d_{T_{v(e_{n-1})}}(x_{n-1}^{e_{n-1}}, g_{n-1} x_n^{\overline{e_n}}) + d_{T_{v(e_n)}}(x_n^{e_n}, g_n y)$$

where the infimum is taken on all paths $(g_0, e_1, g_1, \cdots, e_n, g_n)$ representing $g^{-1}h$ from v to w, with $x \in T_v$ and $y \in T_w$, and on all x_i in T_{e_i}, unless $g = h$ and $x, y \in T_v$ in which case, we set

$$\delta((x, g), (y, h)) = d_{T_v}(x, y).$$

The pseudodistance δ is, by construction, invariant under the action of π_1 on $F \times \pi_1$. Note that the infimum defining δ is always achieved. First, one may consider only geodesic paths $(g_0, e_1, g_1, \cdots, e_n, g_n)$. Then let $g_0 x_1^{\overline{e_1}}$ be the closest point of x in $g_0(\mathrm{Im}(T_{\overline{e_1}} \to T_{v(\overline{e_1})}))$ (this is possible since the subtrees are closed). Let $g_1 x_2^{\overline{e_2}}$ be the closest point of $x_1^{e_1}$ in $g_1(\mathrm{Im}(T_{\overline{e_2}} \to T_{v(\overline{e_2})}))$ and so on. To give a hint why this is indeed the minimum, let $z \in T_{v(e)}, t \in T_{v(\overline{e})}, g \in G_{v(\overline{e})}$, let u^e be the closest point of z

in $\mathrm{Im}(T_e \to T_{v(e)})$ and $gv^{\overline{e}}$ be the closest point of t in $g(\mathrm{Im}(T_{\overline{e}} \to T_{v(\overline{e})}))$. Then for every $w \in [u, v]$,

$$d(t, gw^{\overline{e}}) + d(w^e, z) = d(t, gu^{\overline{e}}) + d(u^e, z).$$

By invariance of d_{T_v} under G_v, δ induces a pseudodistance d_T on T, which is indeed a metric, by the above argument. In particular, (T, d) is the metric space naturally associated to the pseudometric space $(F \times \pi_1, \delta)$, that is $T = (F \times \pi_1)/ \sim'$ where $\sim'$ is the equivalence relation given by $x \sim' y$ if and only if $\delta(x, y) = 0$. Hence the action of π_1 on T is isometric.

We claim that T is indeed an $\mathbb{R}$-tree. Since Γ is connected and since $\delta((x^e, g), (x^{\overline{e}}, g)) = 0$, the metric space T is obviously connected. Hence (see for instance [AB]) one has only to check that any four points $u_i = (x_i, g_i)$ satisfy the 0-hyperbolicity inequality :

$$\delta(u_1, u_2) + \delta(u_3, u_4) \leq \max\{\delta(u_1, u_3) + \delta(u_2, u_4), \delta(u_1, u_4) + \delta(u_2, u_3)\}.$$

This is clear if the g_i's are equal, since the T_v's are $\mathbb{R}$-trees. In general, one argue by induction on the diameter of the finite subtree which is the convex hull of the $g_i G_{v_i}$'s in the Bass-Serre tree, with $x_i \in T_{v_i}$.

The T_v's naturally embed in T by the map induced by $x \mapsto (x, 1)$.

If T' is any $\mathbb{R}$-tree satisfying (1) (2) (3) in 2.11, then there is a map from T to T' obtained by sending the standard T_v's in T onto the standard T_v's in T', then extending by equivariance. That this map is continuous follows from the definition of the pseudodistance δ. It is a morphism since any segment $[x, y]$ in T contains a segment $[x, z]$ in the image of some T_v by an element of π_1.

The last two assertions are immediate. □

In particular if a group G acts on an $\mathbb{R}$-tree T, and if G is isomorphic to the fundamental group of a graph of groups, let T_v (resp. T_e) be the minimal subtree of T invariant by the vertex group G_v (resp. the edge group G_e), and $T_e \to T_{v(e)}$ be inclusion, then there is a morphism from the fundamental $\mathbb{R}$-tree of the graph of $\mathbb{R}$-trees to T.

Definition 2.13 *An action of a group G on an $\mathbb{R}$-tree T is said to be decomposable if there is a decomposition of G such that the above morphism is an isometry onto the minimal subtree.*

As G. Levitt as asked, one may wonder when there exists a decomposition of an action on an $\mathbb{R}$-tree which is defined purely in dynamical terms. See [Lev4] for some answers.

Let $X_1, \cdots, X_m$ be connected systems of isometries on multi-intervals $D_1, \cdots, D_m$, satisfying $\overset{\circ}{D}_p \cap \overset{\circ}{D}_q = \emptyset$ for $1 \leq p < q \leq m$, no component of D_p being a single point. Let X be a connected system of isometries on the multi-interval $D = \bigcup_{p=1}^m D_p$ with $X = X_0 \cup (\bigcup_{p=1}^m X_p)$ where X_0 is a finite set of singletons. Fix a lift of D in $\tilde{\Sigma}(X)$. Every x in D has a unique lift $\tilde{x}$ in this lift. Every $\Sigma(X_i) \subset \Sigma(X)$ with $i \geq 1$ has a unique lift $\tilde{\Sigma}(X_i) \subset \tilde{\Sigma}(X)$ containing $\tilde{a}$ where a is the first point in D_i (for the order on ℝ). The inclusion $\Sigma(X_i) \subset \Sigma(X)$ induces a monomorphism $\pi_1 \Sigma(X_i) \to \pi_1 \Sigma(X)$. Since $\overline{\mathcal{L}}(X_i) \subset \overline{\mathcal{L}}(X)$, one gets a map $G(X_i) \to G(X)$ which is a monomorphism according to Proposition 2.8. Hence the map $\tilde{\Sigma}(X_i) \to \Sigma(X_i)$ induced by $\tilde{\Sigma}(X) \to \Sigma(X)$ is the covering associated to $G(X_i)$. So that one gets a map $T(X_i) \to T(X)$. Assume that X (hence X_i) is without reflection, so that $T(X)$ and $T(X_i)$ are the spaces of leaves of $\tilde{\Sigma}(X)$ and $\tilde{\Sigma}(X_i)$. Note that by definition of the covering group, every leaf of $\Sigma(X_i)$ lifts to $\tilde{\Sigma}(X_i)$. Furthermore, a singleton in X_0 between two points of D_i yields a non trivial element of the covering group $G(X)$, mapping $\tilde{\Sigma}(X_i)$ disjointly from itself, if and only if the two points are in distinct X_i-orbits. So the map $T(X_i) \to T(X)$ is injective, hence an equivariant isometric embedding.

Among all singletons in X_0, choose $x_1^- \mapsto x_1^+, \cdots, x_q^- \mapsto x_q^+$ a maximal subset such that x_j^- and x_j^+ are in distinct orbits under $(\bigcup_{p=1}^m X_p) \cup \{x_q^- \mapsto x_q^+, \cdots, x_{j-1}^- \mapsto x_{j-1}^+\}$ for $j = 1 \cdots q$. Assume x_j^- belongs to $D_{p_-(j)}$ and x_j^+ to $D_{p_+(j)}$, making arbitrary choices when x_j^-, x_j^+ belongs to different D_p's. Among all intersection points in $\bigcup_{p \neq p'} D_p \cap D_{p'}$, choose $\{x_{q+1}, \cdots, x_\ell\}$ a maximal subset such that the similar property holds when splitting at all intersection points. Assume x_i for $i = q+1 \cdots \ell$ belongs to $D_{p_-(i)} \cup D_{p_+(i)}$ with $p_-(i) < p_+(i)$. For $j = 1 \cdots q$ and $i = q+1 \cdots \ell$, choose lifts $\tilde{x}_i^-, \tilde{x}_i^+, \tilde{x}_j^-, \tilde{x}_j^+$ of x_i, x_i, x_j^-, x_j^+ in respectively $\tilde{\Sigma}(X_{p_-(i)}), \tilde{\Sigma}(X_{p_+(i)})$, $\tilde{\Sigma}(X_{p_-(j)}), \tilde{\Sigma}(X_{p_+(j)})$; let $\bar{x}_i^-, \bar{x}_i^+, \bar{x}_j^-, \bar{x}_j^+$ be the images of $\tilde{x}_i^-, \tilde{x}_i^+, \tilde{x}_j^-, \tilde{x}_j^+$ by the canonical projections in $T(X_{p_-(i)}), T(X_{p_+(i)}), T(X_{p_-(j)}), T(X_{p_+(j)})$.

Let Γ be the graph having $X_1, \cdots, X_m$ as vertices and the $\pm i$'s with $i \in \{1, \cdots, \ell\}$ as edges from $X_{p_\mp(i)}$ to $X_{p_\pm(i)}$, the opposite of the edge $\pm i$ being $\mp i$ Let $(\Gamma, \mathcal{T})$ be the following graph of ℝ-trees. The ℝ-tree of the vertex X_i is $T(X_i)$ with its action of $G(X_i)$. Every edge ℝ-tree is a point $\{t_{\pm i}\}$ (its group is trivial), the embedding of an edge ℝ-tree into its endpoint ℝ-tree being given by $t_{\pm i} \to \bar{x}_i^\pm$. Up to automorphisms of the fundamental group of the underlying graphs of groups and equivariant isometries of the fundamental ℝ-trees of the graphs of ℝ-trees, the results do not depend on the choices. The choices are made so that one gets an upper bound on the number of edges of Γ by the rank of $G(X)$, and one has only to glue along points.

Proposition 2.14 *With the above notations, if X is without inversions, $T(X)$ is isomorphic to the fundamental $\mathbb{R}$-tree of $(\Gamma, \mathcal{T})$.*

Proof. Since a splitting preserves the associated group and the associated $\mathbb{R}$-tree (see Proposition 2.6), we may suppose that $D_p \cap D_q = \emptyset$, for every $1 \leq p < q \leq m$, and that if $X_0 = \{\varphi_i : \{a_i\} \to \{b_i\}\}_{i=1,\cdots,k}$, then a_i, b_i are not in the interior of $D = \bigcup_{p=1}^m D_p$.

The result then follows from three obvious remarks. Let X' be a connected system of isometries obtained from X by adding a singleton $\varphi : \{a\} \to \{b\}$.

- If X is connected, and a, b are in the same X-orbit, then according to the fact that two systems of isometries having the same orbits have equivalent generated closed pseudogroup and to Proposition 2.6, $G(X)$ and $G(X')$ are isomorphic and $T(X), T(X')$ equivariantly isometric ;

- If X is connected, and a, b are not in the same X-orbit, then $G(X')$ is isomorphic to the free product $G(X) * \mathbb{Z}$ (HNN extension with trivial edge group), and $T(X')$ is equivariantly isometric to $T(X)*_{\{\tilde{a}=\tilde{b}\}}$;

- If X is not connected, and X_1, X_2 are its two connected components, then $G(X')$ is isomorphic to the free product $G(X_1) * G(X_2)$ (amalgamated product with trivial edge group) and $T(X')$ is equivariantly isometric to $T(X_1) *_{\{x\}} T(X_2)$. $\square$

In particular the decomposition of X into minimal components induces a decomposition of $T(X)$ as the fundamental $\mathbb{R}$-tree of a graph of $\mathbb{R}$-trees, with vertex $\mathbb{R}$-trees the $T(X_i)$ for X_i the minimal components of T, and some compact intervals, one of length ℓ for every connected finite family of orbits of length ℓ (see [GLP1] for notations). This decomposition is slightly finer than the one in [Lev4], since when one adds a singleton between two connected minimal systems of isometries, then the action on the associated tree is minimal, hence the decomposition in [Lev4] is trivial. But the decomposition in [Lev4] holds not only for actions of the form $G(X)$ on $T(X)$, but also for non geometric actions.

2.6   Descriptions of $G(X)$ and $T(X)$

The main result of [GLP1] is a proof of the following theorem.

Theorem 2.15 *(E. Rips) Let G be a finitely generated group acting freely on an $\mathbb{R}$-tree. Then G is a free product of free abelian groups and surface groups.* $\square$

A corresponding analysis for the ℝ-tree itself remains to be done. See partial results in for instance [GLP2][BF2] [Lev4] [GL].

If $g \in G$, the translation length of g on T is $\ell_T(g) = \inf\{d(x, gx)/x \in T\}$. Recall from [Sha1] [Sha2] or [Mor1] that $\ell_T(g) = d$ if and only if there is $x \in X$ such that gx is the midpoint of $[x, g^2x]$, and $d = d(x, gx)$. A sequence of ℝ-trees T_n with a minimal action of G *converges strongly* (see [GS]) to an ℝ-tree T with an action of G if there is a morphism ϕ_n from T_n to T and ψ_n from T_n to T_{n+1} such that $\phi_{n+1} \circ \psi_n = \phi_n$, and if, for every $g \in G$, $\ell_{T_n}(g)$ is eventually constant and equal to $\ell_T(g)$. It is easy to see that this definition is equivalent to the original one in [GS].

A sequence of metric G-spaces Y_n converges to a metric G-space Y for the *equivariant Gromov topology* if for every K finite subset of Y, every P finite subset of G containing the identity and every $\epsilon > 0$, for n big enough, there is a finite subset K_n of Y_n, in bijection with K by $x \mapsto x_n$, such that for every x, y in K and g in P,

$$|d(gx, y) - d(gx_n, y_n)| \leq \epsilon.$$

For further informations on this topology, see [Pau1][Pau2], where a slight mistake in the definition took place, as pointed out to the author by R. Skora, all results remaining true.

It is proved in [Pau1] that for ℝ-trees, the convergence for the equivariant Gromov topology implies the convergence of the translation lengths, the converse being true if the actions are minimal the translation lengths are not the absolute values of a morphism from G to ℝ.

Theorem 2.16 *([GLP2]) Let G be a finitely generated group acting freely and minimally on an ℝ-tree T. For every finite subtree K in T, $X(K)$ is without reflection. If K_n is an increasing sequence of finite subtrees with union T, then $G(X(K_n))$ is isomorphic to G for n big enough and $T(X(K_n))$ tends for the equivariant Gromov topology and for the strong convergence towards T.* □

Let us look at the *branching number $b(T)$* of an action on an ℝ-tree T. By definition, $b(T)$ is the number of orbits of branch points in T. One has the following result :

Theorem 2.17 *(R. Jiang [Jia]) The branching number of a minimal free action of a finitely generated group G on an ℝ-tree is bounded by $3r$ where r is the rank of $G/[G, G]$.*

Proof. Let us give an alternate proof of this theorem. Recall the following results : the branching number of a minimal action of a free group of rank n on an ℝ-tree is bounded by $2n - 2$ (see [Jia], and [GL] for a simpler proof, giving also the result for very small actions in M. Cohen-M. Lustig's sense, and more informations on the rank of the stabilizers of branch points). The branching number of a minimal free action of a surface group of Euler characteristic χ on an $\mathbb{R}$-tree is bounded by $2|\chi|$. This follows from R. Skora's result [Sko1] that a free action of a surface group is dual to a measured foliation in [FLP]'s sense, where the branching points correspond precisely to the singularities of the foliation in the universal covering. A simple counting argument (see [FLP] page 75), assuming that all singularities are generic (i.e. the number of separatrix is 3 at every singularity, which is possible by Whitehead moves), yields that the number of generic singularities on the surface is $2|\chi|$ (the actual branching number may be smaller if there is a connection between two singularities). Recall that an abelian free group action on $\mathbb{R}$ has 0 branch point !

Let us first prove Theorem 2.17 for free actions of the form $G(X)$ on $T(X)$. The proof in [GLP1] of Theorem 2.15 shows that the $G(X_i)$ are either free, free abelian, or surface groups. The number of edges in the graph defined in Proposition 2.14 is bounded in terms of the rank of the group $G(X)$ (each one increasing strictly the rank of the group, see the proof of 2.14). Hence one easily gets that the branching number of $T(X)$ is bounded uniformly in terms of the rank r of $G(X)$. An easy computation gives the (non optimal) bound $3r$.

Since any free action is a strong limit of actions of the form $T(X)$ (Theorem 2.16), one gets the result. $\square$

3   Applications

One of the most striking application of Rips' Theorem 2.15 for free actions is the following one. R. Kenyon [Ken] defines a group G of parametrization-free, non-backtracking paths in the plane, up to translation, with product given by concatenating translates, in the following way. The set G is the set of continuous maps $f : [0,1] \rightarrow \mathbb{R}^2$ with $f(0) = (0,0)$, up to the following equivalence. With D the closed unit disk in $\mathbb{R}^2$, we say $f \sim g$ if there is a continuous map $T : D \rightarrow \mathbb{R}^2$ such that:

- $T(e^{i\pi t}) = f(t)$ for $0 \leq t \leq 1$,

- $T(e^{-i\pi t}) = g(t)$ for $0 \leq t \leq 1$,

- D has a lamination by chords such that T is constant on the leaves and on the connected complementary regions

The group structure is induced by the following operation on paths: the product $[f] \cdot [g]$ is the class of the path

$$f \cdot g(t) = \begin{cases} f(2t) & 0 \leq t \leq 1/2 \\ g(2t-1) + f(t) & 1/2 \leq t \leq 1 \end{cases}$$

The identity $[e]$ is given by the constant path $e(t) = (0,0)$, and the inverse of $[f]$ is $[f^{-1}]$, defined by the path $f^{-1}(t) = f(1-t) - f(1)$. See [Ken] for the needed verifications.

Theorem 3.1 *(R. Kenyon [Ken]) A finitely generated subgroup of G is a free product of (finitely many) free abelian groups and surface groups (except for the closed non-orientable surfaces of Euler caracteristic ≥ -1). Moreover, any such free product is isomorphic to a subgroup of G.* $\square$

The first assertion is proved by constructing a free action of a given finitely generated subgroup of G on an ℝ-tree. The existence part is proved by using pseudo-Anosov homeomorphisms on surfaces.

References

[AB]   R. Alperin, H. Bass, *Length functions of group actions on Λ-trees*, in "Combinatorial group theory and topology", S. Gersten, J. Stallings eds, (1987), 265-378.

[Ada]   S. Adams, *Trees and amenable equivalence relations*, Erg. Th. Dyn. Syst. **10** (1990), 1–14.

[AL]   P. Arnoux, G. Levitt, *Sur l'unique ergodicité des 1-formes fermées singulières*, Inv. Math. **84** (1986), 141–156.

[AY]   P. Arnoux, J.-C. Yoccoz, *Construction de difféomorphismes pseudo-Anosov*, C. R. Acad. Sc. Paris **292** (1981) 75–78.

[BF1]   M. Bestvina, M. Feighn, *Stable actions of groups on real trees*, preprint (Feb. 1992).

[BF2]   M. Bestvina, M. Feighn, *Limits in outer space*, preprint (Oct. 1992).

[CL]   M. Cohen, M. Lustig, *Very small group actions on ℝ-trees and Dehn twist automorphisms*, preprint.

[FLP]   A. Fathi, F. Laudenbach, V. Poenaru, *Travaux de Thurston sur les surfaces*, Astérisque **66-67**, Soc. Math. France (1979).

[Gab]   D. Gaboriau, *Dynamique des systèmes d'isométries et actions de groupes sur les arbres réels*, Thèse, Toulouse, June 1993.

[GL]   D. Gaboriau, G. Levitt, *The rank of actions on ℝ-trees*, preprint, Toulouse (Feb. 1993).

[GLP1] D. Gaboriau, G. Levitt, F. Paulin, *Pseudogroup of isometries of ℝ and Rips' theorem on free actions on ℝ-trees*, to appear in Israel J. Math.

[GLP2] D. Gaboriau, G. Levitt, F. Paulin, *Pseudogroup of isometries of ℝ and Rips' theorem on free actions on ℝ-trees, Part II : Reconstructions of ℝ-trees*, preprint.

[GS]   H. Gillet, P. Shalen, *Dendrology of groups in low $\mathbb{Q}$-ranks*, J. Diff. Geom. **32** (1990), 605-712.

[Gus]   P. Gusmao, *Groupes et feuilletages de codimension* 1, Thèse, Toulouse, June 1993.

[Hae1]   A. Haefliger, *Groupoïdes d'holonomie et classifiants*, in "Structures transverses des feuilletages", Astérisque **116** , Soc. Math. Fran. (1984), 70-97.

[Hae2]   A. Haefliger, *Pseudogroups of local isometries*, in "Proc. Vth Coll. in Differential Geometry", ed. L.A. Cordero, Research Notes in Math **131**, Pitman (1985), 174-197.

[Ima]   H. Imanishi, *On codimension one foliations defined by closed one forms with singularities*, J. Math. Kyoto Univ. **19** (1979), 285–291.

[Jia]   R. Jiang, *Number of orbits of branch points of $\mathbb{R}$-trees*, Trans. AMS. **335** (1993), 341-368.

[Ken]   R. Kenyon, *A group of paths in the plane*, preprint Grenoble (1992).

[Lev1]   G. Levitt, *Groupe fondamental de l'espace des feuilles dans les feuilletages sans holonomie*, Jour. Diff. Geom. **31** (1990), 711–761.

[Lev2]   G. Levitt, *La dynamique des pseudogroupes de rotations*, preprint, University Toulouse III, (1990), to appear in Invent. Math..

[Lev3]   G. Levitt, *Constructing free actions on $\mathbb{R}$-trees*, Duke Math. J., **69** (1993), 615-633.

[Lev4]   G. Levitt, *Graphs of actions on $\mathbb{R}$-trees*, preprint, University Toulouse III, (1991).

[Mor1]   J. Morgan, *Λ-trees and their applications*, Bull. A.M.S. **26** (1992), 87-112.

[Mor2]   J. Morgan, *Notes on Rips' lectures at Columbia University, October 1991*, manuscript.

[MS1]   J. Morgan, P. Shalen, *Valuations, trees and degeneration of hyperbolic structures II, III*, Ann. Math. **122** (1988), 403-519.

[MO]   J. Morgan, J.-P. Otal, *Relative growth rate of closed geodesics on a surface under varying hyperbolic structures*, Comm. Math. Helv. **68** (1993), 171-208.

[Pau1]   F. Paulin, *The Gromov topology on $\mathbb{R}$-trees*, Topology and its App. **32** (1989), 197-221.

[Pau2]   F. Paulin, *Topologie de Gromov équivariante, structures hyperboliques et arbres réels*, Invent. Math. **94** (1988), 53-80.

[Pau3]   F. Paulin, *Dégénérescences algébriques de représentations hyperboliques*, to appear in the proceedings of the "Colloque sur les variétés de dimension 3" (Luminy 1989).

[Rim1] F. Rimlinger, *Free actions on ℝ-trees*, Trans. AMS. **332** (1992), 315-331.

[Rim2] F. Rimlinger, *ℝ-trees and normalization of pseudogroups*, to appear in Experimental Mathematics.

[Rim3] F. Rimlinger, *Two complexes with similar foliations*, preprint (1992).

[Sha1] P. Shalen, *Dendrology of groups : an introduction*, Essays in group theory (S.M. Gersten ed.), M.S.R.I Pub. **8**, Springer Verlag, 1987.

[Sha2] P. Shalen, *Dendrology and its applications*, Group theory from a geometrical viewpoint (E. Ghys, A. Haefliger, A. Verjovsky eds.), World Scientific, 1991.

[Sal]   E. Salem, *Riemannian foliations and pseudogroups of isometries*, appendix in "Riemannian Foliation", P. Molino, Progress in Math. **73** , Birkhäuser (1988), 265-296.

[Ser]   J.-P. Serre, *Arbres, amalgames, SL_2*, Astérisque **46** , Soc. Math. France (1983).

[Sko1] R. Skora, *Splitings of surfaces*, Bull. Amer. Math. Soc. **23** (1990), 85-90.

[Sko2] R. Skora, *Combination theorems for actions on ℝ-trees*, preprint 1989.

[SW]   G.P. Scott, C.T.C. Wall, *Topological methods in group theory*, in "Homological group theory", C.T.C. Wall ed., Lond. Math. Soc. Lect. Notes **36**, Cambridge Univ. Press (1979) 137-203.

Unité de Mathématiques Pures et Appliquées, C.N.R.S. UMR 128
Ecole Normale Supérieure de Lyon
allée d'Italie, 69364 LYON CEDEX 07, FRANCE

e-mail address : paulin@umpa.ens-lyon.fr

Contemporary Mathematics
Volume **164**, 1994

ON A TANGLE PRESENTATION OF THE
MAPPING CLASS GROUPS OF SURFACES

SERGEI MATVEEV AND MICHAEL POLYAK

ABSTRACT. A simple presentation of a surface mapping class group by framed tangles is introduced. This leads to a new relation between surgery and Heegard decompositions for 3-manifolds. As an application of our technique we prove that any 3-manifold of Heegard genus two may be obtained by surgery on a framed arborescent link in S^3. An equivalence of two types of quantum invariants for 3-manifolds is shown. The tangle presentation of the mapping class group enables one to provide as well new proofs of some well-known results, e.g. J. Birman's theorem on the existence of special Heegard diagrams and R. Kirby's theorem.

1. Introduction

We establish here a simple geometrical presentation of the mapping class group of surfaces with marked points by framed tangles. For the case of genus zero surface with one puncture this restricts to a natural presentation of the mapping class group of a disc with n marked points by geometrical n-braids. It is, as well, similar in spirit to a presentation for the group of cobordisms of surfaces by ribbon graphs introduced in [RT1]. Our approach leads to a new relation between surgery on links in S^3 and Heegard decompositions for 3-manifolds. This, in particular, enables us to construct explicitly a surgery presentation of 3-manifold from its Heegard decomposition and vice versa. Using this technique we prove that any 3-manifold of Heegard genus two may be obtained by surgery on a framed arborescent link in S^3. We show also an equivalence of two different constructions for 3-manifold invariants- the one based on surgery [RT2] and another of [Ko] defined via Heegard diagrams. We introduce the notion of stable equivalence for tangles and obtain, as a consequence, a new proof (similar to [L]) of R. Kirby's theorem [Ki].

Both authors are grateful to M. Farber for his friendly support. Second author wishes also to thank V. Turaev for stimulating discussions.

1991 Mathematics Subject Classification: Primary 57M60, 57R65, 57S05

The second author is supported by Rotshild Fellowship

The final version of this paper will be submitted for publication elsewhere

2. Mapping class groups and admissible tangles

2.1.   Let $\Sigma_{g,j,n}$ be an orientable surface of genus g obtained from a closed surface by removing j open disks and n points. The mapping class group $M_{g,j,n}$ of $\Sigma_{g,j,n}$ is the group of isotopy classes of orientation-preserving self homeomorphisms of $\Sigma_{g,j,n}$ fixing the boundary pointwise. In our treatment we will (for technical reasons) pay special attension to $M_{g,1} = M_{g,1,0}$ and then formulate similar results for $M_{g,0} = M_{g,0,0}$ and $M_{g,1,n}$. To give a presentation for $M_{g,1}$ we introduce the notion of framed admissible $2g$-tangles.

2.2.   Let Y_{2g} be a set of g pairs $\{i^-, i^+\}$ of points $i^\pm = (0, i \pm 1/4), 1 \le i \le g$ on the xy-plane $\mathbb{R}^2$. By $2g$-*tangle* ξ we mean a tangle in $\mathbb{R}^2 \times [0,1]$ with the boundary $\partial\xi = Y_{2g} \times \{0,1\}$. A *framing* of ξ is a trivialization of its normal bundle such that its restriction to the boundary $\partial\xi$ is induced by the standard xy-structure of $\mathbb{R}^2$. One than defines the product $\xi \cdot \zeta$ of two $2g$-tangles in a usual way by gluing the top of ζ with the bottom of ξ and rescaling.

By a diagram of $2g$-tangle we mean its projection to $\mathbb{R}^1 \times [0,1]$ with indicated over and underpasses at the crossings such that the framing is orthogonal to the plane of projection. Any tangle can be represented in this way and two diagrams represent the same tangle iff they are regularly isotopic. A change of framing by ± 1 corresponds to making a positive (negative) kink on the strand of the diagram; we will encode it by a small white (black) circle, see Figure 1.

Geometrical $2g$-braids are a particular class of $2g$-tangles. Moreover, any framed link in S^3 may be considered (after isotopy into $\mathbb{R}^2 \times [0,1]$) as a 0-tangle. Another example important for our further purposes is given by $2g$-tangles $\beta_i^{\pm 1}$, $1 \le i \le g$ shown in Figure 2.

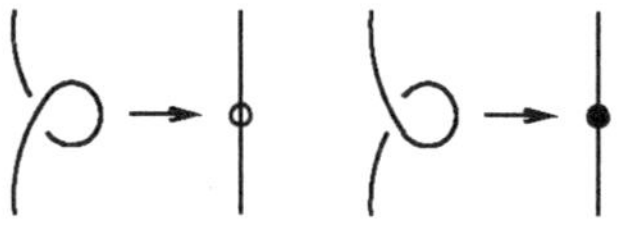

Figure 1    Figure 2

DEFINITION .   A $2g$-braid is *admissible*, if for any i a pair of strings with the lower ends at $i^\pm \times \{0\}$ has the upper ends at $j^\pm \times \{1\}$ for some j. A fraimed $2g$-tangle is called *admissible*, if it can be written as a finite product of admissible braids and tangles $\beta_i^{\pm 1}$, $1 \le i \le g$.

2.3.   An equivalence relation on $2g$-tangles introduced below can be considered as Kirby calculus for tangles. Let us denote by M_L 3-manifold obtained by Dehn surgery on a framed link L in S^3. The following is a well-known fact:

KIRBY THEOREM [Ki]. *Given two framed links L_1, L_2 in S^3 one can pass from L_1 to L_2 by a sequence of moves O_1, O_2 iff M_{L_1} is homeomorphic to M_{L_2}.*

We define here similar moves on framed tangles (see Figures 3,4):

K_1:   Add or delete an unlinked and unknotted closed component with ± 1 framing.

K_2:   Let ξ_0 be a closed component of a tangle ξ. Replace any other component $\xi_1 \subset \xi$ by $\xi_1 \#_b \widetilde{\xi_0}$, where $\#_b$ is a band connected sum and $\widetilde{\xi_0}$ is obtained by pushing ξ_0 off itself along the framing.

K_3:   Let ξ_0 be a closed, 0-framed component of ξ bounding an embedded disk $D \in \mathbb{R}^2 \times [0,1]$ which intersects with $\xi \setminus \xi_0$ in exactly two points belonging to different components ξ_1, ξ_2 of ξ. Suppose also that either at least one of the components ξ_1, ξ_2 is closed or $\partial \xi_1 \subset \mathbb{R}^2 \times \{0\}$, $\partial \xi_2 \subset \mathbb{R}^2 \times \{1\}$. Replace then $\xi_0 \cup \xi_1 \cup \xi_2$ by $\xi_1 \#_b \xi_2$, where the band b intersects D along the middle line of b.

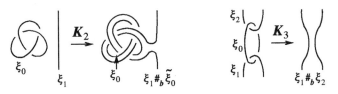

Figure 3   Figure 4

DEFINITION .   Framed $2n$-tangles ξ, ζ are called *K-equivalent* $\xi \underset{K}{\sim} \zeta$, if one can be obtained from another by a sequence of moves K_1, K_2, $K_3^{\pm 1}$.

Remark. (i) $K_1 = O_1$; (ii) $K_2 = O_2$ if ξ_1 in the definition of K_2 is closed; (iii) K_3 may be expressed via K_1, K_2 if one of the components ξ_1, ξ_2 in the definition of K_3 is closed, so the equivalence relation $\underset{K}{\sim}$ for links coincides with the equivalence $\underset{\partial}{\sim}$ of Kirby [Ki].

Denote by T_{2g} the set of K-equivalence classes of admissible $2g$-tangles. One can show that for admissible $2g$-tangles the multiplication agree with the equivalence relation $\underset{K}{\sim}$, thus defining a multiplication on T_{2g}. It is an easy exercise to check that the reflection of $\xi \in T_{2g}$ with respect to the plane $\mathbb{R}^2 \times \{1/2\}$ produces the inverse tangle $\xi^{-1} \in T_{2g}$, so we obtain

PROPOSITION 2.1. T_{2g} *is a group.*

Denote by α_i, δ_i, ϵ_i and β_i, $1 \leq i \leq g$ $2g$-tangles depicted in Figures 5, 2 (we assume that $\alpha_1 = \delta_1$).

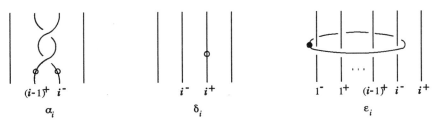

Figure 5

EXERCISE 2.2 [FR]. Let ζ be obtained from ξ by deletion of $+1$ (-1) framed unknotted component and making a full left (right) twist on the strings linked with it. Then $\zeta \underset{K}{\sim} \xi$. In particular, $2g$-tangle ϵ_i, $1 \leq i \leq g$ is K-equivalent to an admissible $2g$-tangle (e.g. for i=1 $\epsilon_1 \underset{K}{\sim} \delta_1$).

EXERCISE 2.3. By a sequence of moves K_1, K_2, $K_3^{\pm 1}$ the white circle of δ_i can be shifted from the string i^+ to i^-.

3. The tangle presentation of $M_{g,1}$, $M_{g,0}$

3.1. Let us return to the mapping class group $M_{g,1}$. Denote by a_i, b_i, d_i, e_i, $1 \leq i \leq n$ right-hand Dehn twists with respect to curves shown in Figure 6.

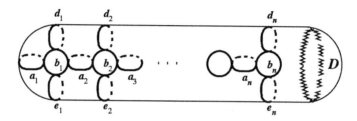

Figure 6

We are ready now to state the following

THEOREM 3.1. *There exists an isomorphism $\phi : T_{2g} \to M_{g,1}$ which maps α_i, β_i, δ_i, ϵ_i to a_i, b_i, d_i, e_i $(1 \leq i \leq g)$ respectively.*

Consider now the case of the mapping class group $M_{g,0}$ of a closed surface obtained by capping $\Sigma_{g,1}$ with a disk. Denote by T'_{2g} the quotient of T_{2g} by relations $\tau = 1$, $\epsilon_g = \delta_g$, where τ is the full twist on $2g$ strands. These relations are shown in Figures 7, 8.

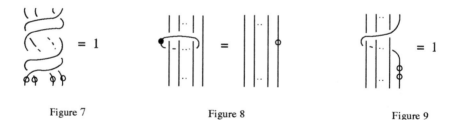

Figure 7   Figure 8   Figure 9

EXERCISE 3.2. T'_{2g} may be obtained as well by imposing relations $\tau = 1$, $\nu = 1$ depicted in Figures 7, 9.

The isomorphism of Theorem 3.1 induces the isomorphism of the quotient groups, so we obtain

PROPOSITION 3.3. *There exists an isomorphism $\phi' : T'_{2g} \to M_{g,0}$ which maps α_i, β_i, δ_i, ϵ_i to a_i, b_i, d_i, e_i $(1 \leq i \leq g)$ respectively.*

3.2.   The proof of Theorem 3.1 is based on the explicit construction of ϕ discussed below.

Let ξ be an admissible $2g$-tangle. We may assume that $\xi \subset B \times [0,1]$ for some large enough disk $B \subset \mathbb{R}^2$. For each i, $1 \le i \le g$ attach to $B \times [0,1]$ a handle $N(A_i) \subset \mathbb{R}^3$ of index one so that $N(A_i) \cap B \times [0,1]$ is the regular neighbourhood of $(i^- \cup i^+) \times \{1\}$ as shown in Figure 10. To close ξ from above (similarly to closure of a $2g$-plat) we add to ξ cores A_i of the handles: $\widetilde{\xi} = \xi \cup (\cup_i A_i)$. Denote by ξ' the union of all unclosed components of $\widetilde{\xi}$ (it consists of g arcs with the ends on $B \times \{0\}$).

DEFINITION.   The framed link $\overline{\xi} = \widetilde{\xi} \setminus \xi'$ is called *the closure* of $2g$-tangle ξ.

Cut out from the solid handlebody $H = B \times [0,1] \cup \left(\cup_i N(A_i) \right) \subset \mathbb{R}^3$ a tubular neighbourhood $N(\xi')$ of ξ'. We obtain 3-manifold $H_\xi = H \setminus \left(N(\xi') \right)$ with framed link $\overline{\xi}$ inside it, see Figure 10.

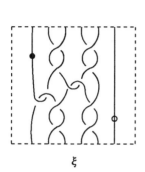
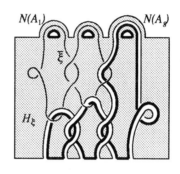

Figure 10

Denote by M_ξ 3-manifold obtained from H_ξ by surgery on the link $\overline{\xi}$. Considering separately the case of admissible braids and the case of $\beta_i^{\pm 1}$, $1 \le i \le g$ one can show that for any admissible $2g$-tangle ξ the manifold M_ξ is homeomorphic to $\Sigma_{n,1} \times [0,1]$ and the product structure induced on M_ξ is an extension of the natural product structure on $\partial B \times [0,1] \subset \partial H_\xi = \partial M_\xi$.

Note now that $\partial M_\xi = \partial H_\xi$ can be decomposed as $\partial M_\xi = \Sigma_{0\xi} \cup \partial B \times [0,1] \cup \Sigma_{1\xi}$, $j = 0,1$ where $\Sigma_{j\xi} \cap \partial B \times [0,1] = \partial B \times \{j\}$ and $\Sigma_{j\xi}$, $j = 0,1$ admits a natural identification $\kappa_j : \Sigma_{g,1} \to \Sigma_{j\xi}$ with standard genus g surface $\Sigma_{g,1}$. Let $p_\xi : \Sigma_{0\xi} \to \Sigma_{1\xi}$ be the restriction on $\Sigma_{0\xi}$ of the direct product projection $p : \partial M_\xi \to \Sigma_{1\xi}$. It remains to define the homeomorphism $\phi(\xi) : \Sigma_{n,1} \to \Sigma_{n,1}$ by $\phi(\xi) = \kappa_1^{-1} p_\xi \kappa_0$ and to notice that the map $\xi \mapsto \phi(\xi)$ determines correctly defined homomorphism $\phi : T_{2g} \to M_{g,1}$.

Remark 3.4. The identification $\kappa_j : \Sigma_{g,1} \to \Sigma_{j\xi}$ above can be constructed explicitly in the following manner: each of the surfaces $\Sigma_{j\xi}$ is the disk $B \times \{j\}$ with g handles ($\Sigma_{g,1}$ can be presented so as well- see Figure 11). The map κ_j maps B with g holes identically to $B \times \{j\}$ with g holes and the curves d_i and b_i are mapped to the meridians and "longitudes" of the corresponding handles (for better illustration we thickened images of b_i in Figure 10).

Figure 11   Figure 12

EXERCISE 3.5.   The homeomorphisms $\phi(\beta_i^{\pm 1})$ are isotopic to the twists along the curve b_i in positive and negative directions respectively (see Figure 12).

One easily obtains from the construction of ϕ that tangles α_i, β_i, δ_i, ϵ_i are mapped to a_i, b_i, d_i, e_i, $1 \leq i \leq g$ respectively. The proof of the fact that Ker $\phi = 1$ is based on straightforward but lengthy technical verification of the fact that the inverse map defined on the generators a_i, b_i, e_i of the mapping class group $M_{g,1}$ by $a_i \mapsto \alpha_i$, $b_i \mapsto \beta_i$, $e_i \mapsto \epsilon_i$ determines correctly defined homomorphism $\phi^{-1} : M_{g,1} \to T_{2g}$. B. Wajnryb's presentation of $M_{g,1}$ [W] is used to show that the relations of the mapping class group under this assignment hold for admissible $2g$-tangles; Figure 13 illustrates verification of the relation $\alpha_i \beta_i \alpha_i = \beta_i \alpha_i \beta_i$.

$$\beta_i \alpha_i \beta_i =$$ 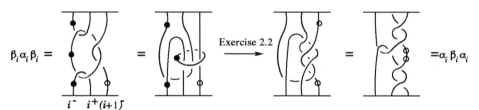 $$= \alpha_i \beta_i \alpha_i$$

Figure 13

4. The tangle presentation of $M_{g,1,n}$

In this section we will consider the mapping class group $M_{g,1,n}$ of a surface $\Sigma_{g,1,n}$ with n marked points p_i, $1 \leq i \leq n$. Without loss of generality one may assume that marked points p_i, $1 \leq i \leq n$ are contained in a disk $D \subset \Sigma_{g,1}$. Denote by s_i, $1 \leq i \leq n-1$ standard generators of the mapping class group of D with n marked points (i.e. generators of the classical braid group B_n) and let a_{g+1} be a Dehn twist with respect to the corresponding curve, see Figure 14.

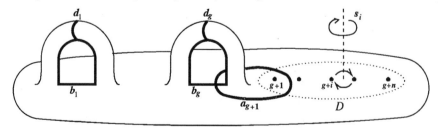

Figure 14

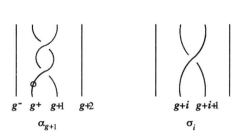

g^- g^+ g+1 g+2 $\qquad$ g+i g+i+1

α_{g+1} $\qquad\qquad$ σ_i

Figure 15

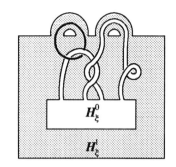

H_ξ^0

H_ξ^1

Figure 16

To handle the case of $M_{g,1,n}$ we modify the definition of admissible $2g$-tangles adding n unframed strings. More precisely, let $Y_{0,n}$ be a set of points $(0, g+i)$, $1 \le i \le n$ and $Y_{2g,n} = Y_{2g} \cup Y_{0,n}$. A $(2g+n)$-tangle ξ is a tangle in $\mathbb{R}^2 \times [0,1]$ with the boundary $\partial\xi = Y_{2g,n} \times \{0,1\}$. Denote by ξ' the union of all components of ξ with the ends on $Y_{0,n}$. We will further assume that tangle $\xi \setminus \xi'$ is framed, while ξ' is unframed. One can consider $(2g+n)$-braids with framed first $2g$ strings as a particular class of $(2g+n)$-tangles; $(2g+n)$-braid ξ is called *admissible*, if its first $2g$ strings $\xi \setminus \xi'$ form an admissible $2g$-braid. Similarly to the definition of Section 2, $(2g+n)$-tangle is *admissible* if it can be written as a finite product of admissible braids and tangles $\beta^{\pm 1}$, $1 \le i \le g$ (tangles $\beta^{\pm 1}$ may be considered as $(2g+n)$-tangles by addition of n trivial strings). Let also α_{g+1} and σ_i, $1 \le i \le n-1$ be $(2g+n)$-tangles depicted in Figure 15. We denote by $T_{2g,n}$ the group of $(2g+n)$-tangles up to K-equivalence. One may then formulate a theorem similar to Theorem 3.1:

PROPOSITION 4.1. *There exists an isomorphism $\phi : T_{2g,n} \to M_{g,1,n}$ which maps α_k, β_i, δ_i, ϵ_i, σ_j to a_k $(1 \le k \le g+1)$, b_i, d_i, e_i $(1 \le i \le g)$ and s_j $(1 \le j \le n-1)$ respectively.*

Remark 4.1. In case of $g = 0$ the group $T_{0,1,n}$ is, in fact, isomorphic to the geometrical braid group B_n and we obtain well-known presentation of the mapping class group $M_{0,1,n}$ of a disk with n marked points by geometrical n-braids.

5. From Heegard diagrams to surgery and back

5.1. Denote by H_g the standard handlebody of genus g in $\mathbb{R}^3 \subset S^3$ and let D be a disk in ∂H_g. By a Heegard diagram of 3-manifold M we mean a homeomorphism $h : \partial H_g \to \partial H_g$ such that $\overline{(S^3 \setminus H_g)} \cup_h H_g = M$. We may identify the group $M_{g,1}$ with the mapping class group of ∂H_g modulo D and to choose for h a representative $h \subset M_{g,1}$. Let $\xi = \phi^{-1}(h)$ be an admissible $2g$-tangle in T_{2g} corresponding to h.

Recall that the framed link $\bar\xi \subset H_\xi \subset S^3$ is obtained from ξ by closing with n small semicircles from above and removing the lower strings. Denote by $M_{\bar\xi}$ a manifold obtained from S^3 by surgery on $\bar\xi$.

THEOREM 5.1. *Manifolds $M_{\bar\xi}$ and $\overline{(S^3 \setminus H_n)} \cup_{\phi(\xi)} H_n$ are homeomorphic.*

Idea of the proof is based on the construction of ϕ described in Section 3, so we will use notatations and ideas of this section furter on. Let $B_1 \subset \mathbb{R}^2$ be a disk

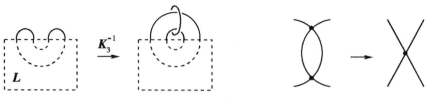

Figure 17   Figure 18

contained in B. Denote by H_ξ^0 solid handlebody $B_1 \times [-1/2, 0] \cup N(\xi')$ and by H_ξ^1 the handlebody $B \times [-1, 1] \cup \left(\cup_{i=1}^n N(A_i) \right)$, see Figure 16. Similarily to Section 3, 3-manifold obtained from $\overline{H_\xi^1 \setminus H_\xi^0}$ by surgery on $\overline{\xi}$ is homeomorphic to $\partial H_\xi^0 \times [0, 1]$. Let $p_\xi : \partial H_\xi^0 \to \partial H_\xi^1$ be the restriction of the direct product projection. Clearly $M_{\overline{\xi}}$ is homeomorphic to $\overline{S^3 \setminus H_\xi^1} \cup_{p_\xi} H_\xi^0$. Homeomorphisms $\kappa_0 : \Sigma_{n,1} \to \Sigma_{0\xi}$ and $\kappa_1 : \Sigma_{n,1} \to \Sigma_{1\xi}$ defined in Section 3 can be extended to homeomorphisms of solid handlebodies $\tilde{\kappa}_0 : H_n \to H_\xi^0$ and $\tilde{\kappa}_1 : \overline{S^3 \setminus H_n} \to \overline{S^3 \setminus H_\xi^1}$ respectively. A homeomorphism $f : \overline{S^3 \setminus H_n} \cup_{p_\xi} H_n \to \overline{S^3 \setminus H_\xi^1} \cup_{p_\xi} H_\xi^0 \approx M_{\overline{\xi}}$ may be defined now by $f(x) = \tilde{\kappa}_0(x)$ for $x \in H_n$ and by $f(x) = \tilde{\kappa}_1(x)$ for $x \in \overline{S^3 \setminus H_n}$.

5.2.   Let us construct now a Heegard diagram of 3-manifold M starting from any surgery presentation $M = M_L$, where L is a framed link in S^3.

Recall ([B1], [RS]) that g-component link $L \subset \mathbb{R}^3$ is said to be represented by pure $2g$-plat if it admits a diagram with only g local maxima (with respect to the projection on z-axis). Note also, that each pure $2g$-plat may be obtained as a plat closure of pure $2g$-braid. Denote by $m(L)$ the number of local maxima and by $g(L)$ the number of components of L. Suppose that $d(L) = m(L) - g(L) > 0$; then there exists a component of L with at least two local maxima. Using the move K_3^{-1} we may pass to a link with $d = d(L) - 1$ as shown in Figure 17. After $d(L)$ steps we will obtain a pure $2g$-plat L', which is a plat closure of some pure $2g$-braid η. Then $\xi = \eta \pi_g$, where $\pi_g = \prod_{i=1}^g \delta_i \beta_i \delta_i$, satisfies the following

THEOREM 5.2. *For any framed link $L \subset S^3$ there exist g and an admissible $2g$-tangle ξ, such that $L \underset{K}{\sim} \overline{\xi}$. Tangle ξ may be chosen in the form $\xi = \eta \pi_g$, where $\pi_g = \prod_{i=1}^g \delta_i \beta_i \delta_i$ and η is a pure $2g$-braid.*

For $h = \phi(\xi)$ we obtain now a Heegard diagram $\overline{S^3 \setminus H_n} \cup_h H_n$ of M in view of Theorem 5.1.

6. Some applications

6.1.   Let us briefly recall the notion of arborescent link (see [3]).

DEFINITION.   A link $L \subset \mathbb{R}^3$ is called *arborescent*, if it admits a projection which, if considered as multigraph, can be reduced to one-vertex multigraph (i.e. to figure eight) by a sequence of local moves shown in Figure 18.

Let ξ be an admissible 4-tangle presented as a product of α_1, α_2, β_1, β_2, δ_2 and let $\overline{\xi}$ be its closure. One may easily deduce that $\overline{\xi}$ is arborescent, so by Theorems 3.1, 5.1 one obtains

PROPOSITION 6.1. *Any 3-manifold of Heegard genus two may be obtained from* S^3 *by surgery on a framed arborescent link.*

6.2.   Reshetikhin and Turaev [RT2] established invariants τ_M of 3-manifolds using surgery, Kirby calculus and representation theory of $U_q(sl_2)$ in roots of unity. Kohno [Ko] defined a similar invariant (refined later by Capell et.al. [CLM]) using a very different approach, involving Heegard decomposition of a manifold, barion graphs and Kniznik-Zamolodchikov equation or SU(2)-level k conformal field theory. Following [CLM] we will denote this invariant by $I(M)$. Passing from Heegard decompositions to surgery as described in Section 5 (and using quantum group interpretation of the WZW models of RCFT discussed in [Ko],[AGS]) one may show that these two approaches are in fact equivalent (see also [CLM]):

PROPOSITION 6.2. *Invariants τ_M and $I(M)$ coincide up to a phase factor.*

6.3.   By Theorem 5.2 we obtain that any 3-manifold M has a Heegard diagram $\overline{S^3 \setminus H_g} \cup_h H_g$, where $h : \partial H_g \to \partial H_g$ has the form $h = \phi(\eta \pi_g)$ for some pure 2g-braid η and $\pi_g = \prod_{i=1}^g \delta_i \beta_i \delta_i$. Homeomorphism $p_g = \phi(\pi_g)$ permutes meridians d_i, $1 \le i \le g$ with the corresponding longitudes b_i , so it can be considered as homeomorphism $p_g : \overline{S^3 \setminus H_g} \to H_g$.

We conclude that M is homeomorhic to $H_g \cup_u H_g$ for $u = p_g^{-1}h$. Note that $\phi(\eta)$ preserves meridians (η is a pure braid), hence $u = p_g^{-1}h = p_g^{-1}\phi(\eta)p_g$ preserves longitudes and we obtain the following version of J. Birman's theorem on the existence of special Heegard diagrams [B2], [RS]:

PROPOSITION 6.3. *Any closed orientable 3-manifold can be obtained by pasting together two copies of standard handlebody $H_g \subset S^3$ via a homeomorphism u : $\partial H_g \to \partial H_g$ which is fixed on all longitudes b_i, $1 \le i \le g$.*

7. Proof of the Kirby theorem

We start with some preliminaries. Let ξ_0 be a closed, 0-framed component of ξ which bounds an embedded disk $D \subset \mathbb{R}^2 \times [0,1]$. Suppose that $D \cap (\xi \setminus \xi_0)$ consists of exactly one point lying on some closed component $\xi_1 \subset \xi$. We define a move K_4 on ξ_0 to be a deletion of components ξ_0 and ξ_1.

EXERCISE 7.1.   The move K_4 may be expressed via K_1 and K_2. (Hint: due to the presence of ξ_0 we may change overcrossings of ξ_1 with itself and with other components of ξ to undercrossings by K_2, hence we can unlink and unknot ξ_1 so that ξ_0 and ξ_1 will form the Hopf link aside from the other components. We may as well make the framing of ξ_1 to be +1. It remains to unlink ξ_0 and ξ_1 by K_2 and delete them by K_1).

Let us verify how an application of moves $K = K_1$, K_2 or K_3 changes the closure $\overline{\xi}$ of an admissible 2g-tagle ξ. If ζ is obtained from ξ by an application of K, $\overline{\zeta}$ is either obtained from $\overline{\xi}$ by the same move, or $\overline{\zeta} = \overline{\xi}$, or (for K_3 move involving a string ending on $Y_{2g} \times \{0\}$) is obtained from $\overline{\xi}$ by K_4, so we have the following

LEMMA 7.2. *Let ξ, ζ be K-equivalent admissible 2g-tangles. Then the corresponding links $\overline{\xi}$, $\overline{\zeta}$ are K-equivalent.*

Moreover, as we are going to show below, multiplication of ξ by a tangle corresponding to a homeomorphism extendable to the inner or outer handlebody also does not change K-equivalence class of $\overline{\xi}$.

Denote by $TI_{2g} \subset T_{2g}$ a subgroup of admissible $2g$-tangles corresponding to a subgroup I_g of homeomorphisms $\partial H_g \to \partial H_g$ which are extendable to H_g. It is known (see [S], [MF]) that the subgroup I_g is generated by $t_1 = b_1 a_2 a_1 b_1 \in I_g$ (so called handle slide) and homeomorphisms of ∂H_n which preserve the union $\cup_{i=1}^n d_i$ of meridians of H_n. It is easy to verify that any homeomorphism of ∂H_g which preserve the union of meridians may be written as $\phi(\eta)$ for some admissible $2g$-braid η (and vice versa), so the corresponding subgroup TI_{2g} is generated by $\tau_1 = \beta_1 \alpha_2 \alpha_1 \beta_1$ and admissible $2g$-braids. One immediately deduce that $\overline{\xi}$ may be obtained from $\overline{\xi\tau_1}$ by K_4 and $\overline{\xi\eta}$ coincides with $\overline{\xi}$ for any admissible $2g$-braid η, so the following holds:

LEMMA 7.3. *Let γ_I be an arbitrary tangle in the subgroup TI_{2g}. Then the links $\overline{\xi}$ and $\overline{\xi\gamma_I}$ are K-equivalent for any admissible $2n$-tangle ξ.*

In a similar way, denote by $TO_{2g} \subset T_{2g}$ a subgroup of admissible $2g$-tangles corresponding to homeomorphisms of ∂H_g which are extendable to $\overline{S^3 \setminus H_g}$. Recall (Section 6.3) that homeomorphism $p_g = \phi(\pi_g)$ for $\pi_g = \prod_{i=1}^g \delta_i \beta_i \delta_i$ permutes each meridian d_i with the corresponding longitude b_i, so it can be considered as homeomorphism $p_g : \overline{S^3 \setminus H_g} \to H_g$. Therefore for any $\gamma_O \in TO_{2g}$ an admissible $2g$-tangle $\pi_g \gamma_O \pi_g^{-1}$ belongs to TI_{2g}, so $\gamma_O = \pi_n^{-1} \gamma_I \pi_n$ for some $\gamma_I \in TI_{2g}$. Considering separately the cases when γ_I is either an admissible $2g$-braid or τ_1, one can pass from $\overline{\gamma_O \xi}$ to $\overline{\xi}$ by n and $n+1$ applications of K_4 respectively, so we have the following

LEMMA 7.4. *Let γ_O be an arbitrary tangle in the subgroup TO_{2g}. Then the links $\overline{\xi}$ and $\overline{\gamma_O \xi}$ are K-equivalent for any admissible $2n$-tangle ξ.*

Note that addition of $2r$ new vertical strings to each admissible $2g$-tangle ξ defines a natural embedding $i_{g,g+r} : T_{2g} \to T_{2(g+r)}$. Obviously, $\overline{i_{g,g+r}(\xi)} = \overline{\xi}$.

DEFINITION . Admissible tangles $\xi \in T_{2m}$, $\zeta \in T_{2n}$ are called *stably equivalent* ($\xi \underset{st}{\sim} \zeta$), if $i_{m,N}(\xi) = \gamma_O i_{n,N}(\zeta) \gamma_I$ for some $N \geq m, n$ and $\gamma_O \in TO_{2N}$, $\gamma_I \in TI_{2N}$.

We are in a position to prove the Kirby theorem. Let L_1, L_2 in S^3 be two links such that $M_{L_1} = M_{L_2}$. By Theorem 5.2 there exist admissible tangles $\xi \in T_{2m}$, $\zeta \in T_{2n}$ for some m, n such that $L_1 \underset{K}{\sim} \overline{\xi}$, $L_2 \underset{K}{\sim} \overline{\zeta}$. It follows from Theorem 5.1 that $\overline{S^3 \setminus H_m} \cup_{\phi(\xi)} H_m = \overline{(S^3 \cdot \setminus H_n)} \cup_{\phi(\zeta)} H_n$. By the Reidemeister-Singer theorem [R] any two Heegard diagrams of the same 3-manifold are stably equivalent. Applying this theorem to tangles via the isomorphism ϕ^{-1} of Theorem 3.1 we obtain that tangles ξ and ζ are stably equivalent. The closure of stably equivalent tangles are K-equivalent by Lemmas 7.2-7.4, so $L_1 \underset{K}{\sim} \overline{\xi} \underset{K}{\sim} \overline{\zeta} \underset{K}{\sim} L_2$ which finishes the proof.

REFERENCES

[AGS] L. Alvarez-Gaumé, C. Gomez, G. Sierra, *Quantum group interpretation of some conformal field theories*, Physics letters B **220**:1 (1989), 142-152.

[B1] J. Birman, *Braids, links and mapping class groups*, Ann. Math. Studies **82** (1975), 3-227.

[B2] J. Birman, *Special Heegard splittings for closed oriented 3-manifolds*, Topology **17** (1978), 157-166.

[CLM] S. Capell, R. Lee, E. Miller, *Invariants of 3-manifolds from conformal field theory*, preprint.

[C] A. Caudron, *Classification des noeuds et des enlacements*, Notes de recherche 76-80, Publ. Math. D'Orsay **82-04**, 350p.

[FR] R. Fenn, C. Rourke, *on Kirby's calculus of links*, Topology **18** (1979), 1-15.

[Ki] R. C. Kirby, *A calculus for framed links in S^3*, Invent. Math **45** (1978), 35-56.

[Ko] T. Kohno, *Topological invariants for 3-manifolds using representation of mapping class groups*, preprint.

[L] Ning Lu, *A simple proof of the fundamental theorem of Kirby calculus on links*, Trans. Amer. Math. Soc. **331** (1992), 143-156.

[MF] S. Matveev, A. Fomenko, Algorithmical and computer methods in 3-manifold topology, Moscow University Press (1991), 300p. (Russian).

[R] K. Reidemeister, *Zur dreidimensionalen Topologie*, Ann. Math. Sem. Univ. Hamburg **9** (1983), 184-194.

[RS] E. Rego, E.C. de Sa, *Special Heegard diagrams and the Kirby calculus*, Topology and its Appl. **37** (1990), 11-24.

[RT1] N. Yu Reshetikhin, V. G. Turaev, *Ribbon graphs and their invariants derived from quantum groups*, Commun. Math. Phys. **127** (1990), 1-26.

[RT2] N. Yu Reshetikhin, V. G. Turaev, *Invariants of 3-manifolds via link polynomials and quantum groups*, Invent. Math. **103** (1991), 547-597.

[S] S. Suzuki, *On homeomorphisms of a 3-dimensional handlebodies*, Can. J. Math **29** (1977), no.4, 111-129.

[W] B. Wajnryb, *A simple presentation for the mapping class group of an orientable surface*, Israel J. Math. **45** (1983), 157-174.

DEPARTMENT OF MATHEMATICS, CHELYABINSK STATE UNIVERSITY, CHELYABINSK 454136 RUSSIA
E-mail address: matveev@cgu.chel.su

SCHOOL OF MATHEMATICAL SCIENCES, TEL AVIV UNIVERSITY, RAMAT-AVIV 69978 ISRAEL
Current address: DEPARTMENT OF MATHEMATICS, UNIVERSITY OF CALIFORNIA, BERKELEY CA 94720
E-mail address: polyak@math.berkeley.edu

Contemporary Mathematics
Volume **164**, 1994

Thin position for 3-manifolds

MARTIN SCHARLEMANN and ABIGAIL THOMPSON

ABSTRACT. We define thin position for 3-manifolds, and examine its relation to Heegaard genus and essential surfaces in the manifold. We show that if the width of a manifold is smaller than its Heegaard genus then the manifold contains an essential surface of genus less than the Heegaard genus.

The concept of thin position for knots in the 3-sphere was developed by D. Gabai [2]. Its relation to bridge position for knots and to incompressible meridinal planar surfaces in the knot complement is discussed in [3]. We define an analogous notion of thin position for 3-manifolds, and discuss its relation to Heegaard splittings, Heegaard genus, and incompressible surfaces.

For illustrative purposes we will restrict to connected orientable closed manifolds. In principle the arguments here could be extended not only to non-orientable closed manifolds, but in fact to all compact 3-manifolds, merely by regarding a compact 3-manifold as a cobordism from one (possibly empty) collection of

1991 Mathematics Subject Classification. Primary 57M25.
Both authors are supported in part by a grant from the National Science Foundation.
The second author is supported in part by a grant from the Alfred P. Sloan Foundation.
This paper is in final form and no version of it will be submitted for publication elsewhere.

boundary components to the collection of remaining components. More elaborate generalizations are also possible, in which the sequential ordering of the intermediate surfaces, analogous to the ordering of vertices in a linear graph, is replaced by a more complicated configuration, analogous to vertices in a more complicated graph.

Any closed orientable 3-manifold M can be constructed as follows: begin with some 0-handles, add some 1-handles, then some 2-handles, then some more 1-handles, then some more 2-handles, etc., and conclude by adding some 3-handles. Of course M can be built less elaborately: in the previous description, all the 1-handles can be added at once, followed by all the 2-handles. This corresponds to a Heegaard splitting of the manifold; the 0- and 1-handles comprise one handlebody of the Heegaard splitting, the 2- and 3-handles the other. The idea of thin position is to build the manifold as first described, with a succession of 1-handles and 2-handles chosen to keep the boundaries of the intermediate steps as simple as possible.

DEFINITIONS: For M a connected, closed, orientable 3-manifold, let $M = b_0 \cup N_1 \cup T_1 \cup N_2 \cup T_2 \cup ... \cup N_k \cup T_k \cup b_3$, where b_0 is a collection of 0-handles, b_3 is a collection of 3-handles, and for each i, N_i is a collection of 1-handles and T_i is a collection of 2-handles. We think of building M in steps, starting with b_0, then adding N_1, then T_1, etc.
Let S_i, $1 \leq i \leq k$, be the surface obtained from $\partial[b_0 \cup N_1 \cup T_1 \cup N_2 \cup T_2 \cup ... \cup N_i]$ by deleting all spheres bounding 0- or 3-handles in the decomposition. Let F_i, $1 \leq i \leq k-1$, be the surface obtained from $\partial[b_0 \cup N_1 \cup T_1 \cup N_2 \cup T_2 \cup ... \cup T_i]$ by similarly deleting all such spheres.
Let W_i = (collar of F_{i-1}) $\cup N_i \cup T_i$ together with every 0- and 3-handle incident to N_i or T_i. W_i is divided by a copy of S_i into two compression bodies: $\overline{N}_i$ = (0-handles) $\cup$ (collar of F_{i-1}) $\cup N_i$ and $\overline{T}_i$ = (collar of S_i) $\cup T_i \cup$ (3-handles). Thus S_i describes a Heegaard splitting of W_i into compression bodies $\overline{N}_i$ and $\overline{T}_i$.

Our goal will be to find a decomposition for which the S_i are as simple as possible. To that end, for S a closed connected orientable surface other than the sphere, define the complexity $c(S) = 1 - \chi(S) = 2\text{genus}(S) - 1$. Define $c(S^2) = 0$. For S not necessarily connected define $c(S) = \sum\{c(S') \mid S'$ a component of $S\}$. The complexity is then always decreased by 2-surgery along an essential circle and is unchanged by 2-surgery along an inessential circle.

Define the <u>width of the decomposition</u> of M to be the set of integers $\{c(S_i) \mid 1 \leq i \leq k\}$. There is a natural way to order finite multi-sets of integers: arrange each multi-set of integers in monotonically non-increasing order, then compare the ordered multi-sets lexicographically. For example the multi-set $\{3, 3, 5, 3, 2, 1\}$ is less than the multi-set $\{2, 2, 5, 3, 4\}$ since [5, 3, 3, 3, 2, 1] precedes [5, 4, 3, 2, 2] in lexicographic order. Define the <u>width</u> w(M) of M to be the minimal width over all decompositions, using the above ordering on finite multi-sets of integers. A given decomposition is <u>thin</u> if the width of the decomposition is the width of M.

EXAMPLE: Suppose M is the connected sum of two lens spaces. Then M has a Heegaard splitting of genus two. The width of this decomposition is $\{c(\text{genus two surface})\} = \{3\}$. But the decomposition can be re-arranged to a thin decomposition: (0-handle) $\cup$ (1-handle) $\cup$ (2-handle) $\cup$ (1-handle) $\cup$ (2-handle) $\cup$ (3-handle), which has width $\{1, 1\}$. This is then the width of M.

NOTE: It is straightforward to show in general that width is "additive" in the obvious manner under the connect-sum operation: $w(M_1 \# M_2) = w(M_1) \cup w(M_2)$.

There are some obvious ways in which a decomposition of M might be thinned: If a component of an F_i is an inessential sphere , then the ball it bounds contains some 1- and 2-handles, since we removed from F_i any sphere which merely bounds a 0- or 3-handle. Replace the ball with either a 0- or a 3-handle, depending on which side of F_i the ball lies on. This will thin the decomposition. We conclude:

RULE 1: *In a thin decomposition, any sphere component of any F_i is essential.*

If there is a component of an F_{i-1} to which no 1-handle is attached, but some 2-handles in T_i are (so the component persists into S_i), the decomposition would be thinner if those 2-handles were considered to be part of T_{i-1}. Similarly, if there is a component of an F_{i-1} to which 1-handles in N_i are attached, but no 2-handles in T_i, the decomposition would be thinner if those 1-handles were regarded as part of N_{i+1} instead. We conclude:

RULE 2: *In a thin decomposition, each component of F_{i-1} either persists into F_i, or has handles from both N_i and T_i attached to it.*

There is a less obvious way to thin a decomposition: Recall that S_i describes a Heegaard splitting of W_i. This Heegaard splitting is called <u>weakly reducible</u> [1] if there are essential disks D_N in $\overline{N_i}$ and D_T in $\overline{T_i}$ such that $\partial D_N \cap \partial D_T = \varnothing$ in S_i. In particular, if W_i contains a 1-handle in one component and a 2-handle in another, then W_i is automatically weakly reducible.

RULE 3: *In a thin decomposition, no W_i is weakly reducible.*

PROOF: Suppose W_i is weakly reducible. Remove a neighborhood of D_N from $\overline{N}$, converting it into a compression body N' with one fewer 1-handle or one more component. Then attach a 2-handle with core D_T to N'. Next attach the 1-handle dual to D_N, followed by the remainder of the 2-handles. This replaces the original decomposition in W_i, which yields the integer $\{c(S_i)\}$, by a decomposition with two surfaces S_{i_-} and S_{i_+}, obtained from S_i by compressing along ∂D_N and ∂D_T respectively. Since $c(S_{i_\pm}) < c(S_i)$ the new decomposition is thinner.

As an immediate corollary we have:

RULE 4: *In a thin decomposition, all handles in N_i and T_i are incident to the same component of S_i, called the* active *component of S_i.*

PROOF: Otherwise W_i would contain two components each with both 1- and 2-handles, so W_i would be weakly reducible.

EXAMPLE: If S is the splitting surface of a weakly reducible Heegaard splitting of M, then width(M) < $\{c(S)\}$.

LEMMA: *If $\partial W_i = F_{i-1} \cup F_i$ is compressible in W_i, then W_i is weakly reducible.*

PROOF: See [1].

RULE 5: *In a thin decomposition, each component of each F_i is incompressible in M.*

PROOF: Let D be a compressing disk for a component of F_i. Let $F = \cup F_i$. By an innermost disk argument on $D \cap F$ we can assume $D \cap F = \partial D \subset F_i$, so D lies entirely inside either W_i or W_{i+1}, say the former. The lemma implies that W_i is weakly reducible, contradicting Rule 3.

DEFINITION: A separating surface S in a 3-manifold M is weakly incompressible (see [3]) if any two compressing disks for S on opposite sides of S intersect along their boundary.

RULE 6: *In a thin decomposition of a 3-manifold, each surface S_i is weakly incompressible.*

PROOF: By rule 5, F is incompressible, so we can

assume any compressing disk for S_i lies in W_i. Hence if S_i is not weakly incompressible, W_i is weakly reducible, contradicting rule 3.

RULE 7: *If M is irreducible and not a Lens space, then in a thin decomposition no component of any S_i is a torus. In particular no number in w(M) is smaller than 3.*

PROOF: Suppose C is a torus component of some S_i. The decomposition can't be a Heegaard splitting, since M is not a Lens space. So assume that k > 1 and, with no loss of generality, assume that C is the active component of S_i. Let W be the component of W_i containing C. Then $\partial W \subset F$ is a non-empty (since S_i is not a Heegaard splitting) collection of spheres. Rule 1 says that such a sphere component of F would be essential, which is impossible since M is irreducible.

COROLLARY: *Let g be the Heegaard genus of M. If M is irreducible and contains no incompressible surfaces of genus < g, then a minimal genus Heegaard splitting for M is a thin decomposition. Hence w(M) = {2g - 1}.*

In particular, this is true if M is irreducible and non-Haken.

COROLLARY: *Let Σ be a homotopy 3-sphere. If w(Σ) < {5} then Σ = S^3.*

PROOF: We may as well take Σ to be irreducible, and it clearly contains no incompressible surfaces, so if g is the Heegaard genus of Σ, w(Σ) = {2g-1}. If 2g - 1 < 5, then g $\leq$ 2, which forces Σ to be standard.
More generally:

COROLLARY: *If w(M) < {5} then M is a 2-fold cover of $\#(S^1 \times S^2)$ branched along a link.*

PROOF: Choose a thin decomposition of M and let S =

US_i and $F = UF_i$. Then each component of $M - (S \cup F)$ is a (possibly punctured) compression body H with genus$(\partial_+H) \le 2$. On each such compression body there is an obvious involution (rotation by π about the illustrated axis) with 1-dimensional fixed point set. The quotient is a possibly punctured 3-ball. It's well-known that this involution commutes with the mapping class group of the boundary surfaces, so it extends across the gluing maps when the compression bodies are glued together along S and F. The quotient of M is obtained from punctured 3-balls by identifying boundary components, so it's a possibly trivial (= S^3) connected sum of ($S^1 \times S^2$)'s.

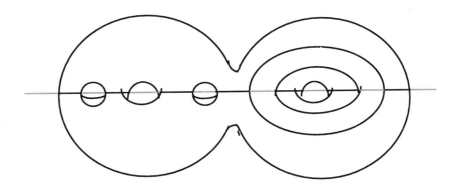

EXAMPLE: Let Q be a genus $g > 0$ orientable surface and $M = Q \times S^1$. We will show that M has width $\{3, 3, ..., 3\}$, with $2g$ copies of 3.

Step 1: A decomposition of this width exists: First observe that (annulus) $\times S^1$ and (pair of pants) $\times S^1$ each admit a genus 2 Heegaard decomposition in which precisely two boundary components lie on one of the compression bodies. But Q can be constructed by successively attaching $2g - 2$ pairs of pants to an annulus, and then topping off with a final annulus. Upon crossing this picture with S^1, the Heegaard splittings of each piece amalgamate to give the required decomposition of (surface) $\times S^1$.

Step 2: We show that no thinner decomposition exists.

Let $w(M)$ be the set of integers from a thin decomposition. It follows from rule 7 and step 1 that the only integer in $w(M)$ is 3 - say $w(M)$ consists of r copies of 3. Again by rule 7 each S_i, $1 \leq i \leq r$ must then be a single genus two surface. It follows that each W_i is connected and that $b_0 \cup N_1$ is a genus two handlebody. No F_i can be a sphere, so each W_i, $1 < i \leq r$ contains a single 1-handle. Hence M can be constructed with $(r+1)$ 1-handles. But $H_1(M)$ has rank $2g + 1$, so we conclude $r \geq 2g$, as required.

REFERENCES

[1] Casson, A. and Gordon, C. McA., Reducing Heegaard splittings, Topology and its applications, 27 (1987), 275-283.

[2] Gabai, D., Foliations and the topology of 3-manifolds III, J. Diff. Geometry, 26 (1987), 479-536.

[3] Thompson, A., Thin position and bridge number for knots in the 3-sphere, preprint.

MATHEMATICS DEPARTMENT, U.C. SANTA BARBARA, SANTA BARBARA, CALIFORNIA 93106
E-mail address: mgscharl@math.ucsb.edu

MATHEMATICS DEPARTMENT, U.C. DAVIS, DAVIS, CALIFORNIA, 95616
E-mail address: thompson@math.ucdavis.edu

Contemporary Mathematics
Volume **164**, 1994

On Embeddings of a Graph into R^3

KOUKI TANIYAMA

§1. Introduction

Let G be a finite graph. Then G can be embedded, in various ways, into the three-dimensional Euclidean space R^3. The placement problem of G in R^3 used to be considered as a branch or an application of knot theory. On the other hand a knot can be regarded as a spatial graph. Therefore there must be a theory of spatial graphs that generalizes knot theory. Some results on knots under ambient isotopy have been generalized in [8], [18], [26], [17], [16], [23], [9] etc, but many recent link invariants are not yet generalized to spatial graphs except trivalent graphs or rigid vertex graphs cf. [7], [25], [15], [6] etc.

On the other hand there are some other important equivalence relations on knots and links. They are, among others, cobordism, isotopy and link homotopy [2], [12], [11]. These concepts have some advantages over ambient isotopy. For example (1) the cobordism classes of knots form an abelian group under connected sum of knots, and (2) recently links were classified up to link homotopy in [5].

In [20], as a generalization of these concepts for links, the author defined eight equivalence relations on the set of embeddings of a graph into R^3 as follows.

Definitions.

Let $f, g : G \to R^3$ be spatial embeddings of a graph G. Let $I = [0, 1]$ be the unit closed interval. We say that a map $\Phi : G \times I \to R^3 \times I$ is

(a) *level preserving* if there is a map $\phi_t : G \to R^3$ for each $t \in I$ such that $\Phi(x, t) = (\phi_t(x), t)$ for all $x \in G$, $t \in I$.

1991 Mathematics Subject Classification. Primary 57M25. Secondary 57M15, 05C10.

The detailed version of this paper has been submitted for publication elsewhere.

(b) *locally flat* if each point of the image of Φ has a neighbourhood N such that the pair $(N, N \cap \Phi(G \times I))$ is homeomorphic to the standard disk pair (D^4, D^2) or $(D^3 \times I, X_n \times I)$ for some non-negative integer n where (D^3, X_n) is shown in Fig. 1.1.

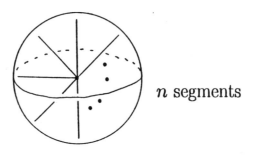

n segments

Fig. 1.1

(c) *between f and g* if there is a real number $\epsilon > 0$ such that $\Phi(x, t) = (f(x), t)$ for all $x \in G$, $0 \leq t \leq \epsilon$ and $\Phi(x, t) = (g(x), t)$ for all $x \in G$, $1 - \epsilon \leq t \leq 1$.

We say that f and g are

(1) *ambient isotopic* if there is a continuous family $h_t : R^3 \to R^3$, $0 \leq t \leq 1$ of self-homeomorphisms such that $h_0 = id_{R^3}$ and $h_1 \circ f = g$. Or equivalently, there is a level preserving locally flat embedding $\Phi : G \times I \to R^3 \times I$ between f and g.

(2) *cobordant* if there is a locally flat embedding $\Phi : G \times I \to R^3 \times I$ between f and g.

(3) *isotopic* if there is a level preserving embedding $\Phi : G \times I \to R^3 \times I$ between f and g.

(4) *I-equivalent* if there is an embedding $\Phi : G \times I \to R^3 \times I$ between f and g.

(5) *(link) homotopic* if g is obtained from f by a series of self-crossing changes (Fig. 1.2) and ambient isotopy.

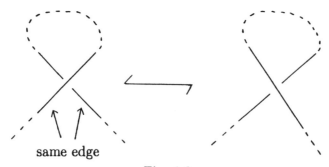

same edge

Fig. 1.2

(6) *weakly (link) homotopic* if g is obtained from f by a series of crossing changes of adjacent edges (Fig. 1.3) and ambient isotopy (a self-crossing change of a loop is considered as a crossing change of adjacent edges).

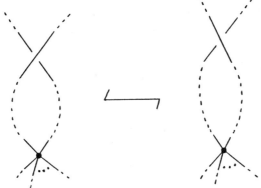

Fig. 1.3

(7) *homologous* if there is a locally flat embedding
$\Phi : (G \times I)\#\bigcup_{i=1}^{n} S_i \to R^3 \times I$ between f and g where n is a natural number, S_i is a closed orientable surface and $\#$ means the connected sum. More precisely, there is an edge e of G for each i such that S_i is attached to the open disk $\text{int}(e \times I)$ by the usual connected sum of surfaces.

(8) Z_2-*homologous* if there is a locally flat embedding
$\Phi : (G \times I)\#\bigcup_{i=1}^{n} S_i \to R^3 \times I$ between f and g where n is a natural number and S_i is a closed (possibly non-orientable) surface.

Thus there are many approaches to spatial graphs. For example the author observed in [19] that the cobordism classes of spatial embeddings of the theta curve form a group under vertex connected sum. Miyazaki showed that this group is not abelian [13].

In [20] the author proved the following fundamental relations between these concepts. This is a generalization of the corresponding results for links [1], [3], [4].

Theorem 1.1.

$$(1) \nearrow^{(2)}_{\searrow}{}_{(3)}^{\nearrow \searrow} (4) \to (5) \to (6) \to (7) \to (8).$$

Thus we are able to approach the ambient isotopy classification problem, which is a main goal of spatial graph theory, step by step. In §3 we will state the homology classification.

§2. Unknotting theorems

Let X and Y be topological spaces. Suppose that X is embeddable into Y. The first question is whether or not there are different embeddings of X into Y. Many unknotting theorems, like those of Jordan-Schönflies, Brown, Zeeman etc., are known as basic theorems of topology. In our case, if G is a forest then it is clear that any two embeddings of G into R^3 are ambient isotopic. Conversely if G is not a forest then G contains a cycle and hence there are many different embeddings of G up to ambient isotopy because the cycle can be knotted in various ways. In order to generalize this observation we make the following definition.

Let $\mathcal{A}$ be one of (1)~(8). A graph G is called *unique up to $\mathcal{A}$-equivalence* if any two embeddings of G into R^3 are $\mathcal{A}$-equivalent.

In [20] the author proved the following results.

Theorem 2.1.

For a graph G, the following conditions are mutually equivalent:

(i) *G is unique up to ambient isotopy.*

(ii) *G is unique up to cobordism.*

(iii) *G is a forest.*

(iv) *G does not contain any subdivision of the loop (Fig.2.1 (a)).*

Theorem 2.2.

For a graph G, the following conditions are mutually equivalent:

(v) *G is unique up to isotopy.*

(vi) *G is unique up to I-equivalence.*

(vii) *G is unique up to link homotopy.*

(viii) *There is a vertex v of G such that $G - v$ is a forest.*

(ix) *G does not contain any subdivision of the graphs Fig.2.1 (b), (c) and (d).*

Theorem 2.3.

For a graph G, the following conditions are mutually equivalent:

(x) *G is unique up to weak link homotopy.*

(xi) *G is unique up to homology.*

(xii) *G is unique up to Z_2-homology.*

(xiii) *G is a planar graph which does not contain disjoint cycles.*

(xiv) *G does not contain any subdivision of the graphs Fig.2.1 (b),(e)* and (f).

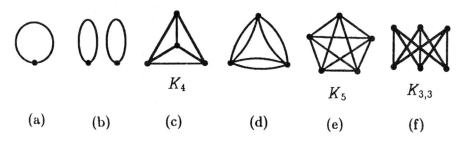

Fig. 2.1

These results show how intrinsic properties of a graph affect extrinsic properties of the spatial embeddings of the graph.

§3. Homology classification

In the summer of 1990, J. Simon gave a lecture in Tokyo. In the lecture he defined an integer invariant for spatial embeddings of the complete graph K_5 and the complete bipartite graph $K_{3,3}$ respectively. See [20] for the definitions. After his lecture the author generalized his invariant for an arbitrary graph G along his original idea in the lecture. The definition was purely combinatorial, cf. §2 of [21]. Recently the author found that this invariant coincides with an invariant defined by W. T. Wu in 1960 [24] in an intrinsic manner as follows.

For a topological space X let $C_2(X)$ be the configuration space of ordered pair of (distinct) points in X. Let σ be the involution on $C_2(X)$ that exchanges the order of the two points, i.e. $\sigma(x,y) = (y,x)$. Let $f : G \to R^3$ be an embedding. Let $f^2 : C_2(G) \to C_2(R^3)$ be the map defined by $f^2(x,y) = (f(x), f(y))$. Then f^2 induces a homomorphism

$$(f^2)^\# : H^2(C_2(R^3), \sigma) \to H^2(C_2(G), \sigma)$$

where $H^2(C_2(X), \sigma)$ denotes the skew-symmetric second cohomology of the pair $(C_2(X), \sigma)$. Namely $H^2(C_2(X), \sigma)$ is the second cohomology of the subcomplex $A_*(C_2(X), \sigma)$ of the singular chain complex $A_*(C_2(X))$ defined by $A_*(C_2(X), \sigma) = \{a \in A_*(C_2(X)) \mid \sigma(a) = -a\}$. It is known that $H^2(C_2(R^3), \sigma)$ is an infinite cyclic group, see [24] or [21]. Let τ be a fixed generator of $H^2(C_2(R^3), \sigma)$. Then Wu defined an invariant of f by $(f^2)^\#(\tau)$. We will denote this element of $H^2(C_2(G), \sigma)$ by $\mathcal{L}(f)$. See [21] for more details.

Wu actually defined his invariant for embeddings of a k-dimensional simplicial complex into R^j. As far as the author knows nothing is known in our case $k = 1$, $j = 3$. It seems to the author that Wu's work is not well-known to knot theorists, but Wu's invariant is a most fundamental geometric invariant in spatial graph theory. In fact it generalizes linking number, and it has clear geometric meaning as well as being explicitly calculable from any regular diagram of the embedding [21]. It is notable that the extended Reidemeister move V of spatial graphs corresponds to a 2-coboundary of $C_2(G)$, as suggested by the schematic picture Fig. 3.1.

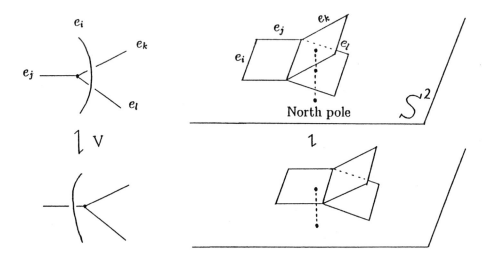

Fig. 3.1

Moreover the author proved the following in [21].

Theorem 3.1.
   Two embeddings f and g of a finite graph G into R^3 are homologous if and only if $\mathcal{L}(f) = \mathcal{L}(g)$.

In other words Wu's invariant is a complete homology invariant. In [21] the author defined $\mathcal{L}_2(f)$ as a modulo 2 reduction of $\mathcal{L}(f)$ and showed that $\mathcal{L}_2(f)$ is a complete Z_2-homology invariant.
   Finally we would like to propose an invariant as follows.
   Let X be a topological space and $X^n = \underbrace{X \times \cdots \times X}_{n}$. Let $1 \leq i < j \leq n$ and let $\sigma_{ij} : X^n \to X^n$ be the involution defined by

$$\sigma_{ij}(x_1, \cdots, x_i, \cdots, x_j, \cdots, x_n) = (x_1, \cdots, x_j, \cdots, x_i, \cdots, x_n).$$

We will denote any restriction of σ_{ij} by σ_{ij} also. Let $C_n(X)$ be the configuration space of ordered n-tuples of distinct points in X. Namely $C_n(X)$ is a subspace of X^n defined by

$$C_n(X) = \{(x_1, \cdots, x_n) \in X^n \mid x_i \neq x_j \text{ if } i \neq j\}.$$

Let Y be another topological space and $f : X \to Y$ be a continuous map. Then a continuous map $f^n : X^n \to Y^n$ is defined by $f^n(x_1, \cdots, x_n) = (f(x_1), \cdots, f(x_n))$. We will denote any restriction of f^n by f^n also. We note that f^n is *equivariant* i.e. $\sigma_{ij} \circ f^n = f^n \circ \sigma_{ij}$ for any $1 \leq i < j \leq n$. Now suppose that f is injective. Then $f^n(C_n(X)) \subset C_n(Y)$ and therefore we have an equivariant map $f^n : C_n(X) \to C_n(Y)$. Then the following proposition is of fundamental importance.

Proposition 3.2.
If two embeddings $f, g : X \to Y$ are isotopic then the two maps $f^n, g^n : C_n(X) \to C_n(Y)$ are equivariantly homotopic. That is, there is a homotopy $h_t : C_n(X) \to C_n(Y)$, $0 \leq t \leq 1$, such that $h_0 = f^n$, $h_1 = g^n$ and h_t is equivariant for each t.

It should be mentioned that even if f and g are homotopic embeddings, f^n and g^n are not neccessarily homotopic because only injections lift to configuration spaces. Hence the equivariant homotopy class of f^n is possibly a nontrivial isotopy invariant of f in many cases. We remark here that this invariant formally generalizes the n-component link homotopy invariant defined by Koschorke in [10].

This is a very natural invariant. Yet as far as the author knows there has been no general study made of this invariant.

We will study link homotopy and weak link homotopy in [22] and [14].

References

[1] A. Casson: Link cobordism and Milnor's invariant, Bull. London Math. Soc., 7, (1975), 39-40.

[2] R. H. Fox and J. W. Milnor: Singularities of 2-spheres in the 4-sphere, Osaka Math. J., 3, (1966), 257-267.

[3] C. H. Giffen: Link concordance implies link homotopy, Math. Scand., 45, (1979), 243-254.

[4] D. L. Goldsmith: Concordance implies homotopy for classical links in M^3, Comment. Math. Helvetici, 54, (1979), 347-355.

[5] N. Habegger and X. -S. Lin: The classification of links up to link homotopy, J. A.M.S., 3, (1990), 389-419.

[6] D. Jonish and K. C. Millett: Isotopy invariants of graphs, Trans. A.M.S., 327, (1991), 655-702.

[7] L. Kauffman: Invariants of graphs in three-space, Trans. A.M.S., 311, (1989), 697-710.

[8] S. Kinoshita: On elementary ideals of polyhedra in the 3-sphere, Pacific J. Math., 42, (1972), 89-98.

[9] K. Kobayashi: Standard spatial graphs, Hokkaido Math. J., 21, (1992), 117-140.

[10] U. Koschorke: Higher order homotopy invariants for higher dimensional link maps, Lecture Notes in Math. 1172, Springer-Verlag, (1985), 116-129.

[11] J. Milnor: Link groups, Ann. Math., 59, (1954), 177-195.

[12] J. Milnor: Isotopy of links, Algebraic Geometry and Topology; A Symposium in honor of S. Lefshetz, ed. Fox, Spencer and Tucker, Princeton University Press, (1957), 280-306.

[13] K. Miyazaki: The theta-curve cobordism group is not abelian, to appear in Tokyo J. Math..

[14] T. Motohashi and K. Taniyama: Homology and weak link homotopy of graphs in R^3, in preparation.

[15] N. Yu. Reshetikhin and V. G. Turaev: Ribbon graphs and their invariants derived from quantum groups, Comm. Math. Phys., 127, (1990), 1-26.

[16] M. Scharlemann and A. Thompson: Detecting unknotted graphs in 3-space, J. Diff. Geom., 34, (1991), 539-560.

[17] J. Simon and K. Wolcott: Minimally knotted graphs in S^3, Topology appl., 37, (1990), 163-180.

[18] S. Suzuki: A prime decomposition theorem for a graph in the 3-sphere, Topology and computer science, edited by S. Suzuki, Kinokuniya, Tokyo, (1987), 259-276.

[19] K. Taniyama: Cobordism of theta curves in S^3, Math. Proc. Camb. Phil. Soc., 113, (1993), 97-106.

[20] K. Taniyama: Cobordism, homotopy and homology of graphs in R^3, preprint.

[21] K. Taniyama: Homology classification of spatial embeddings of a graph, preprint.

[22] K. Taniyama: Link homotopy invariants of graphs in R^3, to appear in Rev. Mat. Univ. Complut. Madrid.

[23] A. Thompson: A polynomial invariant of graphs in 3-manifolds, Topology, 31, (1992), 657-665.

[24] W. T. Wu: On the isotopy of complexes in a eucledian space I, Science Sinica, 9, (1960), 21-46.

[25] S. Yamada: An invariant of spatial graphs, J. Graph Theory, 13, (1989), 537-551.

[26] D. N. Yetter: Category theoretic representations of knotted graphs in S^3, Adv. in Math., 77, (1989), 137-155.

Department of Mathematics, College of Arts and Sciences
Tokyo Woman's Christian University
Zempukuji, Suginami-ku, Tokyo, 167, Japan.

Recent Titles in This Series

(*Continued from the front of this publication*)

(See the AMS catalog for earlier titles)